Human Factors and Ergonomics in Sport

Human Factors and Ergonomics in Sport

Applications and Future Directions

Edited by
Paul M. Salmon, Scott McLean, Clare Dallat,
Neil Mansfield, Colin Solomon and Adam Hulme

CRC Press
Taylor & Francis Group
Boca Raton London New York

CRC Press is an imprint of the
Taylor & Francis Group, an **informa** business

First edition published 2021
by CRC Press
6000 Broken Sound Parkway NW, Suite 300, Boca Raton, FL 33487-2742

and by CRC Press
2 Park Square, Milton Park, Abingdon, Oxon, OX14 4RN

© 2021 Taylor & Francis Group, LLC

CRC Press is an imprint of Taylor & Francis Group, LLC

Library of Congress Cataloging-in-Publication Data

Names: Salmon, Paul M., editor.
Title: Human Factors and Ergonomics in Sport : Applications and Future
    Directions / edited by Paul M. Salmon, Scott McLean, Clare
    Dallat, Neil Mansfield, Colin Solomon, Adam Hulme.
Description: Boca Raton, FL : CRC Press, 2020. | Includes bibliographical
    references and index.
Identifiers: LCCN 2020019008 (print) | LCCN 2020019009 (ebook) | ISBN
    9781138481633 (hardback) | ISBN
    9781351060073 ebook ebook
Subjects: LCSH: Sports--Physiological aspects. | Human engineering.
Classification: LCC RC1235 .H86 2020 (print) | LCC RC1235 (ebook) | DDC
    612/.044--dc23
LC record available at https://lccn.loc.gov/2020019008
LC ebook record available at https://lccn.loc.gov/2020019009

ISBN: 978-1-138-48163-3 (hbk)
ISBN: 978-0-367-52944-4 (pbk)
ISBN: 978-1-351-06007-3 (ebk)

Typeset in Times
by Deanta Global Publishing Services, Chennai, India

Contents

SECTION I Introduction to HFE in Sport

SECTION II Physical HFE Applications

SECTION III   Cognitive HFE Applications

SECTION IV   Systems HFE Applications

SECTION V Future Applications of HFE in Sport

Preface

Eluid Kipchoge crosses the finishing line and jubilantly beats his chest in celebration. He has just covered 26.2 miles (42.2km) in one hour 59 minutes and 40 seconds, making him the first person in history to complete a marathon in under two hours.

Serena Williams falls to the ground in celebration after beating her sister Venus to win the 2017 Australian Open for a record 7th time. At the time of writing Williams has won 23 Grand Slam singles titles, the most by any man or woman in the Open era.

Kylian Mbappe receives the ball five yards outside of the penalty area and drives it past Croatia's goalkeeper, becoming only the second teenager to score in the FIFA World Cup final. His goal gives France an unassailable 4–2 lead against Croatia, sealing their place as the 2018 football World Champions.

David Beckham is stretchered off the pitch after tearing his left Achilles tendon during an Italian Series A match between AC Milan and Chievo at the San Siro stadium in Milan, Italy. The injury rules him out of the 2010 World Cup where he was set to become the first player to represent England in four consecutive World Cups.

It is the opening day of the 2007/2008 English football season and Leeds United Football Club have just lost to opponents Tranmere Rovers FC. In less than a decade, gross financial mismanagement has seen the club go from challenging for the Premier League title and European Champions League to playing in the third tier of English football for the first time in their history.

Jules Bianchi loses control of his Marussia Formula One car in wet conditions and collides with a recovery vehicle during the 2014 Japanese Grand Prix at Suzuka. Suffering major head trauma, Bianchi is placed in an induced coma, tragically succumbing to his injuries just over ten months later.

One of the biggest doping scandals in sport occurred in men's cycling when Lance Armstrong admitted to doping throughout all of his seven Tour de France wins. Armstrong was subsequently stripped of his titles and banned for life from all sanctioned cycling events. Many others were implicated and the case continues to cast a dark shadow over the sport at all levels of competition.

Former USA gymnastics team doctor Larry Nassar is sentenced to 175 years in prison after pleading guilty to seven counts of sexual assault of minors. Days later he is sentenced to a further 40 to 125 years in prison after pleading guilty to three additional counts of sexual assault. Scores of women, including many former gymnasts, came forward with allegations of sexual abuse during his time as USA gymnastics team doctor.

The stories above are real events that have occurred in elite sport over the past three decades. Some represent the amazing feats that can be achieved when athletes, teams, and sports systems perform optimally (Kipchoge, Williams, Mbappe). Some encompass the high-profile failures that occur when injuries are sustained or when athletes, teams, and clubs fail to perform at their full potential (Beckham, Leeds United). Finally, some demonstrate the tragic and catastrophic outcomes that can arise when sports systems fail (Bianchi, Armstrong, Nassar).

Human Factors and Ergonomics (HFE), the discipline that attempts to optimise human health and well-being, can potentially be used to facilitate more of
the amazing feats and to minimise the failures. Whilst the idea that HFE can play
a key role in optimising sports performance is not new (e.g. Reilly & Lees, 1984;
Reilly & Ussher, 1988), it is an area of work that has received relatively less attention than others since HFE was recognised formally as a discipline just after the
Second World War. Thankfully, we are now in a period of new interest, where HFE
theories and methods are increasingly being used to help understand and optimise
sports systems (Salmon & Macquet, 2019). This relates not only to the enhancement
of performance but also to injury prevention, sports equipment design, performance
analysis, and to the prevention of wider sports systems issues such as doping and
corruption.

It is our view that the requirement for HFE input in sport is increasing. This is
in part because sports systems are becoming increasingly technology centric. For
example, the 2019/2020 English Premier League season has seen the challenging
introduction of the Video Assistant Referee (VAR), and similar systems have been in
place in other sports such as cricket and tennis since the turn of the century. In recent
years there has also been widespread recognition that sports systems are inherently
complex in nature (Davids et al., 2013; Bittencourt et al., 2016; Hulme et al., 2018;
McGarry et al., 2002; McLean et al., 2017; Mooney et al., 2017). Consequently,
methods that can cope with such complexity are required. Without such methods, it
is not possible to fully understand the complex set of interacting factors influencing
behaviour, and thus it is not possible to introduce appropriate and effective interventions designed to optimise performance. The stakes are also getting higher, with the
substantial sums of money available through prize money, sponsorship, ticket sales,
and merchandising continuing to increase. Recent estimates suggest that sport now
represents a $75bn a year industry in the US (QARA, 2019), a £20bn-a-year industry
in the UK (Cave, 2015), and between $480bn to $620bn globally (QARA, 2019). As
technology use, complexity, and financial rewards continue to grow, the requirement for HFE in sports research and practice is likely to increase also (Salmon &
Macquet, 2019). Such input is required to support technology design and integration
as well as performance analysis and optimisation efforts.

This book was proposed by the editors in response to this exciting new era for
HFE in sport. The proposal was developed alongside the creation of a Human
Factors and Ergonomics in Sport and Outdoor Recreation conference as part of the
Applied Human Factors and Ergonomics (AHFE) conference in Las Vegas in 2015
(see Salmon & Macquet, 2016), as well as the publication of two special issues of
HFE journals on HFE in sport and outdoor recreation (Salmon, 2017; Salmon &
Macquet, 2019). Building on the conference and special issues, the aim of this book
is to provide a platform for researchers and practitioners to communicate contemporary sports HFE applications, to showcase the issues being tackled, to facilitate
cross-disciplinary interaction between HFE and the sports sciences, and to inspire
the sports science and HFE communities to pursue further applications of HFE in
sport. The intention is also to demonstrate the contribution and potential role of HFE
in future sports research and practice. The book is also intended to be educational,
providing useful references and guidance relating to HFE theory and methods. It is

hoped that this will help researchers and practitioners identify potential applications and ultimately apply HFE theory and methods in their own areas of speciality. The book is thus aimed both at HFE researchers and practitioners and sports scientists as well as those working in sport at all levels.

This book includes contributions from HFE and sports science researchers and practitioners and covers a diverse set of topics ranging from injury prevention and management, sports equipment design, training and coaching, talent identification, athlete decision making and cognitive workload, teamwork, performance analysis, and sports system modelling and analysis. These applications cover a range of sports, including football, running, cycling, cricket, rugby, equestrian sports, parasport, basketball, Formula One, and Australian Rules Football. Although the list of sports is by no means exhaustive in this regard, the book offers insights that can be applied to many other sporting areas.

HOW TO READ THIS BOOK

We expect that some readers will be familiar with the HFE theories and methods discussed in this book while for others there will be much new information. We have asked contributors to achieve a balance in the level of detail provided and where possible refer the novice reader to other texts that they may find useful for further guidance. For example, where HFE methods are applied, reference is given to other texts and articles that provide in-depth guidance on how to apply them in practice.

Readers will no doubt differ in specific interests; however, it is recommended that Chapters 1 and 2 be read by every reader, regardless of field or experience. Following this, the choice of chapters should be made based on the readers' interest in specific theories and methods, areas of work (e.g. sports equipment design versus injury prevention and management), and sporting contexts.

This book is divided into five main sections:

I. Introduction to HFE in Sport
- Chapter 1 (Salmon et al.) introduces HFE and provides an overview of previous HFE applications in sport. In Chapter 2, Hulme et al. discuss how sports systems are complex in nature, and thus require a systems HFE approach.

II. Physical HFE Applications
- Chapters 3 to 8 include a series of studies and commentary articles covering physical HFE applications. These include chapters focussed on injury prevention, coaching, and sports equipment design.

III. Cognitive HFE Applications
- Chapters 9 to 13 include a series of studies and commentary articles covering cognitive HFE applications. These include chapters focussed on decision making, situation awareness, teamwork, cognitive workload, and constraints-based training.

IV. Systems HFE Applications
- Chapters 14 to 19 include a series of studies and commentary articles covering systems HFE applications. These include chapters focussed on

injury prevention and management, performance analysis, sports system analysis, and talent identification.

V. Future Applications of HFE in Sport

- Chapter 20 provides and overview of the conclusions and recommendations arising from the information presented in this book and identifies future research directions for HFE in sport.

REFERENCES

Bittencourt, N. F., Meeuwisse, W. H., Mendonça, L. D., Nettel-Aguiree, A., Ocarino, J. M., and Fonseca, S. T. (2016). Complex systems approach for sports injuries: Moving from risk factor identification to injury pattern recognition-narrative review and new concept. *British Journal of Sports Medicine*, 50, 1309–1314.

Cave, A. (2015). Discover the potential of sport: a £20 billion industry. *The Telegraph*, https://www.telegraph.co.uk/investing/business-of-sport/potential-of-sport-20billion-industry/, accessed 10th June 2020.

Davids, K., Hristovski, R., Araujo, D., Serre, N. B., Button, C., and Passos, P., eds. (2013). *Complex Systems in Sports*. Routledge Research in Sport and Exercise Science.

Hulme, A., Thompson, J., Nielsen, R. O., Read, G. J. M., and Salmon, P. M. (2018). Formalising the complex systems approach: Using agent-based modelling to simulate sports injury aetiology and prevention. *British Journal of Sports Medicine*, 53, 560–569.

McGarry, T., Anderson, D. I., Wallace, S., Hughes, M., and Franks, I. M. (2002). Sport competition as a dynamical selforganizing system. *Journal of Sports Sciences*, 20, 771–781.

McLean, S., Soloman, C., Gorman, A., and Salmon, P. M. (2017). What's in a game? A systems approach to enhancing performance analysis in football. *Plos One*, 12(2), 1–15.

Mooney, M., Charlton, P. C., Soltanzadeh, S., and Drew, M. K. (2017). Who 'owns' the injury or illness? Who 'owns' performance? Applying systems thinking to integrate health and performance in elite sport. *British Journal of Sports Medicine*, 51, 1054–1055.

QARA. (2019). Sports Industry Insights, https://medium.com/qara/sports-industry-report-3244bd253b8, accessed 10th June 2020.

Reilly, T., and Lees, A. (1984). Exercise and sports equipment: some ergonomics aspects. *Applied Ergonomics*, 15(4), 259–279.

Reilly, T., and Ussher, M. (1988). Sport, leisure and ergonomics. *Ergonomics*, 31(11), 1497–1500.

Salmon, P. M. (2017). Ergonomics issues in sport and outdoor recreation. *Theoretical Issues in Ergonomics Science*, 18(4), 299–305.

Salmon, P. M., and Macquet, A. C. (2016). *Advances in Human Factors in Sports and Outdoor Recreation*. Proceedings of the AHFE 2016 International Conference on Human Factors in Sports and Outdoor Recreation, July 27–31, 2016, Florida, USA.

Salmon, P. M., and Macquet, A. C. (2019). Ergonomics in Sport and Outdoor Recreation: From individuals and their equipment to complex systems and their frailties. *Applied Ergonomics*, 80, 209–213.

Editors

Paul M. Salmon is a Professor in Human Factors and is the director of the Centre for Human Factors and Sociotechnical Systems at the University of the Sunshine Coast. Paul has almost twenty years' experience of applied Human Factors research in areas such as road and rail safety, aviation, defence, sport and outdoor recreation, healthcare, workplace safety, and cybersecurity. His research has focussed on understanding and optimizing human, team, organisational, and system performance through the application of Human Factors theory and methods. He has co-authored 19 books, over 200 peer-reviewed journal articles, and numerous book chapters and conference contributions. Paul's contribution has been recognized through various accolades, including the Chartered Institute for Ergonomics and Human Factor's 2019 William Floyd award and 2008 Presidents Medal, the Human Factors and Ergonomics Society Australia's 2017 Cumming memorial medal, and the International Ergonomics Association's 2018 research impacting practice award.

Scott McLean is a Research Fellow and the theme leader for Sport and Outdoor Recreation at the Centre for Human Factors and Sociotechnical Systems at the University of the Sunshine Coast. Scott has a background in Exercise Science (MSc, BExSc) and earned his PhD applying Human Factors and Ergonomics methods in sport, in which he received the David Ferguson Award from the Human Factors and Ergonomics Society of Australia for the best PhD thesis. His research spans a broad range of domains including sports science, safety science, and systems thinking. During his PhD and current post-doctoral research, Scott has made a number of significant research contributions which have advanced knowledge in the areas of team performance analysis, coaching, sports system modelling, applying Human Factors and Ergonomics in sport, incident reporting systems in outdoor recreation, and complex system modelling of the road safety system. Scott has experience working with and conducting research with industry, i.e. professional sporting clubs and an international football team, government agencies, as well as international collaborators. Scott is also a successful and award-winning football coach, which ensures that his research has a focus on delivering practice implications.

Clare Dallat is the Executive Director of Research and The Outdoor Education Foundation at The Outdoor Education Group, a large not-for-profit educational organisation in Australia. Clare also directs Risk Resolve, a consultancy service that works with organisations across the globe helping them to construct knowledge and confidence to proactively and reactively manage risks to their participants and staff. She has over twenty years of practice in all aspects of the led-outdoor activity domain. Clare earned a PhD in Human Factors from the Centre for Human Factors and Sociotechnical Systems at The University of the Sunshine Coast, and an MSc. in Risk, Crisis and Disaster Management. Her research has focussed on the design and application of Human Factors methods for accident prediction and analysis in led outdoor activities. As an elite level cyclist with an international stage win to her name prior to a career ending hip injury, Clare has a unique insight as both an athlete, and a researcher in the elite sport domain. In 2018, she became the first person outside of North America to win the prestigious Reb Gregg Award for exceptional leadership, innovation and contribution to international wilderness risk management.

Neil Mansfield is a Professor of Human Factors Engineering and is Head of the Department of Engineering at Nottingham Trent University, UK. The Department offers programmes in Sport Engineering and Biomedical Engineering in addition to traditional engineering disciplines. His research focusses on optimising the experience, performance, and wellbeing of product users. He has a particular expertise in designing for dynamic environments including vibration and shock, and design for the most vulnerable. Applications have included optimisation of workspaces across the breadth of transport applications from cars, high-speed boats, trains and aircraft, through to mining, agricultural, and military vehicles, and design of transport systems for premature babies. He has also worked with sports equipment manufacturers and elite athletes to understand and optimise performance whilst minimising risk of injury, including golf, badminton, motorsport, cycling, skeleton, and power boats. Neil was editor of the journal *Ergonomics* for 10 years until 2017 when he became President of the Chartered Institute of Ergonomics and Human Factors. He is organizer of the biannual Comfort Congress.

Colin Solomon is an Associate Professor in Exercise Physiology at the University of the Sunshine Coast. He has twenty-five years' experience in research and teaching in human exercise and respiratory physiology, including specifically pulmonary ventilation, oxygen distribution, inhalation toxicology, and asthma, ranging from systemic physiology to molecular biology. To date, he has authored 40+ peer-reviewed research journal articles and government reports and 40+ peer-reviewed conference presentations. He has received substantial competitive funding for his research. He is an experienced research supervisor of Honours, Masters and PhD students, former Associate Dean of Graduate Studies, and former member of the Executive of the Australian Council of Graduate Research.

Adam Hulme is a Post-Doctoral Research Fellow at the Centre for Human Factors and Sociotechnical Systems at the University of the Sunshine Coast. He has a Bachelor of Science in Sport Exercise Science and an Honours degree in Sports Psychology (The University of Cumbria, Lancaster, England), a Master of Health Promotion (USC, Queensland, Australia), and earned a Doctor of Philosophy in Sports Injury Epidemiology and Human Factors and Ergonomics (HFE) in August 2017 (Federation University Australia, Ballarat, Victoria). His doctoral program was completed at the Australian Centre for Research into Injury in Sport and its Prevention (ACRISP), which is recognised by the International Olympic Committee (IOC) as a world leading research centre. His main areas of expertise and interest include the application of qualitative and quantitative methods and approaches that are grounded in systems theory to enhance the health and safety of athletes in complex sociotechnical sports systems. Adam has published numerous peer-reviewed journal articles that involve the use of traditional epidemiological approaches, HFE models and methods, and computational and simulation modelling.

Contributors

Vanessa Beanland is Senior Lecturer in the Department of Psychology at the University of Otago in Dunedin, New Zealand.

Elise Berber is a Doctor of Philosophy Candidate at the Sunshine Coast Mind and Neuroscience – Thompson Institute, University of the Sunshine Coast in Queensland, Australia.

James Brown, MSc, is a Senior Research Assistant and doctoral student in Human Factors Engineering at the University of Southampton in the UK.

Candice Christie is an Associate Professor in the Department of Human Kinetics and Ergonomics at Rhodes University, Grahamstown, South Africa.

Amanda Clacy is a Post-Doctoral Research Fellow at the Sunshine Coast Mind and Neuroscience – Thompson Institute, University of the Sunshine Coast in Queensland, Australia.

Matthew Clark is an Ergonomics Consultant for Ergonomie Australia Pty Ltd, Sydney, New South Wales, Australia.

Clare Dallat has a PhD in Human Factors and is the Director of Research and Innovation at the Outdoor Education Group.

Karl Dodd is a professional football coach and former professional football player. He is currently the National team head coach of Guam.

Alex Donaldson is a Senior Research Fellow at the Centre for Sport and Social Impact at La Trobe University in Melbourne, Victoria, Australia.

Christina Driver is a PhD candidate with the School of Social Sciences, and Associate Lecturer in Mental Health and Neuroscience for the Thompson Institute, at the University of the Sunshine Coast.

Orito Takemoto-Forsyth is a Bachelor of Sports Studies graduate at The University of the Sunshine Coast in Queensland, Australia.

Adam Gorman is a Senior Lecturer in Skill Acquisition in the School of Exercise and Nutrition Sciences at Queensland University of Technology, Brisbane, Australia.

Shona Halson is an Associate Professor in the School of Behavioural and Health Sciences at Australian Catholic University.

David Haydon is a Sports Engineer with the South Australian Sports Institute in Adelaide, Australia.

Jonathon Headrick is a Lecturer and Researcher in the School of Allied Health Sciences at Griffith University, Gold Coast, Australia.

Glenn Holmes is a PhD Researcher at the Sunshine Coast Mind and Neuroscience – Thompson Institute, University of the Sunshine Coast in Queensland, Australia.

Adam Hulme is a Research Fellow at the Centre for Human Factors and Sociotechnical Systems at the University of the Sunshine Coast and holds a PhD in Sports Injury Epidemiology.

David Jenkins is a Professor of Sport and Exercise Science at the University of the Sunshine Coast in Queensland, Australia.

Bridie Kean is a Lecturer in Public Health at the University of the Sunshine Coast in Queensland, Australia. Her PhD research focussed on environmental factors impacting para-athletes and para-studentathletes.

Vincent Kelly is a Senior Lecturer at the School of Exercise and Nutrition Sciences, Queensland University of Technology, Australia; specialising in sport science research for high performance sport.

Ben Lane is a Post-Doctoral Research Fellow at the Centre for Human Factors and Sociotechnical Systems and earned a PhD in Psychology from the University of the Sunshine Coast.

Anne-Claire Macquet is a Researcher in Sports Psychology and Cognitive Ergonomics at the French Institute of Sports in Paris, France.

Laurent Malisoux is a Sports Scientist and Group Leader of the Physical Activity, Sport & Health researcher group at the Luxembourg Institute of Health, Grand-Duchy of Luxembourg.

Neil Mansfield is a Professor of Human Factors Engineering and Head of the Department of Engineering, Nottingham Trent University, UK.

Christopher McCormack is a higher degree research student studying skill acquisition in the School of Health and Sport Sciences at the University of the Sunshine Coast in Queensland, Australia.

Kayla McEwan is a MSc candidate in the Department of Human Kinetics and Ergonomics at Rhodes University, Grahamstown, South Africa.

Scott McLean is a Research Fellow in High Performance Sport and Outdoor Recreation in the Centre for Human Factors and Sociotechnical Systems at the University of the Sunshine Coast in Queensland, Australia.

Timothy Neville is the Umpiring Performance Assessment Manager with the Australian Football League.

Rasmus O. Nielsen is a Post-Doctoral Researcher, Sports Injury Epidemiologist, and Research Coordinator at RUNSAFE, Aarhus University, Denmark.

Erich Petushek is an Assistant Professor in the Department of Cognitive and Learning Sciences at Michigan Technological University, Houghton, Michigan, USA.

Ross Pinder is a Skill Acquisition Specialist with Paralympics Australia, based in Adelaide, Australia.

Gemma J. M. Read is a Senior Research Fellow at the Centre for Human Factors and Sociotechnical Systems at the University of the Sunshine Coast in Queensland, Australia.

Ian Renshaw is an Associate Professor at Queensland University of Technology, Brisbane, Australia. Ian's teaching and research interests are centred on applications of ecological dynamics to sport settings.

Kirsten M. A. Revell, PhD, is a Human Factors Research Fellow in the Transportation Research Group at the University of Southampton, UK.

Suzanna Russell is a PhD candidate in the school of Human Movement and Nutrition Sciences at the University of Queensland, Australia; specialising in athlete performance, fatigue, recovery and well-being.

Paul M. Salmon is a Professor of Human Factors and Director of the Centre for Human Factors and Sociotechnical Systems at the University of the Sunshine Coast in Queensland, Australia.

Colin Solomon is an Associate Professor of Physiology at the University of the Sunshine Coast, Queensland, Australia.

Neville Stanton, PhD, DSc, holds the Chair in Human Factors Engineering in the Faculty of Engineering and the Environment at the University of Southampton in the UK.

Nicholas Stevens is a Landscape Architect and Urban Planner and is the Deputy Director of the Centre for Human Factors and Sociotechnical Systems at the University of the Sunshine Coast.

Daniel Theisen is the co-founder and former Head of the Sports Medicine Research Laboratory at the Luxembourg Institute of Health and has several academic appointments in sports and rehabilitation sciences.

Jason Thompson is a Senior Research Fellow at the Centre for Transport, Health and Urban Design (THUD) at Melbourne University, Australia.

Will Vickery is a Lecturer in the School of Exercise & Nutrition Science, Deakin University, Melbourne, Australia.

Section I

Introduction to HFE in Sport

1 An Introduction to Human Factors and Ergonomics in Sport

Paul M. Salmon, Adam Hulme,
Scott McLean, and Colin Solomon

CONTENTS

1.1 INTRODUCTION

The benefits of sport are widely known and include a range of positive physical, psychological, societal, and economic effects (Khan et al., 2012; Oja et al., 2015; Wankel & Berger, 1990). However, there can also be negative effects, including individual issues such as illness and injury (Brukner & Khan, 2017), drug abuse (Reardon & Creado, 2014), eating and body image disorders (Joy et al., 2016), fatigue and burnout (DiFori et al., 2014), and mental health problems (Hughes & Leavey, 2012). Moreover, when entire sports systems fail, the consequences can be even more catastrophic and can include serious injuries and fatalities (Gabbe et al., 2005; Turk et al., 2008), large-scale financial losses (Lago et al., 2006), and significant adverse effects on individuals and society (Salmon & Macquet, 2019). Such impacts are evidenced in recent sporting tragedies such as the United States gymnastics team sexual abuse scandal, the Port Said stadium disaster, the FIFA corruption scandal, and various fatal incidents in areas such as motorsport, boxing, and equestrian sports.

Human Factors and Ergonomics (HFE) is the scientific discipline that is primarily focussed on optimising human health and well-being. Whilst much of the work of HFE professionals occurs in work systems, it also spans the societal systems that we interact with on a day-to-day basis, such as transport, sports, and outdoor recreation. Accordingly, the role of HFE in optimising sports systems has long been recognised, since the pioneering work of Professor Tom Reilly (e.g. Reilly & Lees, 1984; Reilly & Ussher, 1988), Ergonomist and founding editor of the *Journal of Sports Sciences* (Cable et al., 2009). Though sports HFE applications had occurred previously, it was arguably Reilly who brought HFE to the attention of many sports scientists. Likewise, it was he who raised awareness within HFE of the opportunity to make meaningful contributions in sport (Reilly & Lees, 1984). Since then, a substantial amount of HFE research has been undertaken in sport and it has been suggested that the need for HFE input in sports research and practice is increasing (McLean et al., 2017; Salmon & Macquet, 2019).

The aim of this chapter is to set the scene for the remainder of the book by introducing HFE and providing an overview of previous sports HFE applications. For the reader already familiar with HFE, the intention is to provide an overview of the HFE research undertaken to date in sports settings. For the reader not familiar with HFE or its core theories and methods, the intention is to provide an overview of HFE and how it can be used to understand and optimise sports performance. For both, it is hoped that this chapter will highlight the important role of HFE in sports research and practice as well as help to identify further applications of HFE in sport.

1.2   INTRODUCTION TO HUMAN FACTORS AND ERGONOMICS

HFE is formally defined by the International Ergonomics Association (IEA) as 'the scientific discipline concerned with the understanding of interactions among humans and other elements of a system, and the profession that applies theory, principles, data and methods to design in order to optimise human well-being and overall system performance' (IEA, 2019). Earlier definitions of HFE emphasised 'the relationship between man and his working environment' (Murrell, 1965), the goal of guiding 'technology in the direction of benefiting humanity' (Sanders & McCormick, 1993), and 'the study of how humans accomplish work-related tasks in the context of human-machine systems' (Meister, 1989).

Therefore, HFE attempts to understand and optimise individual, team, organisational, and system performance, both in work and societal systems. This is achieved through the application of a diverse set of HFE theories and methods which enable practitioners to:

a. describe and understand the behaviour of individuals, teams, organisations, and systems; and
b. direct the design and evaluation of products, tools, devices, work and tasks, environments, training programmes, procedures, policy and regulation, and ultimately overall socio-technical systems.

HFE is therefore about understanding and optimising performance to enhance efficiency and productivity, safety, and human health and well-being. To support this, there are various HFE theories and methods relating to different aspects of human, team, and system performance.

1.2.1 HUMAN FACTORS AND ERGONOMICS THEORY

HFE research and practice is based on a diverse set of theoretical models which cover specific components of individual, team, organisational, and system performance. An overview of individual, team, organisational, and system HFE theories is given below.

1.2.1.1 Individual Models

Individual HFE models cover various aspects of physical and cognitive behaviour, including anthropometrics (Pheasant & Haslegrave, 2005), posture and biomechanics (Kumar, 2007), motor skills (Annett, 1994), perception and action (Neisser, 1976), decision making (Klein, 1993), situation awareness (Endsley, 1995a), workload (Hancock & Meshkati, 1988), error (Reason, 1990), distraction (Regan et al., 2008), and attention (Wickens, 1984). These models describe the physical and cognitive processes which contribute to human behaviour, as well factors that might influence behaviour in different contexts. To demonstrate, we discuss Endsley's three-level model of Situation awareness (SA; Endsley, 1995a), arguably one of the most well-known HFE models of all time.

SA is the term used in HFE to describe our understanding of what is going on (Endsley, 1995a). In a recent state of science review, Stanton et al. (2017) describe how SA has been explored in many operational contexts, ranging from defence, transportation, and process control to healthcare, disaster response, and sport. Whilst various theoretical models exist, Endsley's (1995a) 'three-level' model of SA is undoubtedly the most popular. This model describes how individuals develop the SA required to support effective task performance in dynamic environments. According to Endsley's model, SA represents an individual's dynamic mental model of the ongoing situation and incorporates 'the perception of the elements in the environment within a volume of time and space, the comprehension of their meaning and a projection of their status in the near future' (Endsley, 1988). SA is described as comprising three levels: Level 1, perception of the elements in the environment; Level 2, comprehension of their meaning; and Level 3, projection of future system states.

Level 1 SA involves perceiving the status, attributes and dynamics of task-related elements in the surrounding environment (Endsley, 1995a). During Level 2 SA, the individual interprets this information to develop an understanding of its relevance in relation to their task and goals, forming 'a holistic picture of the environment, comprehending the significance of objects and events' (Endsley, 1995a, p. 37). Level 3 SA involves anticipating future task and system states by forecasting the likely behaviour of elements in the environment. This involves the integration of Level 1 and 2 SA and the use of mental models of similar situations to forecast likely events. These mental models play a key role in SA, directing attention to pertinent elements in the environment (Level 1 SA), facilitating the integration of elements to aid comprehension (Level 2 SA), and supporting the generation of future states and behaviours (Level 3 SA) (Endsley, 1995a).

Endsley's model provides a good example of how individual HFE models can be applied to help understand and optimise sports performance. In sport, to date, SA has been applied to the study of sports officiating teams (Neville & Salmon, 2016), hammer throwing (Macquet & Stanton, 2014), rowing (Macquet & Stanton, 2014), cycling (Salmon et al., 2017), squash (Murray et al., 2017), and American football (Pedersen & Cooke, 2006) to name only a few. The three-level model and its associated methods (e.g. Endsley, 1993, 1995b) can be used to determine what information athletes attend to in different match or event scenarios, how they integrate this with their mental models to understand different situations, and how they forecast future events. Likewise, descriptions of expert or elite athlete SA can be used to support the development of coaching exercises designed to enhance athlete SA. Various SA measures are available to support assessments of SA in different contexts, meaning that both the content and accuracy of SA can be assessed. This supports work which aims to evaluate the impact of interventions on athlete and umpire SA, such as new sports equipment, modifications to training regimes, and new tactics. For example, Neville (2017) assessed the effect of introducing radio communication on Australian Rules Football umpires' SA when officiating matches. Whilst we have used SA as an example here, many individual HFE models exist and could be applied to help understand and optimise performance.

1.2.1.2   Team Models

Team level HFE models cover multiple aspects of team behaviour, including teamwork (Salas et al., 2005), team SA (e.g. Salas et al., 1995), communication (Salas et al., 2008) and coordination (Gorman et al., 2006), distributed decision making (Stanton et al., 2010), and team cognition (Salas & Fiore, 2004). These models describe the processes involved in teamwork, as well as the factors that might influence how well a team can perform in different contexts. For example, Salas and colleagues proposed the Big Five model of teamwork based on a review and synthesis of over 130 existing teamwork models (Salas et al., 2005). The Big Five model identifies five key teamwork behaviours that influence the quality of team performance: leadership, mutual performance monitoring, back up behaviour, adaptability, and team orientation. In addition, Salas et al. propose an additional three coordinating mechanisms that are required for effective teamwork: the development of shared mental models; the achievement of mutual trust; and engagement in closed-loop communication (Salas et al., 2005). According to Salas et al. (2005), teams that engage in the five behaviours and the additional coordinating mechanisms are likely to perform better than those that do not.

Given the high number of team sports, it is perhaps not surprising that the analysis of teamwork has been popular in sports science (Carron et al., 2002; Chow & Feltz, 2008; Filho et al., 2014; Fiore & Salas, 2006; Jones et al., 2008; Reimer et al., 2006; McLean et al., 2019). Similarly, many HFE researchers have highlighted the opportunity for applying learnings from the study of elite sports teams in other HFE domains (e.g. Goodwin, 2008; Neville & Salmon, 2016; Salmon & Macquet, 2019). By applying models such as the Big Five in sport, it is possible to identify and examine core teamwork behaviours that are critical to successful performance, such as communication, shared mental models, mutual performance monitoring, and back up behaviours.

1.2.1.3 Organisational and Systems Models

Organisational- and systems-level HFE models are holistic and attempt to describe the behaviour of organisations and/or entire socio-technical systems (e.g. Kleiner, 2006; Leveson, 2004; Rasmussen, 1997) or to provide principles for organisational or system design (Clegg, 2000; Davis et al., 2014). Perhaps the best example of systems models comes from the area of accident causation, where systems thinking-based models are now widely accepted as state-of-the-art (e.g. Dekker, 2011; Leveson, 2004; Perrow, 1984; Rasmussen, 1997). These models are based on the concept that behaviour, safety and accidents are emergent properties arising from non-linear interactions between multiple components across entire socio-technical systems (e.g. Dekker, 2011; Hollnagel, 2004; Leveson, 2004; Rasmussen, 1997). This notion is currently gaining increasing attention in sports science, with many arguing that a systems thinking approach is required to understand aspects of sport such as performance (McLean et al., 2017) and sports injury (Bittencourt et al., 2016; Hulme et al., 2018).

Rasmussen's Risk Management Framework (RMF; Rasmussen, 1997) is arguably the most popular systems HFE model (Salmon et al., 2017). According to RMF, systems comprise various levels (e.g. government, regulators, company, company management, staff, and work), each of which are co-responsible for production and safety. Decisions and actions at all levels of the system interact with one another to produce performance and accidents are thus created by the decisions of all actors, not just the front-line workers in isolation. Moreover, all accidents are caused by multiple contributing factors, not just one decision or action.

A major tenet of systems models is that it is not possible to understand behaviour by reducing a system into its component parts and examining these parts in isolation from one another; rather, it is the interactions between the components across the overall system that are of interest (Ottino, 2003). This requires the assessment of more than the behaviour of the individuals involved in the incident and the immediate circumstances of the event. This view also encompasses factors within the broader organisational, social, or political system in which processes or operations occur. It has been argued that the prevalent approach in many areas of sports science has been based on the study of individual components such as athletes (McLean et al., 2017). Whilst understanding individual athlete behaviour is important, proponents of systems thinking argue that the athlete is just one part of a complex interlinked network of human and technical components that should be investigated together to obtain a detailed understanding of performance and the factors influencing it.

Rasmussen's framework makes a series of assertions regarding accident causation (Cassano-Piche et al., 2009; Rasmussen, 1997), all of which are seemingly relevant when applied to sports performance. Accordingly, we have modified them for the sports context below.

1. Sports performance is an emergent property impacted by the decisions and actions of all actors within a sport system, not just athletes alone.
2. Threats to sports performance are caused by multiple contributing factors, not just a single poor decision or action.

3. Threats to sports performance can result from a lack of poor communication and feedback (or 'vertical integration') across levels of a sport system, not just from deficiencies at one level alone.
4. Lack of vertical integration is caused, in part, by lack of feedback across levels of a sport system.
5. Sport-related behaviours are not static, they migrate over time and under the influence of various pressures such as financial and psychological pressures;
6. Migration occurs at multiple levels of a sport system.
7. Migration of practices cause performance to degrade and erode gradually over time, not all at once. Sub-optimal sports performance is caused by a combination of this migration and a triggering event(s).

When applying organisational and systems HFE models to study sports performance, it is possible to develop in-depth models of overall sports organisations or systems and to use these to identify what factors influence athlete or team performance. For example, Hulme et al. (2019) applied Work Domain Analysis (Naikar, 2013) to develop a model of an elite women's netball organisation and used it to identify the organisational factors that may impact the netball team's performance. Hulme et al. (2019) found various functions that facilitated success, including a strong club ethos, a shared responsibility for performance, and a strong focus on player and staff health and well-being; however, the model also revealed potential negative issues, including organisational priorities not related to playing netball, and additional coach and athlete roles beyond coaching, training, and playing.

1.3 APPLICATIONS OF HUMAN FACTORS AND ERGONOMICS IN SPORT

HFE models can be applied in multiple sports settings. Individual HFE models can be used to explain athlete behaviour and to generate hypotheses regarding key aspects of athlete performance such as injury and physical conditioning, workload, decision making, SA, and error. Team HFE models can be used when studying or working with sports teams to help understand and optimise key aspects of teamwork such as team orientation, shared mental models, communication, situation awareness, and cognition. Finally, organisational and systems HFE models provide a wider view, expanding the focus to sports organisations and entire sport systems to help understand what factors interact to influence the performance of athletes, teams, and organisations.

Most of the popular HFE theoretical models have an associated analysis method that can be used to support experimental or naturalistic studies. The Situation Awareness Global Assessment Technique (SAGAT; Endsley, 1995b) is a measure that is used to assess SA in simulated environments. The SAGAT uses probes administered during freezes in the task to assess how accurate an individual's Level 1, 2, and 3 SA. The Event Analysis of Systemic Teamwork (EAST) has been used extensively to assess teamwork in different domains (Stanton et al., 2018), including to examine the Big Five model behaviours and associated coordinating mechanisms during sports performance (Salmon et al., 2017). EAST uses task, social, and information networks to assess teamwork during tasks, communication, and the information used and shared throughout the team. AcciMap is an accident analysis method that is based on the structure and core tenets of Rasmussen's Risk Management Framework (Svedung & Rasmussen, 2002). AcciMap has been used extensively in HFE both for the analysis of accidents (Salmon et al., 2020) but also to test the assertions made by Rasmussen's RMF in different contexts (e.g. Cassano-Piche et al., 2009; Jenkins et al., 2010; Salmon et al., 2010). Sports applications of RMF and AcciMap include investigating the responsibility for community rugby concussion (Clacy et al., 2019) and concussion in Australian Rules Football (Dawson et al., 2017). In the next section we provide an overview of HFE methods and their previous applications in sport.

1.4 HUMAN FACTORS AND ERGONOMICS METHODS

The HFE professional has access to a diverse and flexible toolkit of HFE methods. According to Stanton et al. (2013), there are now more than 100 structured HFE methods available for designing and evaluating aspects of device, operator, team, organisation, and system performance (see Stanton et al., 2013). These include methods designed to help understand and optimise critical aspects of individual and team behaviour such as perception, decision making (Klein et al., 1989), situation awareness (Endsley et al., 1995b), cognitive workload (Hart & Staveland, 1988), error (Shorrock & Kirwan, 2002), human–machine interaction, communication, and coordination and teamwork (Stanton et al., 2013). More recently methods such as Accimap (Svedung & Rasmussen, 2002), EAST (Stanton et al., 2018), the MacroErgonomic Analysis and Design method (MEAD; Kleiner, 2006), the Functional Resonance Analysis Method (FRAM; Hollnagel, 2012), and Cognitive Work Analysis (CWA; Vicente, 1999) have been developed to support the design and analysis of overall systems. As well as informing the design of devices, tools, and environments, these methods allow the HFE practitioner to contribute to the design of policies, procedures, training, and education programmes, to risk and safety management via activities such as risk assessment, incident reporting, and accident analysis, and to the development of national and international regulatory frameworks.

Ideally, HFE methods are applied throughout the system design lifecycle (Stanton et al., 2013). This includes contributing to the development, evaluation, and iteration of design concepts, supporting implementation via the development of procedures and training programmes, evaluating performance once new designs or interventions are introduced, and analysing adverse events such as accidents and injury-causing incidents. As stated above, the goal is performance optimisation in support of human health and well-being; however, there is also an obvious economic imperative for HFE input given the concomitant increases in usability, productivity, and efficiency.

There are various ways in which HFE methods can be applied. Many of the methods can be applied naturalistically to real world scenarios, during simulations of scenarios, predictively to model performance, or retrospectively during case study analyses. HFE methods can also be applied in isolation, or together as part of a toolkit, or in the form of a 'many model' approach (Kirwan, 1992; Salmon & Read, 2019; Stanton et al., 2013). There is long history of using 'toolkits' comprising of different HFE methods to investigate complex issues (e.g. Kirwan, 1998; Stanton et al, 2013). The aim of using HFE method toolkits is to enhance comprehensiveness by using different methods to assess different features of performance that cannot be covered by one approach independently. For example, the Workload Error Situation awareness Time and Teamwork (WESTT) toolkit method (Houghton et al., 2008) incorporates measures of workload, errors, SA, time, and teamwork. Early instantiations of the EAST method used a combination of methods including task analysis, social network analysis, information networks, coordination assessment, and communications technology assessment, and operator event sequence diagrams to examine tasks, communications, and interactions between agents, distributed situation awareness, and the influence of technology on teamwork. More recently Salmon & Read (2019) outlined a many model systems ergonomics approach where multiple systems analysis methods are used together to provide an in-depth analysis of complex systems.

Fourteen categories of HF methods are presented in Table 1.1, including an overview of each type of method, a representative state-of-the-art HFE method in each category, and a summary of their previous applications (if applicable) in sport from the peer-reviewed literature.

As indicated in Table 1.1, a large body of HFE work has been undertaken in the area of sport; however, there remain HFE methods that are yet to be applied in sport. Broadly speaking, this body of work has focussed on either optimising performance or on injury prevention and management. Within these two areas, however, there are a diverse set of applications covering a wide range of purposes and sports. Since the early pioneering work of Tom Reilly on aspects such as the ergonomics of sports equipment design (Reilly, 1984; Reilly & Lees, 1984), HFE work has expanded to incorporate sports teams (Neville et al., 2017), sports organisations (Hulme et al., 2019), and overall sports systems (Hulme et al., 2017). This has seen HFE research extend from physical HFE (e.g. anthropometrics, physiology, injury, disabilities, sports equipment design) to cognitive HFE (e.g. situation awareness, decision making, error) and now to systems HFE (e.g. injury causation and prevention, organisational modelling) (Salmon & Macquet, 2019). An overview of selected applications in each area is given in the following sections.

1.4.1 Physical Human Factors and Ergonomics Applications

Physical HFE is used to investigate how the individual's anatomy, physiology, anthropometry, and biomechanics functionally interact with the design of physical equipment and structures (Hulme et al., 2019). From a sports perspective, optimising human–artefact, and human–machine interactions can help to facilitate desired performances and attenuate maladaptive physical processes during play (Lake, 2000; Born et al., 2013; Balasubramanian et al., 2014; Hsiao et al., 2015). Physical HFE applications in sport have focussed on issues ranging from sports equipment and clothing design (Lake, 2000; McGhee et al., 2013; Reilly & Lees, 1984; Rochat et al., 2019; Varadarajura & Srinivasan, 2019) to sports injury (e.g. Theberge, 2012), and biomechanics (e.g. Lees et al., 2000).

TABLE 1.1
HFE Methods and Their Use in Sport

Type of method	Description and example method	Example applications in sport (peer-reviewed literature only)
Task analysis	Used to describe tasks, processes, and systems in terms of the goals, sub-goals, tasks, and operations required. Hierarchical Task Analysis (HTA; Annett et al., 1971)	N/A
Cognitive task analysis	Used to elicit and describe the cognitive processes involved during task performance Critical Decision Method (Klein et al., 1989)	Decision making in elite badminton players (Macquet & Fleurance, 2007)
Process charting	Used to diagrammatically represent task performance Operator Sequence Diagrams (OSDs; Stanton et al., 2013)	N/A
Human error identification	Used to identify different human errors that are likely to be made during task performance Systematic Human Error Reduction and Prediction Approach (Embrey, 1986)	Error identification in elite athlete motor control (Hossner et al., 2015)
Situation awareness assessment	Used to assess situation awareness during tasks performance Situation Awareness Global Assessment Technique (SAGAT, Endsley, 1995)	Situation awareness in Australian Rules Football umpire teams (Neville et al., 2018) Situation awareness in hammer throwing and rowing athletes and coaches (Macquet & Stanton, 2014)
Mental workload assessment	Used to assess cognitive task load during task performance NASA Task Load Index (NASA-TLX, Hart, & Staveland, 1988)	Subjective task load in rugby players (Mullen et al., 2019) Mental fatigue in table tennis players (Le Mansec et al., 2018) Perceived exertion during cycling (Vera et al., 2018)
Teamwork assessment	Used to assess key features of teamwork during team task performance such as communication, coordination, team situation awareness, and leadership. Event Analysis of Systemic Teamwork (EAST; Stanton et al., 2013)	Teamwork assessment in elite women's cycling (Salmon et al., 2017)
Interface analysis	Used to evaluate and redesign displays and interfaces. Human Computer Interaction Checklist (Ravden & Johnson, 1989)	N/A

(Continued)

TABLE 1.1 (CONTINUED)
HFE Methods and Their Use in Sport

Type of method	Description and example method	Example applications in sport (peer-reviewed literature only)
Usability evaluation	Used to assess the usability of devices, interfaces, tools and processes. System Usability Scale (Brooke, 1986)	N/A
Performance time prediction	Used to assess or forecast the amount of time required for component tasks. Critical Path Analysis	N/A
Systems analysis	Used to describe and analyse the composition and behaviour of organisations or entire socio-technical systems. Cognitive Work Analysis (CWA; Vicente, 1999)	Performance analysis in elite women's netball (McLean et al., 2019) Analysis of elite women's netball organisation (Hulme et al., 2019) Analysis of football match system (McLean et al., 2017)
System design	Used to support the design or redesign of organisations, systems, or other aspects such as jobs, procedures, training programmes, and regulatory systems. Cognitive Work Analysis Design Toolkit (CWA-DT; Read et al., 2018)	N/A
Risk assessment	Used to assess the risks associated with tasks and identify appropriate risk management strategies. NETworked Hazard Analysis and Risk Management System (NET-HARMS; Dallat et al., 2017)	N/A
Accident analysis	Used to describe and analyse the network of contributory factors involved in accidents and adverse events. AcciMap (Svedung & Rasmussen, 2002)	Concussion in junior and community rugby (Clacy et al., 2017; Holmes et al., 2019)
Computational modelling	Used to simulate the behaviour of agents, organisations, or systems. Agent-Based Modelling (ABM; Bonabeau, 2002)	Running injury prevention (Hulme et al., 2018, 2019)

An example of a physical HFE study is one undertaken by Varadarajura and Srinivasan (2019) to compare physiological thermal comfort when running and during post-running rest periods whilst wearing three different 'body mapping'-based shirts versus a conventional running shirt design. Participants performed a running activity on a treadmill, and skin temperature, heart rate, and skin micro climate was recorded as well as participants' subjective ratings of skin temperature, skin moisture, and overall comfort. At the end of each trial, participants rested for 10 minutes and the same measurements were repeated. Each shirt was subjected to 20 trials. The findings identified key differences between the shirts based on objective and subjective ratings of thermal comfort, with Varadarajura and Srinivasan (2019) concluding that more thermal physiological benefit is achieved during running when wearing body mapping-based shirts versus conventional shirts (Salmon & Macquet, 2019).

1.4.2 Cognitive HFE Applications

Cognitive HFE applications are used to understand the cognitive processes undertaken by athletes and teams during sports performance. In sport, understanding cognitive processes such as perception, decision making, SA, and error can help to support the design of equipment, training, coaching, and tactics to optimise athlete and team cognition.

Cognitive HFE applications have covered a range of areas in different sporting contexts, including decision making (McNeese et al., 2015; Macquet & Fleurance, 2007), SA (Macquet & Stanton, 2014; Macquet et al., 2015; Neville & Salmon, 2016), error (Sanli et al., 2017), teamwork and team cognition (McLean et al., 2018), and sensemaking (e.g. Macquet & Kragba, 2015).

An example cognitive HFE study is that undertaken by Salmon et al. (2017) who used EAST (Stanton et al., 2018) and the Critical Decision Method (CDM; Klein et al., 1989) to examine situation awareness and teamwork in an elite women's cycling team during two elite women's road races. Participants were members of an elite women's Australian NRS cycling team and included five riders, one Director Sportif (DS) and one mechanic. Salmon and colleagues observed the race from within the cycling team's support vehicle and used cameras to record the races and Dictaphones to record team race planning meetings, post-race CDM interviews, and post-race team debriefs. EAST task, social, and information networks were subsequently constructed for each race.

Salmon et al. (2017) reported numerous examples of the Big Five teamwork behaviours (leadership, mutual performance monitoring, back up behaviour, adaptability, team orientation), and three key supporting mechanisms (shared mental models, mutual trust, communication). They concluded that SA is distributed across the cycling team and peloton, and that non-human agents play a critical role in SA development. For example, the EAST analysis of the information used by team members during the race showed the importance of non-human agents, such as the bike-mounted computer and annotated handlebars. SA decrements were identified whereby awareness and understanding of the race plan was not optimal across all team members. The important role of 'SA transactions' in team performance was also demonstrated, including both verbal (e.g. talking to a teammate) and non-verbal (monitoring another riders' behaviour to assess their energy level) exchanges. Based on the study findings it was proposed that performance improvements could be made through improved pre-race planning, including contingency planning, as well as education on the verbal and non-verbal transactions required during different race phases.

1.4.3 Organisational and Systems HFE Applications

As sports systems become increasingly complex, competitive, and technology-centric, there is a greater need for systems HFE applications to consider the performance of athletes, teams, and sports organisations in the context of the wider sport system in which they operate (Hulme et al., 2019). Systems HFE is used to understand and optimise the functioning of sports systems, including their organisational structures, policies, and processes (Hulme et al., 2019). Accordingly, systems HFE applications use sport systems as the primary unit of analysis.

Whilst there has been an increase in systems HFE research in other areas, sports systems HFE applications have only begun to emerge in the past decade. Hulme et al. (2019) conducted a systematic review of systems HFE applications in sport, identifying seven studies undertaken in the areas of women's cycling (Salmon et al., 2017), distance running (Hulme et al., 2017, 2018), football (McLean et al., 2017), Australian Rules Football (AFL) (Dawson et al., 2017), and Rugby Union (Clacy et al., 2017, 2019). The studies were undertaken to support injury prevention and management, and performance analysis and optimisation. The review also discovered that a diverse range of systems HFE models and methods have been applied in sport, including the Systems Theoretic Accident Model and Processes (STAMP; Leveson, 2004), EAST (Stanton et al., 2018), CWA (Vicente, 1999), and Rasmussen's RMF (Rasmussen, 1997).

Further applications of systems HFE methods have been published since Hulme et al.'s (2019) review. For example, McLean et al. (2019) used Work Domain Analysis (WDA; Naikar, 2012), the first phase of CWA, to analyse the structure of elite women's netball matches to identify new performance analysis measures. As described earlier, Hulme et al. (2019) also used WDA to develop a model of an elite women's netball organisation with a view to identify the factors influencing the netball team's performance.

1.5   DISCUSSION

This chapter aimed to introduce HFE and provide an overview of previous sports HFE applications. To summarise:

- There is a long history of HFE applications in sport, with the pioneering work of Tom Reilly providing the impetus for more than 40 years of HFE work areas such as sports product design and injury prevention.
- There are a range of HFE theories that can be used in sports applications, including models of individual (e.g. Endsley, 1995a; Klein, 1998), team (e.g. Salas et al., 1995, 2005), and organisational and system behaviour (e.g. Leveson, 2004; Rasmussen et al., 1997).
- There are 14 categories of HFE methods that can be used to support the design and analysis of sports systems, including task and cognitive task analysis methods, human error identification and risk assessment methods, situation awareness and mental workload assessment methods, interface

design and evaluation methods and teamwork assessment, and systems analysis and design methods.
- Previous and current HFE applications in sport can be categorised into physical HFE (e.g. anthropometrics, physiology, injury, disabilities, sports equipment design), cognitive HFE (e.g. situation awareness, decision making, error), and systems HFE applications (e.g. injury causation and prevention, organisational modelling).

REFERENCES

Annett, J. (1994). The learning of motor skills: Sports science and ergonomics perspectives. *Ergonomics*, 37:1, 5–16.

Annett, J., Duncan, K. D., Stammers, R. B., & Gray, M. J. (1971). Task analysis. Department of Employment Training Information Paper No. 6. London, UK: Her Majesty's Stationary Office (HMSO).

Balasubramanian, V., Jagannath, M., & Adalarasu, K. (2014). Muscle fatigue based evaluation of bicycle design. *Applied Ergonomics*, 45:2, 339–345.

Bittencourt, N. F., Meeuwisse, W. H., Mendonça, L. D., Nettel-Aguiree, A., Ocarino, J. M., & Fonseca, S. T. (2016). Complex systems approach for sports injuries: Moving from risk factor identification to injury pattern recognition-narrative review and new concept. *British Journal of Sports Medicine*, 1309–1314.

Bonabeau, E. (2002). Agent-based modeling: Methods and techniques for simulating human systems. *Proceedings of the National Academy of Sciences of the United States of America*, 99:3 Supplement 3, 7280–7287.

Born, D. P., Sperlich, B., & Holmberg, H. C. (2013). Bringing light into the dark: Effects of compression clothing on performance and recovery. *The International Journal of Sports Physiology and Performance*, 8:1, 4–18.

Brooke, J. (1986). SUS: A "quick and dirty" usability scale. In: P. W. Jordan, B. Thomas, B. A. Weerdmeester, & A. L. McClelland (eds.), *Usability Evaluation in Industry*. London, UK: Taylor and Francis.

Brukner, P., & Khan, K. (2017). *Brukner & Khan's Clinical Sports Medicine: Injuries*. 5th edition, Volume 1. Australia: McGraw-Hill Education.

Cable, T., Sanderson, F., & Nevill, A. (2009). A tribute to Professor Thomas Reilly (1941–2009). *Journal of Sports Sciences*, 27:11, 1107–1108.

Carron, A., Bray, S., & Eys, M. (2002). Team cohesion and team success in sport. *Journal of Sports Sciences*, 20, 119–126.

Cassano-Piche, A. L., Vicente, K. J., & Jamieson, G. A. (2009). A test of Rasmussen's risk management framework in the food safety domain: BSE in the UK. *Theoretical Issues in Ergonomics Science*, 10:4, 283–304. doi: 10.1080/14639220802059232.

Chow, G. M., & Feltz, D. L. (2008). Exploring the relationship between collective efficacy, perceptions of success, and team attributions. *Journal of Sports Sciences*, 26:11, 1179–1189.

Clacy, A., Goode, N., Sharman, R., Lovell, G. P., & Salmon, P. M. (2017). A knock to the system: A new sociotechnical systems approach to sport-related concussion. *Journal of Sports Sciences*, 35:22, 2232–2239. DOI: 10.1080/02640414.2016.1265140

Clacy, A., Goode, N., Sharman, R., Lovell, G. P., & Salmon, P. M. (2019). A systems thinking approach to understanding the identification and treatment of sport-related concussion in community rugby union. *Applied Ergonomics*, 80, 256–264.

Clegg, C. W. (2000). Sociotechnical principles for system design. *Applied Ergonomics*, 31, 463–477.

Davis, M. C., Challenger, R., Jayewardene, D. N. W., & Clegg, C. W. (2014). Advancing socio-technical systems thinking: A call for bravery. *Applied Ergonomics*, 45, 133–220.

Dawson, K., Salmon, P. M., Read, G. J. M., Neville, T., Goode, N., & Clacy, A. (2017). Removing concussed players from the field: The factors influencing decision making around concussion identification and management in Australian Rules Football. Proceedings of the NDM13, University of Bath, UK.

Dekker, S. W. A. (2011). *Drift into Failure*. Surrey, UK: Ashgate Publishing Limited.

DiFiori, J. P., Benjamin, H. J., Brenner, J. S., Gregory, A., Jayanthi N., Landry, G. L., & Luke A. (2014). Overuse injuries and burnout in youth sports: A position statement from the American Medical Society for Sports Medicine. *British Journal of Sports Medicine*, 48, 287–288.

Embrey, D. E. (1986). SHERPA: A systematic human error reduction and prediction approach. Proceedings of the International Topical Meeting on Advances in Human Factors in Nuclear Power Systems, Knoxville, TN; American Nuclear Society, La Grange Park, IL.

Endsley, M. R. (1988). Design and evaluation for situation awareness enhancement. Proceedings of the Human Factors Society 32nd Annual Meeting (pp. 97–101), Human Factors and Ergonomics Society, Santa Monica, CA.

Endsley, M. R. (1993). A survey of situation awareness requirements in air-to-air combat fighters. *International Journal of Aviation Psychology*, 3:2, 157–168.

Endsley, M. R. (1995a). Towards a theory of situation awareness in dynamic systems. *Human Factors*, 37, 32–64.

Endsley, M. R. (1995b). Measurement of situation awareness in dynamic systems. *Human Factors*, 37, 65–84.

Fiore, S. M., & Salas, E. (2006). Team cognition and expert teams: Developing insights from cross-disciplinary analysis of exceptional teams. *International Journal of Sports and Exercise Psychology*, 4, 369–375.

Filho, E., Tenenbaum, G., Yang, Y. (2014). Cohesion, team mental models, and collective efficacy: towards an integrated framework of team dynamics in sport. *Journal of Sports Sciences*, doi:10.1080/02640414.2014.957714

Gabbe B. J., Finch C. F., Cameron P. A., & Williamson O. D. (2005). Incidence of serious injury and death during sport and recreation activities in Victoria, Australia. *British Journal of Sports Medicine*, 39, 573–577.

Goodwin, G. F. (2008). Psychology in sports and the military: Building understanding and collaboration across disciplines. *Military Psychology*, 20:1, S147–S153.

Gorman, J. C., Cooke, N. J., & Winner, J. L. (2006). Measuring team situation awareness in decentralized command and control environments. *Ergonomics*, 49:12–13, 1312–1325.

Hancock, P. A., & Meshkati. N. (1988). *Human Mental Workload*. North Holland: Elsevier Science Publishers.

Hart, S. G., & Staveland, L. E. (1988). Development of NASA-TLX (Task Load Index): Results of empirical and theoretical research. In: P. A. Hancock & N. Meshkati (eds.), *Advances in Psychology, 52. Human Mental Workload* (pp. 139–183). North-Holland: Elsevier Science Publishers.

Hollnagel, E. (2004). *Barriers and Accident Prevention*. Aldershot, UK: Ashgate.

Hollnagel, E. (2012). *FRAM: The Functional Resonance Analysis Method: Modelling Complex Socio-Technical Systems*. Surrey, UK: Ashgate Publishing Limited.

Holmes, G., Clacy, A., Salmon, P. M. (2019). Sports-related concussion management as a control problem: Using STAMP to examine concussion management in community rugby. *Ergonomics*, 10.1080/00140139.2019.1654134

Houghton, R., Baber, C., Cowton, M., Walker, G., & Stanton, N. A. (2008). WESTT (workload, error, situation awareness, time and teamwork): An analytical prototyping system for command and control. *Cognition, Technology and Work*, 10, 199–207.

Hsiao, S. W., Chen, R. Q., & Leng, W. L. (2015). Applying riding-posture optimization on bicycle frame design. *Applied Ergonomics*, 51, 69–79.

Hughes, L., & Leavey, G. (2012). Setting the bar: Athletes and vulnerability to mental illness. *The British Journal of Psychiatry*, 200, 95–96.

Hulme, A., McLean, S., Read, G. J. M., Dallat, C., Bedford, A., & Salmon, P. M. (2019). Sports organisations as complex systems: Using cognitive work analysis to identify the factors influencing performance in an elite netball organisation. *Frontiers in Sports and Active Living*, 1, 56.

Hulme, A., Salmon, P. M., Nielsen, R. O., Read, G. J. M., & Finch, C. F. (2017). Closing Pandora's Box: Adapting a systems ergonomics methodology for better understanding the ecological complexity underpinning the development and prevention of running-related injury. *Theoretical Issues in Ergonomics Science*, 18:4, 338–359.

Hulme, A., Thompson, J., Nielsen, R. O., Read, G. J. M., & Salmon, P. M. (2018). Formalising the complex systems approach: Using agent-based modelling to simulate sports injury aetiology and prevention. *British Journal of Sports Medicine*, 53, 560–569.

Hossner, E.-J., Schiebl, F., & Ulrich Göhner. (2015). A functional approach to movement analysis and error identification in sports and physical education. *Frontiers in Psychology*, (6): 1339.

International Ergonomics Association. (2019). *What Is Ergonomics?* https://www.iea.cc/whats/index.html, accessed on 5 December 2019.

Jenkins, D. P., Salmon, P. M., Stanton, N. A., & Walker, G. H. (2010). A systemic approach to accident analysis: A case study of the Stockwell shooting. *Ergonomics*, 3:1, 1–17.

Jones, N. M. P., James, N., & Mellalieu, S. D. (2008). An objective method for depicting team performance in elite professional rugby union. *Journal of Sports Sciences*, 26:7, 691–700. DOI: 10.1080/02640410701815170

Joy, E., Kussman, A., & Nattiv, A. (2016). 2016 update on eating disorders in athletes: A comprehensive narrative review with a focus on clinical assessment and management. *British Journal of Sports Medicine*, 50, 154–162.

Khan, K. M., Thompson, A. M., Blair, S. N., Sallis, J. F., Powell, K. E., Bull, F. C., & Bauman, A. E. (2012). Sport and exercise as contributors to the health of nations. *Lancet*, 380:9836, 59–64.

Kirwan, B. (1992). Human error identification in human reliability assessment. Part 2: Detailed comparison of techniques. *Applied Ergonomics*, 23, 371–381.

Kirwan, B. (1998). Human error identification techniques for risk assessment of high risk systems – Part 1: Review and evaluation of techniques. *Applied Ergonomics*, 29:3, 157–177.

Klein, G. A. (1993). A recognition-primed decision (RPD) model of rapid decision making. In: G. A. Klein, J. Orasanu, R. Calderwood, & C. E. Zsambok (eds.), *Decision Making in Action: Models and Methods* (pp. 138–147). New York, NY: Ablex Publishing.

Klein, G. A., Calderwood, R., & Macgregor, D. (1989). Critical decision method for eliciting knowledge. *IEEE Transactions on Systems, Man, and Cybernetics*, 19:3, 462–472.

Kleiner, B. M. (2006). Macroergonomics: Analysis and design of work systems. *Applied Ergonomics*, 37:1, 81–89.

Kumar, S. (2007). *Biomechanics in Ergonomics*. 2nd edition. Boca Raton, FL: CRC Press.

Lake, M.J. (2000). Determining the protective function of sports footwear. *Ergonomics* 43 (10): 1610–1621.

Lago, U., Simmons, R., & Szymanski, S. (2006). The financial crisis in European football. *Journal of Sports Economics*, 7:1, 3–12.

Lees, A., Rojas, J., Cepero, M., Soto, V., & Gutierrez, M. (2000). How the free limbs are used by elite high jumpers in generating vertical velocity. *Ergonomics*, 43:10, 1622–1636.

Le Mansec, Y., Pageaux, B., Nordez, A., Dorel, S., & Jubeau, M. (2018). Mental fatigue alters the speed and the accuracy of the ball in table tennis. *Journal of Sports Sciences*, 36:23, 2751–2759. DOI: 10.1080/02640414.2017.1418647

Leveson, N. G. (2004). A new accident model for engineering safer systems. *Safety Science* 42:4, 237–270.

Macquet, A. C., & Fleurance, P. (2007). Naturalistic decision-making in expert badminton players. *Ergonomics*, 50:9, 1433–1450. DOI: 10.1080/00140130701393452

Macquet, A. C., & Kragba, K. (2015). What makes basketball players continue with the planned play or change it? A case study of the relationships between sense-making and decision-making. *Cognition, Technology and Work*, 17:3, 345–353.

Macquet, A. C., & Stanton, N. A. (2014). Do the coach and athlete have the same 'picture' of the situation? Distributed situation awareness in an Elite sport context. *Applied Ergonomics*, 45:3, 724–733.

McGhee, D. E., Steele, J. R., Zealey, W. J., & Takacs, G. J. (2013). Bra–breast forces generated in women with large breasts while standing and during treadmill running: Implications for sports bra design. *Applied Ergonomics*, 44:1, 112–118.

McLean, S., Soloman, C., Gorman, A., & Salmon, P. M. (2017). What's in a game? A systems approach to enhancing performance analysis in football. *PLOS One*, 1–15.

Mclean, S. Salmon, P. M. Gorman, A.D., Wickham, J., Berber, E., Solomon, C. (2018). The effect of playing formation on the passing network characteristics of a professional football team. *Human Movement*, (5): 14–22.

McLean, S., Solomon, C., Gorman, A., & Salmon, P. M. (2019). Integrating communication and passing networks in football using social network analysis. *Science and Medicine in Football*, 3:1, 29–35.

McLean, S., Salmon, P. M., Hulme, A., Read, G. J. M., & Bedford, A. (2019). A systems approach to performance analysis in women's netball: Using work domain analysis to model elite netball performance. *Frontiers in Psychology*, 10, 201.

McNeese, N. J., Cooke, N., Fedele, M. A., & Gray, R. (2015). Theoretical and methodical approaches to studying team cognition in sports. Proceedings of the 6th International Conference on Applied Human Factors and Ergonomics.

Meister, D. (1989). *Conceptual Aspects of Human Factors*. Baltimore, MD: Johns Hopkins University Press.

Mullen, T., Twist, C., & Highton, J. (2019). Stochastic ordering of simulated rugby match activity produces reliable movements and associated measures of subjective task load, cognitive and neuromuscular function. *Journal of Sports Sciences*, 37:21, 2506–2512. DOI: 10.1080/02640414.2019.1646071

Murray, S., James, N., Pers, J., Mandeljc, R., & Vuckovic, G. (2017). Using a situation awareness approach to determine decision making in squash. *Journal of Sports Sciences*, 36:2, 1415–1422.

Murrell, K. F. H. (1965). *Ergonomics: Man in His Working Environment*. London, UK: Chapman Hall.

Naikar, N. (2013). *Work Domain Analysis: Concepts, Guidelines, and Cases*. Boca Raton, FL: CRC Press.

Neisser, U. (1976). *Cognition and Reality: Principles and Implications of Cognitive Psychology*. San Francisco, CA: Freeman.

Neville, T. (2017). Testing and measuring teamwork and distributed situation awareness in Australian rules football field umpiring teams. PhD thesis, University of the Sunshine Coast, Australia.

Neville, T., Salmon, P. M. (2016). Never blame the umpire? A review of situation awareness models and methods for examining the performance of officials in sport. *Ergonomics*, 59:7, 962–975

Oja P., Titze S., Kokko S., Kujala, U. M., Heinonen, A., Kelly, P., Koski, P., & Foster, C. (2015). Health benefits of different sport disciplines for adults: Systematic review of observational and intervention studies with meta-analysis. *The British Journal of Sports Medicine*, 49, 434–440.

Ottino, J. (2003). Complex systems. *AIChE Journal*, 49, 292–299.

Pedersen, H. K., & Cooke, N. J. (2006). From battle plans to football plays: Extending military team cognition to football. International Journal of Sport and Exercise Psychology, 4, 422–446.

Perrow, C. (1984). *Normal Accidents: Living with High-Risk Technologies.* New York: Basic Books.

Pheasant, S., & Haslegrave, C. M. (2005). *Bodyspace: Anthropometry, Ergonomics and the Design of Work.* 3rd edition. Boca Raton, FL: CRC Press.

Rasmussen, J. (1997). Risk management in a dynamic society: A modelling problem. *Safety Science*, 27:2/3, 183–213. doi: 10.1016/S0925-7535(97)00052-0.

Ravden, S. J., & Johnson, G. I. (1989). *Evaluating Usability of Human-Computer Interfaces: A Practical Method.* Chichester, UK: Ellis Horwood Ltd. Proceedings of '90 Conference on Human Factors in Computing Systems.

Read, G. J. M., Salmon, P. M., Lenne, M. G., & Goode, N. A. (2018). A sociotechnical design toolkit for bridging the gap between systems-based analyses and system design. *Human Factors and Ergonomics in Manufacturing and Service Industries*, 28:6, 327–341.

Reardon, C. L., & Creado, S. (2014). Drug abuse in athletes. *Substance Abuse and Rehabilitation*, 5, 95–105.

Reason, J. (1990). *Human Error.* Cambridge, UK: Cambridge University Press.

Regan, M. A., Lee, J. D., & Young, K. L. (2008). *Driver Distraction: Theory, Effects and Mitigation.* Boca Raton, FL: CRC Press.

Reilly, T., & Lees, A. (1984). Exercise and sports equipment: Some ergonomics aspects. *Applied Ergonomics*, 15:4, 259–279.

Reilly, T., & Ussher, M. (1988). Sport, leisure and ergonomics. *Ergonomics*, 31:11, 1497–1500.

Reimer, T., Park, E. S., & Hinsz, V. B. (2006). Shared and coordinated cognition in competitive and dynamic task environments: An information-processing perspective for team sports. *International Journal of Sport and Exercise Psychology*, 4:4, 376–400. DOI: 10.1080/1612197X.2006.9671804

Rochat, N., Hauw, D., & Seifert, L. (2019). Enactments and the design of trail running equipment: An example of carrying systems. *Applied Ergonomics*, 80, 238–247.

Salas, E., Cooke, N. J., & Rosen, M. A. (2008). On teams, teamwork, and team performance: Discoveries and developments. *Human Factors*, 50, 540–547.

Salas, E., & Fiore, S. (2004). *Team Cognition: Understanding the Factors that Drive Process and Performance.* American Psychological Association.

Salas, E., Prince, C., Baker, D. P., & Shrestha, L. (1995). Situation awareness in team performance: Implications for measurement and training. *Human Factors*, 37, 1123–1136.

Salas, E., Sims, D. E., & Burke, C. S. (2005). Is there a big five in teamwork? *Small Group Research*, 36:5, 555–599.

Salmon, P. M., Dallat, C., & Clacy, A. (2017). It's not all about the bike: Distributed situation awareness and teamwork in elite women's cycling teams. Proceedings from Contemporary Ergonomics and Human Factors (pp. 240–248), Staverton Estate, Daventry, Northamptonshire, UK.

Salmon, P. M., Hulme, A., Walker, G. H., Berber, E., Waterson, P., & Stanton, N. A. (2020). The big picture on accident causation: A review, synthesis and meta-analysis of AcciMap studies. *Safety Science*, 126.

Salmon, P. M., & Macquet, A.-C. (2017). *Advances in Human Factors in Sports and Outdoor Recreation.* Springer.

Salmon, P. M., & Macquet, A. C. (2019). Ergonomics in sport and outdoor recreation: From individuals and their equipment to complex systems and their frailties. *Applied Ergonomics*, 80, 209–213.

Salmon, P. M., & Read, G. J. M. (2019). Many-model thinking in systems ergonomics: A case study in road safety. *Ergonomics*, 62:5, 612–628.

Salmon, P. M., Walker, G. H., Read, G. J. M., Goode, N. & Stanton, N. A. (2017). Fitting methods to paradigms: Are ergonomics methods fit for systems thinking? *Ergonomics*, 60:2, 194–205.

Salmon, P. M., Williamson, A., Lenne, M. G., Mitsopoulos, E., & Rudin-Brown, C. M. (2010). Systems-based accident analysis in the led outdoor activity domain: Application and evaluation of a risk management framework. *Ergonomics*, 53:8, 927–939.

Sanders, M. S., & McCormick, E. J. (1993). *Human Factors in Engineering and Design*. 7th edition. McGraw-Hill Book Company.

Sanli, E. A., Slauenwhite, J., & Carnahan, H. (2017). The relationship between error production when performing motor skills in high and low-stakes situations. *Theoretical Issues in Ergonomics Science*, 18:4, 360–369. DOI: 10.1080/1463922X.2016.1269845

Shorrock, S. T., & Kirwan, B. (2002). Development and application of a human error identification tool for air traffic control. *Applied Ergonomics*, 33:4, 319–336. DOI: 10.1016/S0003-6870(02)00010-8

Stanton, N. A., Rafferty, L., Salmon, P. M., Revell, K. M. A., McMaster, R., Caird-Daley, A., & Cooper Chapman, C. (2010). Distributed decision making in multi-helicopter teams: Case study of mission planning and execution from a non-combatant evacuation operation training scenario. *Journal of Cognitive Engineering and Decision Making*, 4:4, 328–353.

Stanton, N. A., Salmon, N. A., & Walker, G. H. (2018). *Systems Thinking in Practice: The Event Analysis of Systemic Teamwork*. Boca Raton, FL: CRC Press.

Stanton, N. A., Salmon, P. M., Rafferty, L., Walker, G. H., Jenkins, D. P., & Baber, C. (2013). *Human Factors Methods: A Practical Guide for Engineering and Design*. 2nd edition. Aldershot, UK: Ashgate.

Svedung, I., & Rasmussen, J. (2002). Graphic representation of accident scenarios: Mapping system structure and the causation of accidents. *Safety Science*, 40:5, 397–417.

Theberge, N. (2012). Studying gender and injuries: A comparative analysis of the literatures on women's injuries in sport and work. *Ergonomics*, 55:2, 183–193. DOI: 10.1080/00140139.2011.592602

Turk, E. E., Riedel, A., & Püeschel, K. (2008). Natural and traumatic sports-related fatalities: A 10-year retrospective study. *British Journal of Sports Medicine*, 42, 604–608.

Varadaraju, R., & Srinivasan, J. (2019). Design of sports clothing for hot environments. *Applied Ergonomics*, 80, 248–255.

Vera, J., Perales, J. C., Rodriguez, R. J., Velez, D. C. (2018). A test-retest assessment of the effects of mental load on ratings of affect, arousal and perceived exertion during submaximal cycling. *Journal of Sports Sciences*, 36:22.

Vicente, K. J. (1999). *Cognitive Work Analysis: Toward Safe, Productive, and Healthy Computer-Based Work*. Mahwah, NJ: Lawrence Erlbaum Associates.

Wankel, L. M., & Berger, B. G. (1990). The psychological and social benefits of sport and physical activity. *Journal of Leisure Research*, 22:2, 167–182. DOI: 10.1080/00222216.1990.11969823

Wickens, C. D. (1984). Processing resources in attention. In: R. Parasuraman & R. Davies (eds.), *Varieties of Attention* (pp. 63–101). New York: Academic Press.

2 Sport as a Complex Socio-Technical System

Adam Hulme, Scott McLean, and Paul M. Salmon

CONTENTS

2.1 INTRODUCTION

To better understand the core concepts and features of complexity described herein, it is useful to provide a definition of what a 'complex system' is. According to Luke and Stamatakis (2012), complex systems exhibit the following properties:

> They are made up of a large number of heterogenous elements; these elements interact with each other; the interactions produce an emergent effect that is different from the effects of the individual elements alone; and, this effect persists over time and adapts to changing circumstances.
>
> **(p. 2)**

The reader is encouraged to consider these key system properties throughout the remainder of this chapter. Its deeper meaning and implication for the study and analysis of 'complex sports systems' will become apparent by way of example and illustration.

2.1.1 A BACKGROUND TO COMPLEXITY SCIENCE

Complexity science is the discipline that attempts to understand and respond to problems that are dynamic and unpredictable, multi-dimensional, and comprise various interrelated components (Salmon & McLean, 2019). Researchers and practitioners who study complexity – and by definition complex problems – focus on the *interactions* among fundamentally different elements within a 'complex system', rather than on the role and contribution of those elements in *isolation* (Ottino, 2003; Batty & Torrens, 2005; Senge, 2006). Indeed, the ability to offer an accurate and comprehensive causal description for the occurrence of complex issues at the biological, individual, and/or societal level is largely dependent on whether a broad or narrow view of a given problem is adopted (Senge, 2006; Sterman, 2006). Accordingly, complexity science proponents will advocate for a *holistic*, or 'systems thinking' perspective over a reductionist one, as doing so is to consider the whole system (or multiple interacting elements of it) as the primary unit of analysis (Ottino, 2003). This affords insight into how the constituent elements of a complex system converge in context of a much greater whole, which is useful when trying to make sense of resilient, persistent, and oftentimes policy resistant problems (e.g. global climate change, non-communicable disease, food security) (Rutter et al., 2017).

The field of complexity science is extensive, involving multiple traditions, disciplines, methods, techniques, and analytical tools. For instance, Castellani's (2018) 'Map of the complexity sciences' (Figure 2.1) spans several decades and visualises the historical progression of five major intellectual traditions: (i) dynamical systems theory; (ii) systems science; (iii) complex systems theory; (iv) cybernetics; and, (v) artificial intelligence. These traditions share a number of philosophical and theoretical similarities, as well as practical commonalities in terms of the approaches used to study and examine complex phenomena. Figure 2.1 indicates that there is no single, unified understanding of what exactly complexity is from an operational standpoint when it is subjected to formal investigation and analysis. That is, researchers have a multitude of highly capable scientific approaches and modelling techniques at their disposal to understand complexity and complex problems so long as a suitable justification for their selection is offered.

2.1.2 EMERGENCE: A KEY CONCEPT IN COMPLEXITY SCIENCE

The concept of emergence is a widely recognised feature in the study of complexity and complex systems (Rickles, Hawe, & Shiell, 2007; Diez Roux, 2011). Definitions of emergence in the academic literature differ based on the discipline of study, whether it be ecology, economics, systems engineering, or biology (Levin, 1998; Markos, 2005; Sheard & Mostashari, 2009; Feller, Khammissa, & Lemmer, 2017; Cunha, Xavier, & de Castro, 2018). For the purpose of this chapter, emergence is defined as:

> Difficult-to-predict properties, behaviours, or outputs that arise from the dependencies, interactions, and relationships among a range of heterogenous living and/or non-living elements within a complex system.
>
> **(Hulme, McLean, Salmon et al., 2019)**

There are various cases of emergent phenomena in everyday systems and work contexts that appear across micro (e.g. molecular/cellular; physical objects/artefacts),

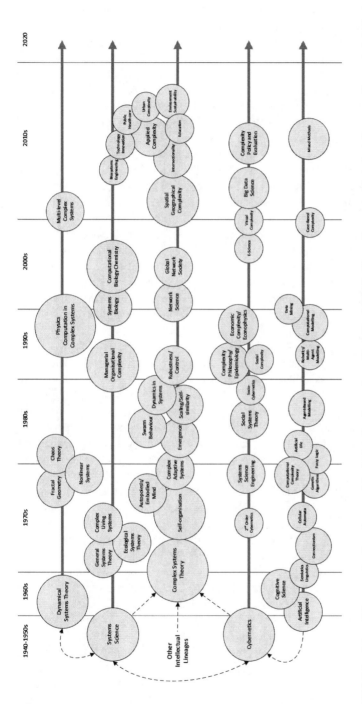

FIGURE 2.1 Map of the complexity sciences. Adapted and redrawn from Castellani (2018). The author notes that, first, there are five main intellectual traditions, including philosophies, methods, analyses, tools and techniques for studying complexity and complex systems. Second, each node along the timeline is roughly positioned at the point it became a major area of study (or was introduced) given that all are either still relevant today, or have seen a resurgence in popularity (e.g. artificial intelligence). The full-colour depiction with associated leading scholars in the corresponding fields can be viewed at https://www.art-sciencefactory.com/complexity-map_feb09.html.

meso (e.g. individuals/group-level; organisational/subsystems), and macroscopic (e.g. societal/political; systems of systems) levels. Examples include the structure and function of the human immune system (Grilo, Caetano, & Rosa, 2002), the evolution and propagation of infectious disease (Marshall et al., 2012), fluctuating prices on the global financial market (Sornette, 2017), the Earth's changing climate and weather system (Rial et al., 2004), the coordination of flight operations in air traffic management (Cook et al., 2015), nuclear power generation (Perrow, 2011), the growth of antimicrobial resistance (Holmes et al., 2016), deforestation and natural mineral resource exhaustion (Walsh et al., 2008), increasing levels of crime and terrorist behaviour (Elliott & Kiel, 2004), the rise in obesity levels (Lee et al., 2017), and even the health, safety, and performance of athletes across all levels of sport (Bittencourt et al., 2016; Mooney et al., 2017). For example, complexity researchers have referred to football (soccer) performance as an emergent phenomenon of a complex 'super system' spanning multiple scales, including the match system (i.e. in this instance micro-level), club system (i.e. meso), and league system (i.e. macro) (Salmon & McLean, 2019). Thus, whilst an analysis of individual player performance would consider isolated elements of football performance such as distance (run) covered, successful passes, tackles, shots, corners, and fouls, an analysis of the football match (sub)system of the broader super system would consider the interactions between components such as players, officials, coaching staff, the crowd, the ball and pitch, and the weather to better understand emergent properties such as goals.

Common to the examples above is the fact that they all involve self-organising processes that continually adapt to, and are affected by, both internal perturbation and outside influence and change (Ladyman, Lambert, & Wiesner, 2013). Self-organisation necessarily implies that the elements within a complex system exhibit memory, awareness, and intelligence, or display behaviours that adhere to the physical laws of nature. For instance, the element hydrogen is an emergent property arising from one proton and one electron particle; a water molecule emerges when two hydrogen and one oxygen atom form a bond; and, larger bodies of water such as rivers and lakes ultimately emerge when many water molecules combine (Galea, Riddle, & Kaplan, 2010). Human consciousness emerges, in part, from the complex interactions that take place among trillions of synapses, billions of neurons, and millions of axons. Consciousness is one of the very things that makes us human, engendering personality, cognition, emotion, and sensory awareness. By extension, the existence of life itself, notwithstanding its mathematical improbability, is perhaps one of the most striking cases of emergence in a complex system of massive ecological proportion (Pearce & Merletti, 2006). The key message is that human consciousness (or life) cannot be explained by the behaviour of individual molecules, neurons, or cells alone; rather, it emerges as a result of many complex interactions that give rise to more sophisticated levels of complexity.

There are two broad forms of emergence in reference to complexity and complex systems. The first is hierarchical, integrative emergence which is generative in the sense that it operates from the 'bottom up' and across multiple levels (Rickles, Hawe, & Shiell, 2007). According to this view, a single subordinate level of organisation within a complex system determines the level above it, and so on. Second, 'top-down causation' describes how properties at the higher levels can influence activities at those found beneath (Ladyman, Lambert, & Wiesner, 2013; Sturmberg, 2018). For example, the aetiology of sports injury has been expressed through a multi-level emergent process involving feedback loops between elements at the molecular, organism, and social levels (Pol et al., 2019). In contrast to a hierarchical, integrative understanding, emergence may also be universal or *multiply realisable* in the sense that there are many diverse ways in which the same (or different) emergent properties can be generated at any level of a complex system (Rickles, Hawe, & Shiell, 2007). As demonstrated in the next section, many different elements and factor configurations within a complex sports system can contribute to the development of athletic injury (Hulme & Finch, 2015). In sum, emergence occurs continually across time and all scales and systems of life. It is therefore a matter of defining the boundary and scope with which it can best be realised and understood in relation to a given problem.

2.2   CHARACTERISTICS OF COMPLEXITY AND COMPLEX SYSTEMS

This next section introduces the main characteristics of complexity by drawing on the aetiology of sports injury as an example. Whilst the example of sports injury has been used throughout, it is possible to apply the same lessons to sports performance, athlete health and illness, as well as extrapolate to the team, club, and wider sports organisational level. This section elaborates in greater detail on the definition of a complex system as provided at the beginning of the chapter (Luke & Stamatakis, 2012). The subsequent characteristics and examples, which have been adapted from Hulme et al. (2019), are logically ordered such that each makes sense in consideration of the one before.

2.2.1   MULTIPLE SYSTEM LEVELS, SCALABLE

Complex systems vary in type, size and scale, from the micro, through to the meso, and macroscopic levels (Simon, 1996; Ottino, 2003; Galea, Riddle, & Kaplan, 2010; Ladyman, Lambert, & Wiesner, 2013; Hulme, McLean, Salmon et al., 2019). For example, the occurrence of sports injury can be conceptualised at the cellular, individual, group, and/or societal level of a complex system. For this reason, attempting to modify an athlete's equipment, biomechanics, or technique in isolation – albeit necessary when the situation calls for it – cannot be relied upon to prevent sports injury. Individual behaviours and deeper underlying personal motivations should be explored, understood, and contextualised within an even greater whole. Sports injury systems are composed of, and are a component of, other usually complex systems (i.e. complex systems are nested, akin to a 'system of systems' approach, e.g. Karwowski, 2012).

2.2.2   DIVERSE RANGE OF AGENTS (E.G. PEOPLE/ORGANISATIONS) AND FACTORS

Complex systems contain many fundamentally different types of agents and factors (i.e. both human and non-human) that interact, both within and across multiple system levels (Cilliers, 1998; Bittencourt et al., 2016; Hulme, McLean, Salmon et al., 2019). For example, the risk of sports injury is influenced by the following:

- The athlete's biological and physiological predisposition (e.g. genetics, physiology, biomechanics, anthropometry) (Nigg, 1985; Malisoux et al., 2017);
- Individual beliefs, norms, training behaviours, and lifestyle practices (e.g. psychology, personality, training workloads, sleep, stress, diet, employment obligations) (Von Rosen et al., 2017; Nielsen et al., 2018; Eckerman et al., 2019) (see Chapter 12 on cognitive load);
- Equipment and technologies (e.g. footwear (see Chapter 5 on running footwear), protective equipment including helmets and guards, performance tracking devices, health monitoring software, 'e-health') (Olivier & Creighton, 2017; Rossi et al., 2018);
- The operating environment (e.g. playing surfaces and terrain, infrastructure, town and city design, geography) (Hardin, Van Den Bogert, & Hamill, 2004; Bertelsen et al., 2017; Nielsen et al., 2018);
- Other athletes and coaches (e.g. sports culture, psychosocial influences, peer influence) (Timpka, Ekstrand, & Svanström, 2006; Hanson et al., 2014);
- Equipment design, and product and manufacturing companies (e.g. footwear production cycles, sports nutrition, advertising) (McIntosh, 2012; Hulme et al., 2017b) (see Chapters 6 and 7 on sports equipment design);
- The medical, healthcare, and pharmaceutical industry (e.g. surveillance/data systems, general practitioners, clinicians, allied health professionals) (Buckler, 1999; Baarveld et al., 2011; Ekegren, Gabbe, & Finch, 2016);
- Educational programmes, training pathways (e.g. medical curriculums, coaching preparation pathways, uptake and adoption of injury prevention interventions) (Donaldson & Finch, 2013; Hulme et al., 2017b);
- Official sponsors and governing bodies (e.g. event promotion, organised competition) (Hulme et al., 2017b);
- Sports safety policies (e.g. club and team rules, regulations) (Finch, 2012; Donaldson, Leggett, & Finch, 2012; Bekker & Finch, 2016); and,
- Wider national economic and taxation systems (e.g. availability and allocation of funding for both recreational and competitive sports programmes) (Green, 2007; Kavetsos & Szymanski, 2009; ASC, 2017).

If the occurrence of sports injury at the individual or population-level is reflective of the spatiotemporal interactions among many heterogeneous living and/or non-living elements, then it is necessary to utilise a complex systems approach that can dynamically capture the combined effect of multi-scale factor interactions.

2.2.3 OPEN BOUNDARIES

Complex systems are 'open systems' with permeable boundaries (Cilliers, 1998). They continually learn and recon Figure 2.2 in response to internal perturbation (i.e. the deviation of a system from its regular or normal state or path) and external influence and change (Ottino, 2003; Batty & Torrens, 2005; Pearce & Merletti, 2006;

Diez Roux, 2011; Ladyman, Lambert, & Wiesner, 2013; Holland, 2014). Athletes and players do not participate in sport within a social or environmental vacuum. Personal routines, lifestyle preferences, professional obligations, and daily stressors can dictate training quality, quantity, and psychological states (Hulme & Finch, 2015; Bolling et al., 2019). The boundary around a complex system is defined according to the purpose and aims of the study, and/or the specific problem that requires solving. From the perspective of sports injury, a complex systems approach might widen boundary to encapsulate the meso (e.g. individual workloads) and macroscopic (e.g. wider socio-cultural and political environments) levels. Indeed, if socio-cultural attitudes shape athletic behaviours at the meso level, then it is fair to say that a focus on macro level determinants is worthy of our time and scientific interest, potentially leading to the greatest differences in the aggregate.

2.2.4 ADAPTIVE AND SELF-ORGANISING

Complex systems continually shift towards and away from acceptable boundaries of safety and performance. Abrupt transitions without adequate adaptation to maintain equilibrium can result in a tipping point, or system failure (Trochim et al., 2006; Cilliers, 1998; Sturmberg, 2018), resulting in injury and/or sub-optimal performance. For example, athletes continuously navigate through a changing set of everyday circumstances to maintain health and well-being, to maximise sports performance, and minimise injury risk. Increasing pressure from peers to become more active or to push current physiological limits, the evolving nature of social norms and expectations, the hype surrounding an innovative footwear or ergogenic product, or the arrival of new community initiative can continually 'pull' the sports injury system in different directions and encourage the migration of system behaviours towards a state of optimal performance or higher injury risk (Hulme, McLean, Salmon et al., 2019). There is no hierarchy of command or single identifiable controller of events, only a sports injury system that is forced to readjust to systemic change, and athletes who are responding accordingly.

2.2.5 COMPLEX BEHAVIOURS AND RELATIONSHIPS

Complex systems exhibit non-linear interactions and feedback loops among their many elements (Willy, Neugebauer, & Gerngroß, 2003; Ottino, 2003; Batty & Torrens, 2005; Diez Roux, 2011; Holland, 2014; Sturmberg, 2018). This means that small causes can reverberate throughout the system leading to a disproportionately larger effect at a future point in time (i.e. otherwise known as 'sensitivity on initial conditions') (Rickles, Hawe, & Shiell, 2007; Dekker, 2016). The following scenario illustrates a series of events across the meso and macroscopic levels of the 'recreational distance running system':

- Technological innovations and automated work systems are encouraging sedentary lifestyles. In response to a chronic disease epidemic, more runners are taking to the streets to improve health, fitness, and overall well-being;

- Other individuals notice this growing trend in running popularity and are motivated to do the same. The effect is therefore *cyclical* and is *reinforced* over time;
- Through a diffusion of ideas and information facilitated by tightly connected social networks, running attracts further interest leading to a growth in fun runs, charity events, organised competitions, and big city marathons, many of which are widely broadcast and publicised;
- Increasing participatory trends mobilise multinational sports corporations to invest in new technologies and footwear products targeting the needs of consumers with increasingly active lifestyles;
- Marketing and promotional materials advertising new products are disseminated through various media channels, encouraging a new wave of potential participants to start running; and
- With accelerated growth in running participation comes a high rate of injury at the population-level, with scientific research and evidence-based healthcare interventions working to 'catch up' and offer suitable resolution.

In response to this hypothetical timeline of events, the question remains: *What is the cause of running injury and what is the key to its prevention?* The inputs and outputs of a complex sports injury system can be difficult to identify. Knowing where exactly to intervene with the appropriate injury prevention intervention and pinpointing the optimal leverage points for effective countermeasures (Meadows, 2008) in the presence of non-linear interactions and feedback can challenge even the most experienced complexity scientists. Chapters 16, 17, 18, and 19 demonstrate that it is possible to investigate and better understand complex behaviours and relationships with systems HFE approaches and computational modelling methods.

2.2.6 EMERGENCE

As already covered, complex systems give rise to emergent phenomena (Cilliers, 1998; Ottino, 2003; Rickles, Hawe, & Shiell, 2007; Diez Roux, 2011; Sturmberg, 2018). The previously listed characteristics can be considered as the inputs, and the emergence of sports injury as one of many potential outputs (e.g. alongside performance, health, safety, illness). Tied to emergence is the concept of non-determinism, which can be thought of as uncertainty despite reliable, stable, and consistent input.

2.3 SIMPLE, COMPLICATED, AND COMPLEX PROBLEMS AND SYSTEMS

Not all seemingly complex systems are, in fact, truly complex in the sense of the word. It is important to delineate between simple, complicated, and complex problems and systems. One of the better overviews highlighting these differences is provided by Glouberman and Zimmerman (2002) (Table 2.1).

With regard to Table 2.1, simple problems such as manufacturing a steering wheel for a Formula One race vehicle carries with it a high chance of success. There is very little deviation (if any) from the expected result, and the manufacturing process

TABLE 2.1

Differences between Simple, Complicated, and Complex Problems

Simple	Complicated	Complex
Manufacturing a steering wheel for a Formula One race vehicle	*Formula One race vehicle*	*Winning a Formula One championship*
Clear instructions for production are essential	Formulae critical and necessary	Formulae have limited application (e.g. strategy, tactics do not necessarily deliver)
Production processes tested and evaluated leading to easy replication	Producing and using one vehicle increases the assurance that the next will deliver similar results	Winning one race provides experience however there is no assurance of subsequent success
Expertise required however appropriate experience increases the success of replication	High levels of expertise in a variety of fields necessary for vehicle design, engineering, operation	Expertise can contribute but is neither necessary nor sufficient to assure success
Production processes, when perfected, guarantee results	Vehicles similar in critical ways (e.g. safety, performance)	Every race/championship is unique and must be understood on its own
The most efficient production lines lead to the best results	High degree of certainty of outcome	Uncertainty of outcome remains

Adapted from Glouberman and Zimmerman (2002) and applied to the Formula One context.

can be mastered through repeat exposure and trial and error. A complicated problem would be the Formula One racing vehicle itself, which essentially comprises a subset of many simple systems that are designed to integrate and function collectively. Although a high level of expertise and knowledge is required to understand the design and engineered specifications of the machine as a whole, it can nevertheless be broken down, diagnosed, repaired, reassembled, and brought back to full working order should a problem be suspected. On the other hand, complex problems such as attempting to figure out how to win a Formula One championship, includes both simple and complicated subsidiary problems, however, it is not reducible to either. Indeed, a desirable outcome such as winning is never guaranteed even with an impressive track record or historical success (Glouberman & Zimmerman, 2002).

Others have taken the idea of complicated and complex problems further, explaining that the number of parts within a system is not the critical issue; rather, it is the *adaptability* of the system and its response to changing conditions that makes it complex (Ottino, 2003). For example, the removal of a single piece of technology from a highly critical and deterministic system, such as a Formula One race vehicle, would not only prevent it from operating but there would be no self-organisation or internal adaptability to circumvent the issue. Conversely, each Formula One race is totally unique; outcomes are unpredictable, drivers have to continually adapt to changing external circumstances such as weather and the behaviours of other drivers, and team managers regularly communicate strategies and instructions that further modify individual and collective behaviours. For this reason, a complicated system can be distinguished from a complex one on the basis of whether or not there is a central controller (Ladyman, Lambert, & Wiesner, 2013). In other words, organisation and control in a complex system is distributed across the system and is locally generated through self-organisation and emergent processes.

In light of the distinction between simple, complicated, and complex concepts, it can be concluded that the complexity of a given sports system is predicted on a description of *where* the boundary is set, and *what* type of problem requires solving. For example, Chapters 3 to 8, in the physical HFE section of this book, touch on topics such as scaling, sports injury, and sports equipment design. Whilst these topics are not focussed on systems as a whole (or complex systems according to the definitions above), the accompanying discussions highlight the importance of designing and optimising equipment, products, and physical systems given that each has a critical role to play *within* broader complex sports systems.

2.4   COMPLEX SOCIO-TECHNICAL SYSTEMS

Socio-technical systems are comprised of both social and technical elements co-engaged in the pursuit of shared goals. The interaction of these social and technical aspects creates emergent properties and the conditions for either successful or unsuccessful system performance (Walker, Stanton, Salmon, & Jenkins, 2008). More applicably, the term socio-technical carries with it specific meaning for the HFE community since it was first coined by Eric Trist and Ken Bamforth in the 1950s (Trist & Bamforth, 1951). Trist and Bamforth worked in the field of organisational development at the London Tavistock Institute of Human Relations. The institute focused on understanding the role of human skill and methods to facilitate teamwork on productivity within English coal mines (Waterson et al., 2015). The goal of the socio-technical systems approach was (and still) is to optimise the relationship between humans (i.e. socio), and the tools, technologies, and techniques (i.e. technical) used during work (Clegg, 2000; Carayon, 2006). This joint optimisation of the socio and technical – as opposed to optimisation the social or technical aspect in isolation – is thus the aim when applying socio-technical systems approaches. The underlying theory is that by linking the socio and technical aspects of work systems together, it is possible to enhance worker productivity, satisfaction, and overall safety at the individual and organisational levels (Waterson et al., 2015).

By the 1960s, the socio-technical systems approach was recognised as a legitimate area of study and formed the basis of many contemporary HFE subdisciplines, including macro-ergonomics (i.e. work system design, human–machine interface optimisation) (Hendrick & Kleiner, 2001) and systems HFE (e.g. Salmon & Read, 2019). In particular, it is systems ergonomics – or systems HFE in context of this book – that widens the scope and more explicitly draws on the characteristics of complex systems as discussed in the previous section, including multi-scale interactions, heterogeneity, open boundaries, self-organisation, non-linearity, and emergence.

One popular HFE area that the socio-technical systems approach has been applied to is occupational risk and organisational safety, as well as the broader field of accident analysis and prevention (Rasmussen, 1997; Leveson, 2004; Kleiner, 2004; Carayon, 2006; Walker et al., 2008; Underwood & Waterson, 2014; Read et al., 2017; Hulme, Stanton et al., 2019). For example, the socio-technical systems approach has played a pivotal role in detracting attention away from the role of human error and failed technologies in relation to disaster causation, in favour of a holistic perspective that considers the entire system and its many complex interactions therein (Stanton et al., 2019).

In a similar way to traditional safety contexts in HFE, be it nuclear power, mining, civil engineering, public health, transportation, defence, healthcare, and manufacturing, both recreational and professional sport occurs within complex socio-technical systems. Scholars in the field of systems HFE have begun to model and elucidate socio-technical sports systems as well as the multiple interactions between their socio (e.g. athletes, teams, coaches, staff) and technical (e.g. equipment, technologies, products, services, and policies) aspects (McLean et al., 2017; Salmon, 2017; Hulme, McLean, Read et al., 2019; McLean et al., 2019; Salmon & McLean, 2019). Although systems HFE applications in sport are showing early promise (Hulme, Thompson et al., 2019), there is a need to continue to design and conduct studies that have the capacity to explore new ways in which the socio and technical aspects of sports systems can be optimised to support the health and performance of athletes and teams in practice. Table 2.2 outlines existing systems HFE methods that are available to the sports scientists and ergonomics researcher. Some of these methods and approaches appear across select chapters in this book, and/or have already been applied in the contexts of both recreational and elite sports performance and injury.

2.5   CONCLUSION

The aim of this chapter was to provide a brief introduction to complexity science, including how sports systems possess the characteristics of complex socio-technical systems. Consideration of different sport systems as complex systems has a number of implications for sports research and practice. First, the acceptance that sport is inherently complex brings with it the argument that sports performance and injury, whether at the athlete, team, match, club, organisation, or league level, cannot be understood by studying components in isolation. Whilst to date the majority of sports science research has attempted to reduce sports performance and injury causation down to the level of the individual athlete, the need for a paradigm shift towards a complex systems approach is once again emphasised. It is our view that systems HFE methods (see Chapters 14, 16, 17, 18, and 19) are important tools with which to initiate this shift. Second, there is a need for research that attempts to model and understand complex sports systems. This includes capturing the many fundamentally different types of elements that comprise them, as well as using computational approaches that have the capacity track and monitor how systems change over time (see Chapter 19). As above, various systems HFE methods exist that can be used to support efforts to model complex systems. These include both quantitative and qualitative modelling approaches such as EAST (Walker et al., 2006; Stanton & Baber,

TABLE 2.2

Systems HFE Methods

Method/approach	Associated methods/ extensions	Background/utility/purpose	Previous applications in sport
The Event Analysis of Systemic Teamwork (EAST) (Walker et al., 2006; Stanton & Baber, 2017)	• Hierarchical Task Analysis (HTA) (Kirwan & Ainsworth, 1992) • Task networks • EAST Broken Links (EAST-BL) (Stanton & Harvey, 2017) • EAST Broken Nodes (EAST-BN) (Stanton et al., 2019)	Model and understand distributed cognition and situation awareness in complex work and organisational systems Produces three network-based representations of a complex system: (i) a task network; (ii) a social network; and, (iii) an information network. A composite task-social-information network shows which (agents) are using what information to undertake and perform certain tasks EAST-BL is a systemic risk assessment and prediction approach that could theoretically be used to predict the risks that could degrade optimal sports performance EAST-BN is an accident and incident analysis approach that can be used to understand how, for example, previous sports performances may have been compromised so essential future learnings can be made	Elite women's road cycling (Salmon, Clacy, & Dallat, 2017) Research is currently underway in applying EAST-BL in the elite cycling race system to predict the risks that could degrade optimal sports performance
Cognitive Work Analysis (CWA) (Vicente, 1999; Jenkins, Stanton, & Walker, 2017)	Work Domain Analysis (WDA) Control task (or activity) analysis Strategies analysis Social Organisation and Co-operation Analysis (SOCA) Worker competencies analysis	A framework and toolkit to model complex socio-technical systems The framework models different types of constraints, resulting in the development of a model describing how work could proceed within a given work system Optimisation of complex systems behaviours and properties, including the appropriate allocation of workers and resources, the design of future systems, process design, training needs analysis and evaluation, and information requirements specification	Football (soccer) performance (WDA) (McLean et al., 2017); elite netball performance (WDA and SOCA) (Hulme, McLean, Read et al., 2019; McLean et al., 2019)

(Continued)

TABLE 2.2 (CONTINUED)
Systems HFE Methods

Method/approach	Associated methods/extensions	Background/utility/purpose	Previous applications in sport
Risk Management Framework (RMF) (Rasmussen, 1997) and the Accident Mapping (AcciMap) technique (Rasmussen & Suedung, 2000)	ActorMap, ImprovisationMaps (ImproMap) (Trotter, Salmon, & Lenne, 2014), ProtectiveMap (ProtectiMap)	Model complex systems as a hierarchy of levels, describe the actors, organisations, and/or entities that reside across those levels. Model multi-linear interactions/relationships among components/factors from across all levels of a complex system, ranging from the equipment and environment level, through to the government, policy, and budgeting levels of a system	Injury in competitive Australian Football League (AFL) (Dawson et al., 2017); concussion in competitive rugby union (Clacy et al., 2019); Injury in the led outdoor activity system (McLean et al., 2020)
The Systems–Theoretic Accident Model and Processes (STAMP) method (Leveson, 2004)	Systems–Theoretic Process Analysis (STPA) (Leveson, 2011), Causal Analysis based on STAMP (CAST) (Leveson, 2011)	Underpinned by 'control theory' to model various types of constraints within a system. STAMP models control and feedback loops and process models linking actors, organisations, and/or entities that reside across system levels. STPA is a systemic risk assessment and prediction approach that is applied to the STAMP control and feedback loops (i.e. forecasting). CAST is an accident and incident analysis approach that is applied to the STAMP control and feedback loops (i.e. retrospective following incident occurrence)	Overuse injury in recreational distance running (Hulme et al. 2017b, Hulme et al., 2017a) (i.e. system modelling). Research is currently underway in applying STPA in the elite cycling race system to predict the risks that could degrade optimal sports performance
The NETworked Hazard Analysis and Risk Management System (NET-HARMS) (Dallat, Salmon, & Goode, 2018)	HTA, Systematic Human Error Reduction and Prediction Approach (SHERPA) (Embrey, 1986), Task networks	Systemic, network-based risk assessment and prediction approach that identified both task risks and 'emergent risks' (i.e. new risks that are created when tasks risks combine and interact)	Identify risks in the led outdoor recreation system (Dallat, Salmon, & Goode, 2018). Research is currently underway in applying NET-HARMS in the elite cycling race system to predict the risks that could degrade optimal sports performance

2017), CWA (Vicente, 1999; Jenkins, Stanton, & Walker, 2017), STAMP (Leveson, 2004, 2011), and network analysis. Whilst this book showcases initial applications of these methods, further applications are encouraged. When applying these HFE methods, emphasis should be placed on attempting to understand how factors interact across sports systems to influence athlete, coach, team, sports club, and sports system behaviour (Salmon & McLean, 2019). It is our view that, once it is accepted that sports systems are complex, it becomes clear that there is much to learn regarding their composition and behaviour, and how they can be optimised. As such, there is a need for future sports research to embrace and investigate this complexity. The discipline of HFE provides theory and methods to support this.

REFERENCES

ASC. (2017). Sport 2030 – National Sport Plan. Australian Government (Australian Sports Commission), accessed 3 February. https://www.sportaus.gov.au/nationalsportplan.

Baarveld, Frank, Visser, Chantal A. N., Kollen, Boudewijn J., & Backx, Frank J. G. (2011). Sports-related injuries in primary health care. *Family Practice*, 28:1, 29–33.

Batty, Michael, & Torrens, Paul M. (2005). Modelling and prediction in a complex world. *Futures*, 37:7, 745–766. doi: 10.1016/j.futures.2004.11.003

Bekker, Sheree, & Finch, Caroline F. (2016). Too much information? A document analysis of sport safety resources from key organisations. *BMJ Open*, 6:5, e010877. doi: 10.1136/bmjopen-2015-010877.

Bertelsen, M. L., Hulme, A., Petersen, J., Brund, R. K., Sørensen, H., Finch, C. F., Parner, E. T., & Nielsenm, R. O. (2017). A framework for the etiology of running-related injuries. *Scandinavian Journal of Medicine and Science in Sports*, 27:11, 1170–1180. doi: 10.1111/sms.12883.

Bittencourt, N. F. N., Meeuwisse, W. H., Mendonça, L. D., Nettel-Aguirre, A., Ocarino, J. M., & Fonseca, S. T. (2016). Complex systems approach for sports injuries: Moving from risk factor identification to injury pattern recognition – narrative review and new concept. *British Journal of Sports Medicine*, 50:21, 1309–1314.

Bolling, Caroline, Mellette, Jay, Roeline Pasman, H., Van Mechelen, Willem, & Verhagen, Evert. (2019). From the safety net to the injury prevention web: Applying systems thinking to unravel injury prevention challenges and opportunities in Cirque du Soleil. *BMJ Open Sport and Exercise Medicine* 5:1, e000492.

Buckler, D. G. (1999). General practitioners' training for, interest in, and knowledge of sports medicine and its organisations. *British Journal of Sports Medicine*, 33:5, 360–363.

Carayon, Pascale. (2006). Human factors of complex sociotechnical systems. *Applied Ergonomics*, 37:4, 525–535. doi: 10.1016/j.apergo.2006.04.011.

Castellani, Brian. (2018). Map of the complexity sciences. Art & Science Factory. https://www.art-sciencefactory.com/complexity-map_feb09.html, accessed on 31 January.

Cilliers, Paul. 1998. *Complexity and Postmodernism: Understanding Complex Systems*. London, UK: Routledge.

Clacy, Amanda, Goode, Natassia, Sharman, Rachael, Lovell, Geoff P., & Salmon, Paul. (2019). A systems approach to understanding the identification and treatment of sport-related concussion in community rugby union. *Applied Ergonomics*, 80:256–264.

Clegg, Chris W. (2000). Sociotechnical principles for system design. *Applied Ergonomics*, 31:5, 463–477. doi: 10.1016/S0003-6870(00)00009-0

Cook, Andrew, Blom, Henk A. P., Lillo, Fabrizio, Mantegna, Rosario Nunzio, Miccichè, Salvatore, Rafael, Vázquez, Damián, Rivas, & Zanin, Massimiliano. (2015). Applying complexity science to air traffic management. *Journal of Air Transport Management*, 42, 149–158. doi: 10.1016/j.jairtraman.2014.09.011.

Cunha, Danilo, Xavier, Rafael, & Nunes de Castro, Leandro. (2018). Bacterial colonies as complex adaptive systems. *Natural Computing*, 17:4, 781–798. doi: 10.1007/s11047-018-9689-7.

Dallat, Clare, Salmon, Paul M., & Goode, Natassia. (2018). Identifying risks and emergent risks across sociotechnical systems: The NETworked hazard analysis and risk management system (NET-HARMS). *Theoretical Issues in Ergonomics Science*, 19:4, 456–482.

Dawson, Katelyn, Salmon, Paul M., G. J. Read, Neville, Timothy, Goode, Natassia, & Clacy, Amanda. (2017). Removing concussed players from the field: the factors influencing decision making around concussion identification and management in Australian Rules Football. *Proceedings of the 13th International Conference on Naturalistic Decision Making*.

Dekker, Sidney. (2016). *Drift into Failure: From Hunting Broken Components to Understanding Complex Systems*. Boca Raton, FL: CRC Press.

Diez Roux, Ana V. (2011). Complex systems thinking and current impasses in health disparities research. *American Journal of Public Health*, 101:9, 1627–1634. doi: 10.2105/AJPH.2011.300149.

Donaldson, Alex, & Finch, Caroline F. (2013). Applying implementation science to sports injury prevention. *British Journal of Sports Medicine*, 47, 473–475.

Donaldson, Alex, Leggett, Susan, & Finch, Caroline F. (2012). Sports policy development and implementation in context: Researching and understanding the perceptions of community end-users. *International Review for the Sociology of Sport*, 47:6, 743–760.

Eckerman, Mattias, Gunnar, Edman, Kjell, Svensson, & Alricsson, Marie. (2019). The relationship between personality traits and muscle injuries in Swedish elite male football players. *Journal of Sport Rehabilitation*, 1:aop, 1–6.

Ekegren, Christina L., Gabbe, Belinda J., & Finch, Caroline F. (2016). Sports injury surveillance systems: a review of methods and data quality. *Sports Medicine*, 46:1, 49–65.

Elliott, Euel, & Kiel, L. Douglas. (2004). A complex systems approach for developing public policy toward terrorism: An agent-based approach. *Chaos, Solitons and Fractals*, 20:1, 63–68. doi: 10.1016/S0960-0779(03)00428-4.

Embrey, D. E. (1986). *SHERPA: A Systematic Human Error Reduction and Prediction Approach*. La Grange Park, IL: American Nuclear Society.

Feller, Liviu, Khammissa, Razia Abdool Gafaar, & Lemmer, Johan. (2017). Biomechanical cell regulatory networks as complex adaptive systems in relation to cancer. *Cancer Cell International*, 17:1, 16. doi: 10.1186/s12935-017-0385-y.

Finch, Caroline F. (2012). Getting sports injury prevention on to public health agendas – addressing the shortfalls in current information sources. *British Journal of Sports Medicine*, 46:1, 70–74.

Galea, Sandro, Riddle, Matthew, & Kaplan, George A. (2010). Causal thinking and complex system approaches in epidemiology. *International Journal of Epidemiology*, 39:1, 97–106. doi: 10.1093/ije/dyp296.

Glouberman, Sholom, & Zimmerman, Brenda. (2002). *Complicated and Complex Systems: What Would Successful Reform of Medicare Look Like?* Vol. 2. Toronto, ON: Commission on the Future of Health Care in Canada.

Green, Mick. (2007). Olympic glory or grassroots development? Sport policy priorities in Australia, Canada and the United Kingdom, 1960–2006. *The International Journal of the History of Sport*, 24:7, 921–953.

Grilo, António, Caetano, Artur, & Rosa, Agostinho. (2002). Immune system simulation through a complex adaptive system model. In: Rajkumar Roy, Mario Köppen, Seppo Ovaska, Takeshi Furuhashi and Frank Hoffmann (eds.), *Soft Computing and Industry: Recent Applications* (pp. 675–698). London, UK: Springer.

Hanson, Dale, Allegrante, John P., Sleet, David A., & Finch, Caroline F. (2014). Research alone is not sufficient to prevent sports injury. *British Journal of Sports Medicine*, 48, 682–684.

Hardin, Elizabeth C., Van Den Bogert, Antonie J., & Hamill, Joseph. (2004). Kinematic adaptations during running: Effects of footwear, surface, and duration. *Medicine and Science in Sports and Exercise*, 36:5, 838–844.

Hendrick, Hal W., & Kleiner, Brian M. (2001). *Macroergonomics: An Introduction to Work System Design*. Santa Monica, CA: Human Factors and Ergonomics Society.

Holland, John H. (2014). *Complexity: A Very Short Introduction*. Oxford, UK: Oxford University Press.

Holmes, Alison H., Moore, Luke S. P., Sundsfjord, Arnfinn, Steinbakk, Martin, Regmi, Sadie, Karkey, Abhilasha, Guerin, Philippe J., & Piddock, Laura J. V. (2016). Understanding the mechanisms and drivers of antimicrobial resistance. *The Lancet*, 387:10014, 176–187. doi: 10.1016/S0140-6736(15)00473-0.

Hulme, Adam, & Finch, C. F. (2015). From monocausality to systems thinking: A complementary and alternative conceptual approach for better understanding the development and prevention of sports injury. *Injury Epidemiology*, 2:1, 31. doi: 10.1186/s40621-015-0064-1.

Hulme, Adam, McLean, Scott, Read, Gemma J. M., Dallat, Clare, Bedford, Anthony, & Salmon, Paul M. (2019). Sports organizations as complex systems: Using cognitive work analysis to identify the factors influencing performance in an elite netball organization. *Frontiers in Sport and Active Living*, 1:56. doi: 10.3389/fspor.2019.00056.

Hulme, Adam, McLean, Scott, Salmon, Paul M., Thompson, Jason, Lane, Ben R., & Nielsen, Rasmus Oestergaard. (2019). Computational methods to model complex systems in sports injury research: Agent-based modelling (ABM) and systems dynamics (SD) modelling. *British Journal of Sports Medicine*, 53:24, 1507. doi: 10.1136/bjsports-2018-100098.

Hulme, Adam, Salmon, P. M., Nielsen, Rasmus Oestergaard, Read, Gemma J. M., & Finch, C. F. (2017a). Closing Pandora's box: Adapting a systems ergonomics methodology for better understanding the ecological complexity underpinning the development and prevention of running-related injury. *Theoretical Issues in Ergonomics Science*, 18:4, 338–359. doi: 10.1080/1463922X.2016.1274455.

Hulme, Adam, Salmon, P. M., Nielsen, Rasmus Oestergaard, Read, Gemma J. M., & Finch, C. F. (2017b). From control to causation: Validating a 'complex systems model' of running-related injury development and prevention. *Applied Ergonomics*, 65, 345–354.

Hulme, Adam, Stanton, Neville A., Walker, Guy H., Waterson, Patrick, & Salmon, Paul M. (2019). What do applications of systems thinking accident analysis methods tell us about accident causation? A systematic review of applications between 1990 and 12018. *Safety Science*, 117, 164–183. doi: 10.1016/j.ssci.2019.04.016.

Hulme, Adam, Thompson, Jason, Plant, Katherine L., Read, Gemma J. M., McLean, Scott, Clacy, Amanda, & Salmon, Paul M. (2019). Applying systems ergonomics methods in sport: A systematic review. *Applied Ergonomics*, 80, 214–225. doi: 10.1016/j.apergo.2018.03.019.

Jenkins, Daniel P., Stanton, Neville A., & Walker, Guy H. (2017). *Cognitive Work Analysis: Coping with Complexity*. Boca, Raton, FL: CRC Press.

Karwowski, Waldemar. (2012). A review of human factors challenges of complex adaptive systems: Discovering and understanding chaos in human performance. *Human Factors*, 54:6, 983–995. doi: 10.1177/0018720812467459.

Kavetsos, Georgios, & Szymanski, Stefan. (2009). From the Olympics to the grassroots: What will London 2012 mean for sport funding and participation in Britain? *Public Policy Research*, 16:3, 192–196. doi: 10.1111/j.1744-540X.2009.00580.x.

Kirwan, Barry, & Ainsworth, Les K. (1992). *A Guide to Task Analysis: The Task Analysis Working Group*. Boca Raton, FL: CRC Press.

Kleiner, B. M. (2004). Macroergonomics as a large work-system transformation technology. *Human Factors and Ergonomics in Manufacturing and Service Industries*, 14:2, 99–115. doi: 10.1002/hfm.10060.

Ladyman, James, Lambert, James, & Wiesner, Karoline. (2013). What is a complex system? *European Journal for Philosophy of Science*, 3:1, 33–67.

Lee, Bruce Y., Bartsch, Sarah M., Mui, Yeeli, Haidari, Leila A., Spiker, Marie L., & Gittelsohn, Joel. (2017). A systems approach to obesity. *Nutrition Reviews*, 75:suppl 1, 94–106. doi: 10.1093/nutrit/nuw049.

Leveson, Nancy. (2004). A new accident model for engineering safer systems. *Safety Science*, 42:4, 237–270.

Leveson, Nancy. (2011). *Engineering a Safer World: Systems Thinking Applied to Safety*. Cambridge, MA: MIT Press.

Levin, Simon A. (1998). Ecosystems and the biosphere as complex adaptive systems. *Ecosystems*, 1:5, 431–436. doi: 10.1007/s100219900037.

Luke, Douglas A., & Stamatakis, Katherine A. (2012). Systems science methods in public health: Dynamics, networks, and agents. *Annual Review of Public Health*, 33, 357–376. doi: 10.1146/annurev-publhealth-031210-101222.

Malisoux, Laurent, Delattre, Nicolas, Urhausen, Axel, & Theisen, Daniel. (2017). Shoe cushioning, body mass and running biomechanics as risk factors for running injury: A study protocol for a randomised controlled trial. *BMJ Open*, 7:8, e017379.

Markose, Sheri M. (2005). Computability and evolutionary complexity: Markets as complex adaptive systems (CAS). *The Economic Journal*, 115:504, F159–F192. doi: 10.1111/j.1468-0297.2005.01000.x.

Marshall, Brandon D. L., Paczkowski, Magdalena M., Seemann, Lars, Tempalski, Barbara, Pouget, Enrique R., Galea, Sandro, & Friedman, Samuel R. (2012). A complex systems approach to evaluate HIV prevention in metropolitan areas: Preliminary implications for combination intervention strategies. *PloS one*, 7:9, e44833–e44833. doi: 10.1371/journal.pone.0044833.

McIntosh, Andrew. (2012). Biomechanical considerations in the design of equipment to prevent sports injury.*Proceedings of the Institution of Mechanical Engineers, Part P: Journal of Sports Engineering Technology*, 226:3–4, 193–199.

McLean, Scott, Finch, Caroline F., Goode, Natassia, Clacy, Amanda, Coventon, Lauren J., & Salmon, Paul M. (2020). Applying a systems thinking lens to injury causation in the outdoors: Evidence collected during 3 years of the understanding and preventing led outdoor accidents data system. *Injury Prevention*.

McLean, Scott, Hulme, Adam, Mooney, Mitchell, Read, Gemma J. M., Bedford, Anthony, & Salmon, Paul M. (2019). A systems approach to performance analysis in women's netball: Using Work domain analysis to model elite netball performance. *Frontiers in Psychology*, 10, 201–201. doi: 10.3389/fpsyg.2019.00201.

McLean, Scott, Salmon, Paul M., Gorman, Adam D., Read, Gemma J. M., & Solomon, Colin. (2017). What's in a game? A systems approach to enhancing performance analysis in football. *PloS one*, 12:2.

Meadows, Donella H. (2008). *Thinking in Systems: A Primer*. Hartford, VT: Chelsea Green Publishing.

Mooney, Mitchell, Charlton, Paula C., Soltanzadeh, Sadjad, & Drew, Michael K. (2017). Who 'owns' the injury or illness? Who 'owns' performance? Applying systems thinking to integrate health and performance in elite sport. *British Journal of Sports Medicine*, 51:14.

Nielsen, Rasmus Oestergaard, Bertelsen, Michael Lejbach, Møller, Merete, Hulme Adam, Windt, Johann, Verhagen, Evert, Mansournia, Mohammad Ali, Casals, Martí, & Parner, Erik Thorlund. (2018). Training load and structure-specific load: Applications for sport injury causality and data analyses. *British Journal of Sports Medicine*,52:16, 1016. doi: 10.1136/bjsports-2017-097838.

Nigg, Benno M. (1985). Biomechanics, load analysis and sports injuries in the lower extremities. *Sports Medicine*, 2, 5, 367–379.

Olivier, Jake, & Creighton, Prudence. (2017). Bicycle injuries and helmet use: A systematic review and meta-analysis.*International Journal of Epidemiology*, 46:1, 278–292.

Ottino, J. M. (2003). Complex systems. *AIChE Journal American Institute of Chemical Engineers Journals*, 49:2, 292–299. doi: 10.1002/aic.690490202.

Pearce, Neil, & Merletti, Franco. (2006). Complexity, simplicity, and epidemiology. *International Journal of Epidemiology*, 35:3, 515–519. doi: 10.1093/ije/dyi322.

Perrow, Charles. (2011). *Normal Accidents: Living with High Risk Technologies – Updated Version*. Princeton, NJ: Princeton University Press.

Pol, Rafel, Hristovski Robert, Medina, Daniel, & Balague, Natalia. (2019). From microscopic to macroscopic sports injuries. Applying the complex dynamic systems approach to sports medicine: A narrative review. *British Journal of Sports Medicine*, 53:19, 1214–1220. doi: 10.1136/bjsports-2016-097395.

Rasmussen, Jens. (1997). Risk management in a dynamic society: A modelling problem. *Safety Science*, 27:2, 183–213. doi: 10.1016/S0925-7535(97)00052-0.

Rasmussen, Jens, & Suedung, Inge. (2000). *Proactive Risk Management in a Dynamic Society*. Swedish Rescue Services Agency.

Read, Gemma J. M., Beanland, Vanessa, Lenné, Michael G., Stanton, Neville A., & Salmon, Paul M. (2017). *Integrating Human Factors Methods and Systems Thinking for Transport Analysis and Design*. Boca Raton, FL: CRC Press.

Rial, José A., Pielke, Roger A., Beniston, Martin, Claussen, Martin, Canadell, Josep, Cox, Peter, Held, Hermann, de Noblet-Ducoudré, Nathalie, Prinn, Ronald, & Reynolds, James. (2004). Nonlinearities, feedbacks and critical thresholds within the Earth's climate system. *Climatic Change*, 65:1–12, 11–38.

Rickles, Dean, Hawe, Penelope, & Shiell, Alan. (2007). A simple guide to chaos and complexity. *Journal of Epidemiology and Community Health*, 61:11, 933–937. doi: 10.1136/jech.2006.054254.

Rossi, Alessio, Pappalardo, Luca, Cintia, Paolo, Iaia, F. Marcello, Fernández, Javier, & Medina, Daniel. (2018). Effective injury forecasting in soccer with GPS training data and machine learning. *PloS one*, 13:7, e0201264.

Rutter, Harry, Savona, Natalie, Glonti, Ketevan, Bibby, Jo, Cummins, Steven, Finegood, Diane T., Greaves, Felix, Harper, Laura, Hawe, Penelope, Moore, Laurence, Petticrew, Mark, Rehfuess, Eva, Shiell, Alan, Thomas, James, & White, Martin. (2017). The need for a complex systems model of evidence for public health. *Lancet* (London, England), 390:10112, 2602–2604. doi: 10.1016/S0140-6736(17)31267-9.

Salmon, Paul M. (2017). Ergonomics issues in sport and outdoor recreation. *Theoretical Issues in Ergonomics Science*, 18:4, 299–305.

Salmon, Paul M., Clacy, A., & Dallat, C. (2017). It's not all about the bike: distributed situation awareness and teamwork in elite women's cycling teams. *Contemporary Ergonomics*, 2017, 240–248.

Salmon, Paul M., & McLean, Scott. (2019). Complexity in the beautiful game: Implications for football research and practice. *Science and Medicine in Football*, 1–6. doi: 10.1080/24733938.2019.1699247.

Salmon, Paul M., & Read, Gemma J. M. (2019). Many model thinking in systems ergonomics: a case study in road safety. *Ergonomics*, 62:5, 612–628.

Senge, Peter M. (2006). *The Fifth Discipline: The Art and Practice of the Learning Organization*. New York, NY: Broadway Business.

Sheard, Sarah A., & Mostashari, Ali. (2009). Principles of complex systems for systems engineering. *Systems Engineering*, 12:4, 295–311. doi: 10.1002/sys.20124.

Simon, Herbert A. (1996). *The Architecture of Complexity*. Cambridge, MA: MIT Press.

Sornette, Didier. (2017). *Why Stock Markets Crash: Critical Events in Complex Financial Systems*. Vol. 49. Princeton, NJ: Princeton University Press.

Stanton, Neville A, & Baber, Chris. (2017). *Modelling Command and Control: Event Analysis of Systemic Teamwork*. Boca Raton, FL: CRC Press.

Stanton, Neville A., & Harvey, Catherine. (2017). Beyond human error taxonomies in assessment of risk in sociotechnical systems: A new paradigm with the EAST 'broken-links' approach. *Ergonomics*, 60:2, 221–233.

Stanton, Neville A., Salmon, Paul M., Walker, Guy H., & Stanton, Maggie. (2019). Models and methods for collision analysis: A comparison study based on the Uber collision with a pedestrian. *Safety Science*, 120, 117–128. doi: 10.1016/j.ssci.2019.06.008.

Sterman, John D. (2006). Learning from evidence in a complex world. *American Journal of Public Health*, 96:3, 505–514. doi: 10.2105/AJPH.2005.066043.

Sturmberg, Joachim P. (2018). Complexity sciences. In: Joachim P. Sturmberg (ed.), *Health System Redesign: How to Make Health Care Person-Centered, Equitable, and Sustainable* (pp. 21–44). Cham: Springer International Publishing.

Timpka, Toomas, Ekstrand, Jan, & Svanström, Leif. (2006). From sports injury prevention to safety promotion in sports. *Sports Medicine*, 36:9, 733–745.

Trist, Eric Lansdown, & Bamforth, Ken W. (1951). Some social and psychological consequences of the longwall method of coal-getting: An examination of the psychological situation and defences of a work group in relation to the social structure and technological content of the work system. *Human Relations*, 4:1, 3–38.

Trochim, William M., Cabrera, Derek A., Milstein, Bobby, Gallagher, Richard S., & Leischow, Scott J. (2006). Practical challenges of systems thinking and modeling in public health. *American Journal of Public Health*, 96:3, 538–546. doi: 10.2105/AJPH.2005.066001.

Trotter, Margaret J., Salmon, Paul M., & Lenne, Michael G. (2014). Impromaps: Applying Rasmussen's risk management framework to improvisation incidents. *Safety Science*, 64, 60–70.

Underwood, Peter, & Waterson, Patrick. (2014). Systems thinking, the Swiss Cheese Model and accident analysis: A comparative systemic analysis of the Grayrigg train derailment using the ATSB, AcciMap and STAMP models. *Accident Analysis and Prevention*, 68, 75–94.

Vicente, Kim J. (1999). *Cognitive Work Analysis: Toward Safe, Productive, and Healthy Computer-Based Work*. Boca Raton, FL: CRC Press.

Von Rosen, P., Frohm, A., Kottorp, A., Friden, C., & Heijne, A. (2017). Too little sleep and an unhealthy diet could increase the risk of sustaining a new injury in adolescent elite athletes. *Scandinavian Journal of Medicine and Science in Sports*, 27:11, 1364–1371.

Walker, Guy H., Gibson, Huw, Stanton, Neville A., Baber, Chris, Salmon, Paul, & Green, Damian. (2006). Event analysis of systemic teamwork (EAST): a novel integration of ergonomics methods to analyse C4i activity. *Ergonomics*, 49:12–13, 1345–1369.

Walker, Guy H., Stanton, Neville A., & Jenkins, Daniel P. (2017). *Command and Control: The Sociotechnical Perspective*. Boca Raton, FL: CRC Press.

Walker, Guy H., Stanton, Neville A., Salmon, Paul M., & Jenkins, Daniel P. (2008). A review of sociotechnical systems theory: A classic concept for new command and control paradigms. *Theoretical Issues in Ergonomics Science*, 9:6, 479–499. doi: 10.1080/14639220701635470.

Walsh, Stephen J., Messina, Joseph P., Mena, Carlos F., Malanson, George P., & Page, Philip H. (2008). Complexity theory, spatial simulation models, and land use dynamics in the Northern Ecuadorian Amazon. *Geoforum*, 39:2, 867–878. doi: 10.1016/j.geoforum.2007.02.011.

Waterson, Patrick, Robertson, Michelle M., Cooke, Nancy J., Militello, Laura, Roth, Emilie, & Stanton, Neville A. (2015). Defining the methodological challenges and opportunities for an effective science of sociotechnical systems and safety. *Ergonomics*, 58:4, 565–599. doi: 10.1080/00140139.2015.1015622.

Willy, Christian, Neugebauer, Edmund A. M., & Gerngroß, Heinz. (2003). The concept of nonlinearity in complex systems. *European Journal of Trauma*, 29:1, 11–22. doi: 10.1007/s00068-003-1248-x.

Section II

Physical HFE Applications

3 Using the Principle of Scaling to Improve Skill Acquisition and the Overall Sporting Experience in Children's Sport

Adam D. Gorman, Ian Renshaw,
Jonathon Headrick, and Christopher J. McCormack

CONTENTS

3.1 INTRODUCTION

The differences between children and adults in characteristics such as strength and body size mean that junior sporting competitions should ideally use equipment, playing areas, and rules that are scaled (i.e. matched) to the size of the child (for reviews, see Buszard, Reid, Masters, & Farrow, 2016; Buszard, Farrow, & Reid, 2020). This is similar to the discipline of physical ergonomics where the general intention is to produce an efficacious match between the physical characteristics of the individual and the characteristics of the environment in which the individual will be operating (Ahram & Karwowski, 2013; Karwowski, 2005). Fortunately, sporting associations have been proactive in terms of modifying competitions for children (see Eime et al., 2015); however, a key concern is that the majority of changes have generally not been based upon an empirical evaluation of children's action capabilities, meaning that the games that children play often bear little resemblance in terms of the techniques and tactics used by their adult counterparts (Buszard et al., 2016; Buszard et al., 2020).

To understand scaling in sport, it is important to appreciate that the behaviours of competitors are co-adaptable and hence intentions, perceptions, and actions are informed by one's own action capabilities in relation to those of team-mates and opponents (Passos, Araújo, Davids, & Shuttleworth, 2008; Renshaw, Davids, Shuttleworth, & Chow, 2009). For example, three-point attempts are common in adult basketball (Mexas, Tsitskaris, Kyriakou, & Garefis, 2005) because the adult has the action capability to shoot the ball the required distance. Typically, because the actions of defenders and attackers are coupled (e.g. Cordovil et al., 2009), defenders will attempt to prevent this scoring opportunity by closing down a potential shooter whenever he/she gains possession at the three-point line. In contrast, a 12-year-old basketball player who is required to play with a full-size ball typically struggles to reach the basket with a shot from the three-point line, meaning that defenders know that the attacker is not an immediate threat and that they can remain in the key (i.e. close to the basket) and deny space to prevent drives to the basket or passes to attackers who are in closer scoring positions. As a consequence, attacking (and defensive) strategies in junior competitions can become quite different to those employed in the adult version, bringing into question the degree to which tactical understanding developed by playing in junior games with adult equipment transfers to the adult game. If one of the overall aims of junior programs is talent development, a key requirement is critiquing junior programmes by considering the emergence of functionality and fidelity (i.e. the junior game is a simulation of the adult game that results in a positive transfer of performance skills) (Pinder, Davids, Renshaw, & Araujo, 2011). In essence, this can be summarised by asking; how functional are children's competitions in terms of enhancing intentions, perceptions, and actions? In this chapter, we will describe how a principled approach to scaling using the theoretical principles of constraints and affordances is required to ensure that junior sporting competitions replicate more of the key characteristics present in adult competitions (see Buszard et al., 2016; Buszard et al., 2020; Gorman, Headrick, Renshaw, McCormack, & Topp, in review). In the section that follows, we discuss scaling in the context of 'backyard games' (a term used to capture the informal practice environments of children in backyards, streets, and local parks; Cannane, 2009) and highlight how organised sports programs could use this concept to enhance children's sporting experiences and potentially also help to address dropout (Renshaw, Oldham, & Bawden, 2012; see also Buszard et al., 2016; Buszard et al., 2020).

3.2 THE IMPORTANCE OF SCALING IN CHILDREN'S SPORTS

Historically, children's play took place via backyard games, with children playing their favourite sports or games for hours in adult-free and unstructured settings (Cannane, 2009; Renshaw et al., 2012; see also Baker, Côté, & Abernethy, 2003). Whilst the motivation for the child was simply about having fun, for those interested in talent development, backyard games have been viewed as a beneficial precursor for skill acquisition (Baker et al., 2003; Renshaw et al., 2012). Backyard games have been described as the ideal foundation for the development of expertise because they allow young players to engage in hours of unstructured practice, providing the environment that enables the development of technical, physical, tactical, mental, emotional, leadership, and social skills needed for later success (Baker et al., 2003; Cannane, 2009; Cooper, 2010; Renshaw & Chappell, 2010; Renshaw & Moy, 2018). One key aspect of children's play is that it provides the freedom to explore and learn sports without the interference of well-meaning adults (Baker et al., 2003; Renshaw et al., 2012). Many examples of how these unique environments have been significant in shaping future champions across sports have been reported (Baker et al., 2003; Cannane, 2009; Machado et al., 2018; Uehara et al., 2018).

Perhaps the most important characteristic of backyard games in terms of scaling is the fact that the games are typically tailored by the children to suit their current individual capabilities and to meet their own basic psychological needs, thereby helping to foster the requisite intrinsic motivation necessary to invest the time and effort to achieve sporting expertise (Cannane, 2009; Devereux, 2001; Machado et al., 2018; Renshaw et al., 2012; Renshaw & Chappell, 2010). For instance, the quintessential backyard cricket game created by children involves a modified ball (typically a tennis ball) and shorter pitch to (a) help ensure the bowler can deliver the ball without allowing it to bounce more than once before reaching the batter, (b) help provide the batter with deliveries that he/she can actually hit, and (c) provide an engaging and enjoyable environment that caters to the capabilities of the participants (see also Cannane, 2009; Devereux, 2001; Elliott, Plunkett, & Alderson, 2005; Harwood, Yeadon, & King, 2018). These qualities also highlight the fundamental differences that exist between backyard games and many organised sports programmes. That is, children have a natural tendency to self-scale their sporting contests to create a more inclusive and engaging experience, which is in stark contrast to the one size fits all game design that comes with the formal organised sporting programmes that typically involve greater rigidity and structure (Renshaw et al., 2012). Often, these games require the use of adult-sized equipment (or near adult-sized), as well as regulation (adult) playing areas and rules. While speculative, it is possible that this divide may be contributing, at least in part, to the drop-out rate of children from organised sport (see Buszard et al., 2016; Buszard et al., 2020). In some instances, there have been reports of large percentages of participants withdrawing from youth sport programs (Gould, 1987; Kelley & Carchia, 2013; see also Carlman, Wagnsson, & Patriksson, 2013; Delorme, Chalabaev, & Raspaud, 2011; Eime et al., 2015; Visek et al., 2015). A review by Crane and Temple (2015) highlighted that two of the major causes of children and youth withdrawing from organised sport was a lack of enjoyment and perceptions of competence. These two factors tend to be self-managed by children when they apply the notion of scaling to create backyard games (Devereux, 2001; Machado et al., 2018; Phillips, Davids, Renshaw, & Portus, 2010).

Unfortunately, anecdotal evidence suggests that the role of the backyard as the foundation for talent development appears to be diminishing over time due to a range of potential factors. One possible environmental constraint is that the loss of appropriate space in urban areas is reducing the opportunities for spontaneous play (Cannane, 2009; Clements, 2004). Additionally, there is a perceived safety issue with parents becoming reluctant to let their children go out and play in unsupervised environments (Clements, 2004). As more and more peak sports bodies invest considerable amounts of time and money into organised junior programmes, it is important for such programmes to consider their overall role in relation to talent development and the promotion of lifelong participation in sport in general. In lieu of the perceived decline in backyard games, we argue that junior sports programmes should attempt to recreate some of the critical elements of the backyard environment by using scaling as a key underpinning philosophy (Renshaw et al., 2012). As we describe in the section that follows, there are numerous positive outcomes that occur when scaling is implemented, including enhanced skill acquisition and greater engagement (Buszard et al., 2016). The principle of scaling is not only simple and relatively cost effective to implement, but it may also create the type of environment that encourages children to maintain their involvement in sport and other physical activities (Buszard et al., 2016; Buszard et al., 2020).

Some of the core principles that programs should consider adopting have been outlined in the literature (e.g. Allender, Cowburn, & Foster, 2006; Crane & Temple, 2015; Visek et al., 2015). First and foremost, the importance of fun is captured as the key requirement to underpin children's ongoing participation in sport (Visek, Mannix, Mann, & Jones, 2017; see also Allender et al., 2006; Crane & Temple, 2015). Program design must therefore ensure that the basic psychological needs of children are met, which means ensuring that activities are inherently fun and rewarding, and that children can demonstrate competence, have autonomy, and feel connected in some way to other individuals (Balish, McLaren, Rainham, & Blanchard, 2014; Crane & Temple, 2015; Deci & Ryan, 2000; Renshaw et al., 2012; Visek et al., 2015; Visek et al., 2017). To ensure these opportunities are available, the capabilities and age appropriate needs of children must be front and centre. Age appropriate has been taken to mean a focus on several factors including (a) viewing enjoyment as a key feature that underpins ongoing motivation, (b) equal opportunity in terms of participation time, (c) development of fundamental motor skills and a reduced focus on winning, (d) early diversification of sporting experiences, (e) age appropriate competitions, and (f) opportunities to take part in informal recreational sports (Allender et al., 2006; Côté & Hancock, 2016; Crane & Temple, 2015; Renshaw et al., 2012; Visek et al., 2017; Witt & Dangi, 2018). These principles emphasise the importance of designing junior programs that take into consideration the physical and psychological developmental stages and capabilities that exist during childhood, which, as we highlight next, can largely be achieved through the mechanism of scaling (Buszard et al., 2016; Buszard et al., 2020).

3.3   RESEARCH INVESTIGATING SCALING IN SPORT

Some of the earliest research into scaling in sport occurred over 50 years ago (e.g. Egstrom, Logan, & Earl, 1960; Wright, 1967), and while further research is required, a critical mass of studies now exists that provide empirical evidence in support of the efficacy of scaling for facilitating children's motor skill acquisition and enhancing their overall sporting experience (Buszard et al., 2016). This section will provide an overview of the specific areas that have been targeted in the research and will summarise some of the key findings.

Equipment modification is one of the most widely studied areas within the domain of scaling (see Buszard et al., 2016; Buszard et al., 2020). The typical approach has been to compare adult-sized sporting equipment, which is typically larger, heavier,

bouncier, and/or higher, to an equivalent scaled-down version in an attempt to identify the changes, if any, that occur in performance, learning, and/or other related factors (e.g. Hammond & Smith, 2006; Limpens, Buszard, Shoemaker, Savelsbergh, &
Reid, 2018; Satern, Messier, & Keller-McNulty, 1989). As highlighted by Buszard et
al. (2016), research in this area has tended to concentrate upon tennis (e.g. Buszard,
Farrow, Reid, & Masters, 2014) and basketball (e.g. Arias, Argudo, & Alonso, 2012a),
with only a handful of other sporting tasks attracting attention (e.g. volleyball: Pellett,
Henschel-Pellett, & Harrison, 1994; overhand throwing: Burton, Greer, & Wiese,
1992). In general, the results have shown that the use of appropriately scaled equipment
tends to enhance children's skill acquisition, relative to using adult-sized equipment
(Buszard et al., 2016). For example, Buszard et al. (2014) found that when girls and
boys aged 6 to 8 years hit low compression tennis balls (75% less compression than an
adult ball) with a small tennis racket (48.3 cm), hitting performance (using a points-
based system) was significantly better compared to when those same children used a
larger racket (68.6 cm) with a standard adult ball. The reduced speed and bounce of the
lower compression ball also enabled the children to produce better hitting technique as
evidenced by a greater number of low-to-high swings and more impacts with the ball
in front and to the side of their body. Similarly, when the size of a basketball is reduced
(relative to an adult ball), children tend to increase their shooting accuracy (Arias,
2012), engage in more one-on-one situations (Arias, Argudo, & Alonso, 2012b), and
perform more dribbles, passes, and pass receptions (Arias, Argudo, & Alonso, 2012c).
Evidence also suggests that a lowered basket height (8 feet) has positive effects on the
self-efficacy of junior basketball players (Chase, Ewing, Lirgg, & George, 1994).

The other common scaling approach is to modify the size of the playing area
(Buszard et al., 2016; Buszard et al., 2020). The general intention is to reduce the
length and/or width of the adult playing surface to a size that better accommodates
for the reduced size and strength of children (Buszard et al., 2016; Buszard et al.,
2020). Research in this domain has largely been confined to cricket (e.g. Elliott et
al., 2005) and tennis (e.g. Timmerman et al., 2015), suggesting that more research
into other motor tasks across more sports is required to extend the existing knowledge base (Buszard et al., 2016). In cricket, reduced pitch lengths have had positive
results for children in terms of enhancing bowling accuracy (Elliott et al., 2005),
reducing the number of deliveries that bounced twice (Harwood et al., 2018), and
generally increasing the overall level of engagement of the players (Harwood et al.,
2018). Shorter pitches have also elicited positive changes in kinematic variables that
have previously been associated with a heightened risk of injury for cricket bowlers
(Elliott et al., 2005).

In tennis, researchers have examined a combination of scaling approaches by
changing equipment *and* playing dimensions (e.g. Farrow & Reid, 2010; Larson &
Guggenheimer, 2013; Lee, Chow, Komar, Tan, & Button, 2014). Timmerman et al.
(2015) reduced both the net height (0.91 m at net centre) and the court size (18.14 m
× 6.28 m) for junior male tennis players and showed that these modifications brought
the speed of the junior game (i.e. time between players' shots) into closer alignment
with that of the adult game. However, scaling of the net, irrespective of court size,
led to a more aggressive playing style with a greater number of winners and more
forced errors. Playing with a scaled net on an adult-sized court was also reported by

the majority of players as being the most enjoyable (see also Farrow & Reid, 2010). As highlighted by the authors, these latter results suggest that the height of the net may be a particularly important scaling variable in tennis.

Rule scaling (e.g. McCormick et al., 2012) and body scaling (see Cordovil et al., 2009) are two additional modifications that can be used to enhance the junior game. For example, rule changes that reduce the number of players (e.g. three-on-three basketball games or small-sided soccer games) tend to increase the number of opportunities for players to engage in the key skills of the sport, relative to full-sided games (Brown, Wisner, & Kontos, 2000; McCormick et al., 2012). Similarly, substitution rules aimed at rotating children through different playing positions provide increased opportunities for children to experience a broader repertoire of key roles throughout a game (Rupnow & Engelhorn, 1989). Other rule changes that have been implemented, but to our knowledge have not attracted empirical research, include the removal of tackling (such as touch football relative to rugby league or rugby union), the removal of the shot clock in junior basketball competitions (to reduce time constraints), and the removal of the offside rule in soccer (to reduce the complexity of the game). Body scaling based upon the physical characteristics (e.g. height, weight) of the participants could be used to encourage certain behaviours to emerge. While we are unaware of any research in this area that has specifically examined scaling for junior sport, a study by Cordovil et al. (2009) provides a good example of the concept. They found that the heights of pairs of players (aged 14 to 19 years) during a one-on-one full-court basketball contest influenced the intentions of the attackers and defenders in terms of their game strategy, suggesting that further research in this area may yield important results specific to the application of body scaling in children's sporting contests.

The notion of scaling is well supported by the extant literature, however, far more research across a broader range of domains is clearly required to help inform the decisions of practitioners, as well as guide in the design of future research endeavours (Buszard et al., 2016; Buszard et al., 2020). In particular, and in the context of the points outlined in this chapter, it is not only important to examine the extent to which scaling is linked to increased enjoyment and engagement in physical activity by children, but it is also necessary to explore the potential link between scaling, backyard games, and dropout (see Buszard et al., 2016; Buszard et al., 2020; Gorman et al., in review; Renshaw et al., 2012). One of the other missing elements is a sound theoretical rationale to provide a framework upon which the notion of scaling can be based (Buszard et al., 2016; Buszard et al., 2020; Gorman et al., in review).

3.4   A THEORETICAL FRAMEWORK FOR SCALING

Founded on ecological dynamics, here we outline a theoretical framework for scaling based on the interrelated concepts of constraints and affordances (Buszard et al., 2016; Buszard et al., 2020; Gorman et al., in review; for overviews of ecological dynamics, see Seifert, Araújo, Komar, & Davids, 2017; Seifert, Button, & Davids, 2013; Seifert & Davids, 2017). Constraints are considered to be the factors that influence or shape behaviour (Davids, Button, & Bennett, 2008; Newell, 1986; Passos et al., 2008). The name itself suggests that constraints only act to limit or restrict behaviour, but under this theoretical framework, constraints are simply the boundaries that underpin movement solutions and can therefore have a detrimental, neutral, or functional impact on the emergent performance skills (Davids et al., 2008; Newell, 1986). There are three general categories of constraints including task (e.g. equipment, playing area, and rules), performer (e.g. height of an individual), and environment (e.g. playing surface) (Davids et al., 2008; Newell, 1986). In general, task and performer constraints are most commonly associated with the types of modifications that occur in the context of scaling (see Buszard et al., 2016).

The second concept from ecological dynamics that provides an underpinning theoretical framework for scaling is the notion of affordances (Gibson, 1986; see also Buszard et al., 2020; Gorman et al., in review). Closely related to constraints, affordances are the action possibilities offered by a given situation for a given individual (Fajen, 2005; Gibson, 1986; Renshaw et al., 2009). The possibilities afforded to an individual are not only dependent upon the specific nature of the environment, but they are also influenced by the anthropometric characteristics and action capabilities of the person (Fajen, 2005; Renshaw et al., 2009; Warren, 1984). In a classic research example, the relationship between an individual's leg length and the height of a stair riser was found to afford the opportunity to climb the stair in a bipedal manner (Warren, 1984; for another example, see also Seifert et al., 2018). This perception of affordances not only highlights the interconnectedness of the individual and the environment, which is an important feature of ecological dynamics (Gibson, 1979; Seifert et al., 2013; Seifert & Davids, 2017), but it also highlights the links between constraints and affordances (see Seifert et al., 2013). The height of the stair is a task constraint which, when changed, influenced the affordances available to the individual (Warren, 1984). When the height of the stair exceeded a certain point relative to the leg length of the individual, bipedal climbing of the stair was perceived to be no longer afforded (Warren, 1984). To use a scaling example from sport, when a cricket pitch is scaled by reducing its length (a task constraint), the bowlers are afforded the opportunity of providing more accurate deliveries to the batter (Elliott et al., 2005; see also Buszard et al., 2016; Buszard et al., 2020; Gorman et al., in review). Put simply, scaling is aimed at changing the constraints in order to change the affordances (Buszard et al., 2016; Buszard et al., 2020; Gorman et al., in review; Renshaw & Chow, 2018).

The interactive relationship between constraints and affordances exemplifies the individual–environment synergy, and the need to view skill acquisition as attunement or adaptation to the environment (Araújo & Davids, 2011; Renshaw & Chow, 2018). When constraints are modified in a sport setting, the concomitant changes that occur in the performance environment necessitate that the performer attunes and adapts to those changes to achieve the desired outcome (Araújo & Davids, 2011; Gibson & Pick, 2000; Renshaw & Chow, 2018). A smaller racket in tennis, a lower hoop height in basketball, and a small-sided game in soccer create situations that require individuals to adapt by perceiving information and changing their actions accordingly, based upon their specific action capabilities (Araújo & Davids, 2011; Gibson & Pick, 2000; Renshaw et al., 2009; Renshaw & Chow, 2018). Thus, when applying the concept of scaling to children in a sports setting, skill acquisition essentially involves the learner exploiting the affordances provided by the constraints of the performance environment (Araújo & Davids, 2011; Buszard et al., 2016; Buszard et al., 2020; Gorman et al., in review Renshaw & Chow, 2018).

Another important feature that requires greater attention in the research literature is the use of a principled approach towards scaling (Buszard et al., 2016; Buszard et al., 2020; Gorman et al., in review). Given that the nature and extent of the influence of a particular constraint and the opportunities it affords is dependent upon the nature of the specific individual (Renshaw et al., 2009), scaling should ideally use a principled approach based upon the characteristics of the target group (Buszard et al., 2016; Buszard et al., 2020; Gorman et al., in review). For instance, an optimally sized tennis court for 10-year-old girls may not necessarily be optimal for 10-year-old boys. In a recent case study in basketball, Gorman et al. (in review) used a principled approach to scale the size of the ball to the hand size of the players. The hand spans and hand lengths of 11-year-old boys were measured and compared to those of adult basketball players to identify the ball size that provided the closest anatomical match. Using ratios of hand dimensions-to-ball dimensions for the junior and adult players, the optimal size of basketball for the junior players was revealed to be a size 3 or a size 4 (even though the current size used by the players was a size 6: adults use a size 7 ball). However, the results from small-sided games (three-on-three) showed that despite the presumed affordances offered by the smaller basketballs (i.e. size 3 and 4), the children reported an overall preference for using larger basketballs during the games. The authors surmised that the players had become accustomed to using the larger basketballs to such an extent that their increased familiarity outweighed, at least initially, the affordances offered by the smaller balls. This suggests that sporting associations may need to introduce scaled equipment at an early stage along the developmental pathway (Gorman et al., in review).

3.5   PRACTICAL APPLICATIONS IN SPORT

In this section, we use a recent project commissioned by Cricket Australia and reported by Farrow et al. (2016) to highlight a range of considerations when conducting scaling research and implementing findings into practice. The project was concerned with the redesign of the junior/developmental pathways (ages 8–14) with the aim of enhancing skill development and participation in cricket. Phase 1 comprised individual batting, bowling, and overarm throwing tests, each incorporating a range of scaled equipment modifications that were specific to each age group. The tests assessed metrics such as maximum distances that participants could hit and throw, along with how accurately participants could hit or bowl a ball. Common to all skills tests, and the lone manipulation in the throwing test, was the scaling of the ball used; specifically, the weight, size, and construction. Ball size/types ranged from a small circumference plastic ball through to a full size men's 156 g leather ball. Along with the ball size/type, scaling in the batting test involved a range of cricket bats comprising different sizes (combined blade and handle length), weight, and construction (e.g. plastic or wood). Bats ranged from a small hollow plastic bat through to a small wooden adult-size bat. Bowling testing included the scaling of pitch length in tandem with the different sizes/types of ball. Pitch length ranged from 14 m through to the standard adult length of 20.12 m. The results from the tests in Phase 1 provided a principled approach for the design of a game which was used in Phase 2 of the project. The size of the field was scaled using the results from the throwing and hitting tests, while the pitch length was scaled based upon bowling accuracy.

Using the findings of Phases 1 and 2, recommendations to Cricket Australia included shorter pitch lengths, smaller boundaries, and scaled down ball sizes and/or types relative to the existing match formats across all age groups. In unison with current community practice and experiential knowledge, these recommendations have since been implemented in junior associations nationally. Match statistics from a pilot season using the revised formats indicated some positive outcomes, including 24% fewer dot-balls (deliveries where the batter did not score a run), 35% fewer wides and no-balls ('unhittable' or illegal deliveries), 13% more balls hit, 43% more runs scored off the bat, and 53% more balls bowled on a good length (affording wicket-taking and run-scoring). Perhaps most importantly, 87% of players reported increased enjoyment, 89% of coaches felt more effective, and 76% of parents perceived more benefits for the children who participated in the pilot season.

The project also highlighted some important considerations when planning or implementing changes in junior sport. First, researchers and practitioners should establish what is possible in terms of the combination of different constraints that can be examined and/or implemented. In the cricket project, it was not possible for the participants to complete the testing scenarios with every conceivable bat–ball combination in a systematic manner, even though this would have provided a more comprehensive analysis. Similarly, the implementation of scaling by a sporting association may require a more gradual and longer-term approach to provide adequate time for players (as well as coaches and parents) to familiarise themselves with the changes (Gorman et al., in review; see also Buszard et al., 2016; Buszard et al., 2020). This latter point is related to another key consideration that was particularly evident in the Cricket Australia project, and also in the basketball project described earlier (Gorman et al., in review), in terms of the willingness and capability of end-users to implement any necessary changes to their sport (see Husak, Poto, & Stein, 1986). During the Cricket Australia project, participants were at times hesitant to use scaled equipment, potentially because (a) they were familiar with and liked using the existing equipment, (b) they were not using the 'real equipment' they associated with the adult version of the sport, and/or (c) the players were not performing as well as they had hoped. Similarly, a number of coaches and parents involved with junior cricket were hesitant to embrace change in the implementation of the revised junior formats because they perceived change to be interfering with the fabric of what cricket 'should look like'. Such beliefs are often based on traditions in the sport and/or the prior experiences of the individual (Moy, Renshaw, & Davids, 2014; see also Husak et al., 1986).

Finally, a key point for those interested in implementing scaling projects in their programmes is the need to understand and involve the community (end-users) to help ensure the programme is adopted successfully. This includes involving the participants, parents, coaches, and administrators who will be responsible for implementing and hopefully benefiting from any recommended changes. During the planning phases, the insights and experiences of the community can be invaluable in identifying research problems and establishing what is realistic for short- and long-term change. Valuing community involvement in a project is also critical in terms of implementing recommendations in pilot programmes and receiving informed feedback from participants. By maintaining community engagement, the end-users also retain some ownership and connection with the changes that are implemented, rather than having the outcomes imposed upon them.

3.6   SUMMARY

There are many benefits to implementing scaling in junior sport competitions including, but not limited to, improved skill acquisition and enhanced levels of self-efficacy (e.g. Buszard et al., 2016; Chase et al., 1994; Harwood et al., 2018). In particular, the increased levels of engagement and enjoyment that occur through scaling (Buszard et al., 2016; Farrow & Reid, 2010) may help to attenuate the number of children who choose to dropout from organised sport programs (Buszard et al., 2016; Buszard et al., 2020). Given that many of these programs aim to 'capture' children as young as possible, and often include so called elite-pathways that typically reduce or eliminate exploratory-based learning opportunities, initiatives that rekindle the spirit and philosophy of backyard games through the mechanism of scaling are likely to be paramount for maintaining children's ongoing involvement in sport (Renshaw et al., 2012; see also Buszard et al., 2016; Buszard et al., 2020). Fun is frequently and ubiquitously cited as one of the main reasons that children choose to remain involved in a given sport (Visek et al., 2017), and this key ingredient is not only a core attribute of the typical backyard game (see Baker et al., 2003; Devereux, 2001), but it is also a factor that can be cultivated through scaling (Buszard et al., 2016; Farrow & Reid, 2010). Further research should be conducted using a much broader range of motor tasks across more sports, and with a specific focus upon the application of a principled approach (Buszard et al., 2016; Buszard et al., 2020; Gorman et al., in review). It is equally incumbent upon researchers and practitioners alike to consider the timeframe over which any scaling initiatives are implemented to ensure that end-users (e.g. participants, parents, coaches) have sufficient time to familiarise themselves with the modifications (Buszard et al., 2016; Buszard et al., 2020; Gorman et al., in review; Husak et al., 1986).

REFERENCES

Ahram, T., & Karwowski, W. (2013). Preface. In: T. Ahram & W. Karwowski (eds.), *Advances in Physical Ergonomics and Safety* (pp. xiii–xiv). Boca Raton, FL: CRC Press.

Allender, S., Cowburn, G., & Foster, C. (2006). Understanding participation in sport and physical activity among children and adults: A review of qualitative studies. *Health Education Research*, 21, 826–835.

Araújo, D., & Davids, K. (2011). What exactly is acquired during skill acquisition? *Journal of Consciousness Studies*, 18, 7–23.

Arias, J. L. (2012). Influence of ball weight on shot accuracy and efficacy among 9–11-year-old male basketball players. *Kinesiology*, 44, 52–59.

Arias, J. L., Argudo, F. M., & Alonso, J. I. (2012a). Effect of basketball mass on shot performance among 9–11 year-old male players. *International Journal of Sports Science and Coaching*, 7, 69–79.

Arias, J. L., Argudo, F. M., & Alonso, J. I. (2012b). Effect of the ball mass on the one-on-one game situation in 9–11 year old boys' basketball. *European Journal of Sport Science*, 12, 225–230.

Arias, J. L., Argudo, F. M., & Alonso, J. I. (2012c). Effect of ball mass on dribble, pass, and pass reception in 9–11-year-old boys' basketball. *Research Quarterly for Exercise and Sport*, 83, 407–412.

Baker, J., Côté, J., & Abernethy, B. (2003). Sport-specific practice and the development of expert decision-making in team ball sports. *Journal of Applied Sport Psychology*, 15, 12–25.

Balish, S. M., McLaren, C., Rainham, D., & Blanchard, C. (2014). Correlates of youth sport attrition: A review and future directions. *Psychology of Sport and Exercise*, 15, 429–439.

Brown, E. W., Wisner, D. M., & Kontos, A. (2000). Comparison of the incidences of selected events performed by youth players in regulation and modified soccer games. *International Journal of Applied Sports Sciences*, 12, 2–21.

Burton, A. W., Greer, N. L., & Wiese, D. M. (1992). Changes in overhand throwing patterns as a function of ball size. *Pediatric Exercise Science*, 4, 50–67.

Buszard, T., Farrow, D., & Reid, M. (2020). Designing junior sport to maximise potential: The knowns, unknowns, and paradoxes of scaling sport. *Frontiers in Psychology*, 10, Article 2878.

Buszard, T., Farrow, D., Reid, M., & Masters, R. S. (2014). Modifying equipment in early skill development: A tennis perspective. *Research Quarterly for Exercise and Sport*, 85, 218–225.

Buszard, T., Reid, M., Masters, R., & Farrow, D. (2016). Scaling the equipment and play area in children's sport to improve motor skill acquisition: A systematic review. *Sports Medicine*, 46, 829–843.

Cannane, S. (2009). *First Tests: Great Australian Cricketers and the Backyards that Made Them*. Sydney, Australia: ABC Books.

Carlman, P., Wagnsson, S., & Patriksson, G. (2013). Causes and consequences of dropping out from organized youth sports. *Swedish Journal of Sport Research*, 2, 26–54.

Chase, M. A., Ewing, M. E., Lirgg, C. D., & George, T. R. (1994). The effects of equipment modification on children's self-efficacy and basketball shooting performance. *Research Quarterly for Exercise and Sport*, 65, 159–168.

Clements, R. (2004). An investigation of the status of outdoor play. *Contemporary Issues in Early Childhood*, 5, 68–80.

Cooper, P. (2010). Play and children. In: L. Kidman & B. J. Lombardo (eds.), *Athlete-Centred Coaching: Developing Decision Makers* (2nd edition; pp. 137–150). Worcester, UK: IPC Print Resources.

Cordovil, R., Araújo, D., Davids, K., Gouveia, L., Barreiros, J., Fernandes, O., & Serpa, S. (2009). The influence of instructions and body-scaling as constraints on decision-making processes in team sports. *European Journal of Sport Science*, 9, 169–179.

Côté, J., & Hancock, D. J. (2016). Evidence-based policies for youth sport programmes. *International Journal of Sport Policy and Politics*, 8, 51–65.

Crane, J., & Temple, V. (2015). A systematic review of dropout from organized sport among children and youth. *European Physical Education Review*, 21, 114–131.

Davids, K., Button, C., & Bennett, S. (2008). *Dynamics of Skill Acquisition: A Constraints-Led Approach*. Champaign, IL: Human Kinetics.

Deci, E. L., & Ryan, R. M. (2000). The "what" and "why" of goal pursuits: Human needs and the self-determination of behavior. *Psychological Inquiry*, 11, 227–268.

Delorme, N., Chalabaev, A., & Raspaud, M. (2011). Relative age is associated with sport dropout: Evidence from youth categories of French basketball. *Scandinavian Journal of Medicine and Science in Sports*, 21, 120–128.

Devereux, E. C. (2001). Backyard versus little league baseball: Some observations on the impoverishment of children's games in contemporary America. In: A. Yiannakis & M. J. Melnick (eds.), *Contemporary Issues in Sociology of Sport* (pp. 63–71). Champaign, IL: Human Kinetics.

Egstrom, G. H., Logan, G. A., & Wallis, E. L. (1960). Acquisition of throwing skill involving projectiles of varying weights. *Research Quarterly. American Association for Health, Physical Education and Recreation*, 31, 420–425.

Eime, R. M., Casey, M. M., Harvey, J. T., Charity, M. J., Young, J. A., & Payne, W. R. (2015). Participation in modified sports programs: A longitudinal study of children's transition to club sport competition. *BMC Public Health*, 15, 1–7.

Elliott, B., Plunkett, D., & Alderson, J. (2005). The effect of altered pitch length on performance and technique in junior fast bowlers. *Journal of Sports Sciences*, 23, 661–667.

Fajen, B. R. (2005). Perceiving possibilities for action: On the necessity of calibration and perceptual learning for the visual guidance of action. *Perception*, 34, 717–740.

Farrow, D., Headrick, J., Rudd, J. R., Renshaw, I., Buszard, T., & Reid, M. (2016). *Examination of the Optimal Equipment Modification and Game Formats for the Foundation Stage of the Australian Cricket Pathway*. Melbourne, Australia: Report Prepared for Cricket Australia.

Farrow, D., & Reid, M. (2010). The effect of equipment scaling on the skill acquisition of beginning tennis players. *Journal of Sports Sciences*, 28, 723–732.

Gibson, E. J., & Pick, A. D. (2000). *An Ecological Approach to Perceptual Learning and Development*. Oxford, UK: Oxford University Press.

Gibson, J. J. (1979). *The Ecological Approach to Visual Perception*. Boston, MA: Houghton Mifflin.

Gibson, J. J. (1986). *The Ecological Approach to Visual Perception*. Hillsdale, NJ: Erlbaum.

Gorman, A. D., Headrick, J., Renshaw, I., McCormack, C. J., & Topp, K. (in review). A principled approach to equipment scaling for children's sport: A case study in basketball.

Gould, D. (1987). Understanding attrition in children's sport. In: D. Gould & M. R. Weiss (eds.), *Advances in Pediatric Sport Sciences: Volume 2, Behavioral Issues* (pp. 401–411). Champaign, IL: Human Kinetics.

Hammond, J., & Smith, C. (2006). Low compression tennis balls and skill development. *Journal of Sports Science and Medicine*, 5, 575–581.

Harwood, M. J., Yeadon, M. R., & King, M. A. (2018). Reducing the pitch length: Effects on junior cricket. *International Journal of Sports Science and Coaching*, 13(6), 1–9.

Husak, W. S., Poto, C. C., & Stein, G. (1986). The women's smaller basketball: Its influence on performance and attitude. *Journal of Physical Education, Recreation and Dance*, 57, 18–26.

Karwowski, W. (2005). Ergonomics and human factors: The paradigms for science, engineering, design, technology and management of human-compatible systems. *Ergonomics*, 48, 436–463.

Kelley, B., & Carchia, C. (2013). Hey, data data – swing! Retrieved from http://www.espn.com/espn/story/_/id/9469252/hidden-demographics-youth-sports-espn-magazine, 1 February 2019.

Larson, E. J., & Guggenheimer, J. D. (2013). The effects of scaling tennis equipment on the forehand groundstroke performance of children. *Journal of Sports Science and Medicine*, 12, 323–331.

Lee, M. C. Y., Chow, J. Y., Komar, J., Tan, C. W. K., & Button, C. (2014). Nonlinear pedagogy: An effective approach to cater for individual differences in learning a sports skill. *PloS One*, 9, 1–13.

Limpens, V., Buszard, T., Shoemaker, E., Savelsbergh, G. J., & Reid, M. (2018). Scaling constraints in junior tennis: The influence of net height on skilled players' match-play performance. *Research Quarterly for Exercise and Sport*, 89, 1–10.

Machado, J. C., Barreira, D., Galatti, L., Chow, J. Y., Garganta, J., & Scaglia, A. J. (2018). Enhancing learning in the context of street football: A case for nonlinear pedagogy. *Physical Education and Sport Pedagogy*, 1–14.

McCormick, B. T., Hannon, J. C., Newton, M., Shultz, B., Miller, N., & Young, W. (2012). Comparison of physical activity in small-sided basketball games versus full-sided games. *International Journal of Sports Science and Coaching*, 7, 689–697.

Mexas, K., Tsitskaris, G., Kyriakou, D., & Garefis, A. (2005). Comparison of effectiveness of organized offences between two different championships in high level basketball. *International Journal of Performance Analysis in Sport*, 5, 72–82.

Moy, B., Renshaw, I., & Davids, K. (2014). Variations in acculturation and Australian physical education teacher education students' receptiveness to an alternative pedagogical approach to games teaching. *Physical Education and Sport Pedagogy*, 19, 349–369.

Newell, K. M. (1986). Constraints on the development of coordination. In: M. G. Wade & H. T. A. Whiting (eds.), *Motor Skill Acquisition in Children: Aspects of Coordination and Control* (pp. 341–360). Dordrecht, the Netherlands: Martinus Nijhoff.

Passos, P., Araújo, D., Davids, K., & Shuttleworth, R. (2008). Manipulating constraints to train decision making in rugby union. *International Journal of Sports Science and Coaching*, 3, 125–140.

Pellett, T. L., Henschel-Pellett, H. A., & Harrison, J. M. (1994). Influence of ball weight on junior high school girls' volleyball performance. *Perceptual and Motor Skills*, 78, 1379–1384.

Phillips, E., Davids, K., Renshaw, I., & Portus, M. (2010). The development of fast bowling experts in Australian cricket. *Talent Development and Excellence*, 2, 137–148.

Pinder, R. A., Davids, K., Renshaw, I., & Araújo, D. (2011). Representative learning design and functionality of research and practice in sport. *Journal of Sport & Exercise Psychology*, 33, 146–155.

Renshaw, I., & Chappell, G. (2010). A constraints-led approach to talent development in cricket. In: L. Kidman & B. J. Lombardo (eds.), *Athlete-Centred Coaching: Developing Decision Makers* (2nd edition; pp. 151–172). Worcester, UK: IPC Print Resources.

Renshaw, I., & Chow, J. (2018). A constraint-led approach to sport and physical education pedagogy. *Physical Education and Sport Pedagogy*, 1–14.

Renshaw, I., Davids, K. W., Shuttleworth, R., & Chow, J. Y. (2009). Insights from ecological psychology and dynamical systems theory can underpin a philosophy of coaching. *International Journal of Sport Psychology*, 40, 580–602.

Renshaw, I., & Moy, B. (2018). A constraint-led approach to coaching and teaching games: Can going back to the future solve the 'they need the basics before they can play a game' argument? *Ágora para la Educación Física y el Deporte*, 20, 1–26.

Renshaw, I., Oldham, A. R., & Bawden, M. (2012). Nonlinear pedagogy underpins intrinsic motivation in sports coaching. *The Open Sports Sciences Journal*, 5, 88–99.

Rupnow, A., & Engelhorn, R. (1989). Effect of substitution rules on player participation in youth softball and baseball. *Journal of Applied Research in Coaching and Athletics*, 4, 233–242.

Satern, M. N., Messier, S. P., & Keller-McNulty, S. (1989). The effect of ball size and basket height on the mechanics of the basketball free throw. *Journal of Human Movement Studies*, 16, 123–137.

Seifert, L., Araújo, D., Komar, J., & Davids, K. (2017). Understanding constraints on sport performance from the complexity sciences paradigm: An ecological dynamics framework. *Human Movement Science*, 56, 178–180.

Seifert, L., Button, C., & Davids, K. (2013). Key properties of expert movement systems in sport: An ecological dynamics perspective. Sports Medicine, 43, 167–178.

Seifert, L., & Davids, K. (2017). Ecological dynamics: A theoretical framework for understanding sport performance, physical education and physical activity. In: P. Bourgine,

P. Collet, & P. Parrend (eds.), *First Complex Systems Digital Campus World e-Conference 2015* (pp. 29–40). Cham, Switzerland: Springer International Publishing.

Seifert, L., Orth, D., Mantel, B., Boulanger, J., Hérault, R., & Dicks, M. (2018). Affordance realisation in climbing: Learning and transfer. *Frontiers in Psychology*, 9, 1–14.

Timmerman, E., De Water, J., Kachel, K., Reid, M., Farrow, D., & Savelsbergh, G. (2015). The effect of equipment scaling on children's sport performance: The case for tennis. *Journal of Sports Sciences*, 33, 1093–1100.

Uehara, L. A., Button, C., Araújo, D., Renshaw, I., Davids, K., & Falcous, M. (2018). The role of informal, unstructured practice in developing football expertise: The case of Brazilian Pelada. *Journal of Expertise*, 1, 162–180.

Visek, A. J., Achrati, S. M., Mannix, H. M., McDonnell, K., Harris, B. S., & DiPietro, L. (2015). The fun integration theory: Towards sustaining children and adolescents sport participation. *Journal of Physical Activity and Health*, 12, 424–433.

Visek, A. J., Mannix, H., Mann, D., & Jones, C. (2017). Integrating fun in young athletes' sport experiences. In: C. J. Knight, C. G. Harwood, & D. Gould (eds.), *Sport Psychology for Young Athletes* (pp. 68–80). London, UK: Routledge.

Warren, W. H. (1984). Perceiving affordances: Visual guidance of stair climbing. *Journal of Experimental Psychology: Human Perception and Performance*, 10, 683–703.

Witt, P. A., & Dangi, T. B. (2018). Why children/youth drop out of sports. *Journal of Park and Recreation Administration*, 36, 191–199.

Wright, E. J. (1967). Effects of light and heavy equipment on acquisition of sports-type skills by young children. *Research Quarterly. American Association for Health, Physical Education and Recreation*, 38, 705–714.

4 Injury Prevention in Sport – Design with Implementation in Mind

Erich Petushek and Alex Donaldson

CONTENTS

4.1 SPORT INJURY EPIDEMIOLOGY AND PREVENTION FRAMEWORKS

The physical activity inherent in sporting activities is vital to fostering long-term health and well-being in youth/adolescents, with approximately 60 million school-aged children (6–18 years) participating in sports/physical activity annually in the United States alone (Physical Activity Council, 2019). However, sport participation can have negative consequences, including traumatic injury (Avraham et al., 2019; Poulsen et al., 2019). In fact, sports injuries are the second most common and costly cause of severe injury in paediatric populations totalling nearly $2 billion in health care costs annually in the United States (Avraham et al., 2019). Injuries are more common in some sports compared to others (see Figure 4.1), and vary by body region (see Figure 4.2) and level of play (Comstock & Pierpoint, 2019; Hootman et al., 2007).

Epidemiological data is often the first step in many injury prevention research frameworks/processes. A model proposed by van Mechelen in 1992, termed the 'sequence of prevention', has served as a simplified guiding framework for injury prevention research (Figure 4.3) (van Mechelen et al., 1992).

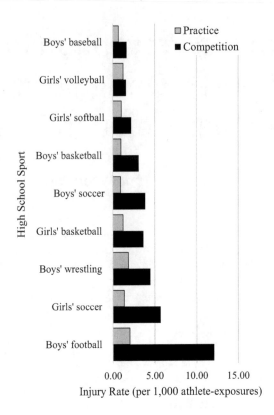

FIGURE 4.1   Normalised injury rates across various high school sports. Data extracted from Comstock and Pierpoint (2019).

4.2   ESTABLISHING THE EXTENT OF THE INJURY PROBLEM (STEP 1)

The first step is establishing the extent of the injury problem, often by using epidemiological methods to quantify the incidence and severity of the injuries. The second step is to explore the risk factors and mechanisms of the injury, and the third step is to introduce preventive measures. The fourth and final step is then to assess the effectiveness of these prevention measures by repeating Step 1. This framework has guided many research investigations over the previous decades, but many have proposed more in-depth models and frameworks as the various stages in the original framework are likely underspecified (Finch, 2006; Hulme & Finch, 2015; Meeuwisse, 1994). For example, once research has shown efficacy for an injury prevention strategy, this doesn't mean that it will be effective; thus further research around the implementation of prevention strategies is warranted (Finch, 2006). Because 'sports-injury prevention requires above all behavioural change' (Verhagen & van Mechelen, 2010), it is our view that the field of HFE can provide novel tools, methods, and strategies to progress the field. The aim of this chapter is to explore how HFE methods have been used to develop tools and methods to augment Steps 2 (risk assessment) and 3

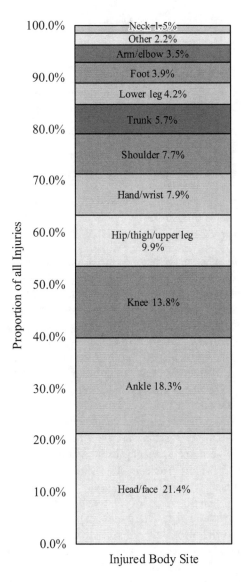

FIGURE 4.2 High school sport injury proportion by body site. Data extracted from Comstock and Pierpoint (2019).

(implementing prevention strategies) of the 'sequence of prevention'. Moreover, the following sections will showcase how HFE methods can support the design of instruments, training, and decision support tools to reduce sport-related injuries.

4.3 OPTIMISING RISK ASSESSMENT (STEP 2)

Biomechanical and athlete movement characteristics have been the focus for much of the work of examining mechanisms and risk factors of injury (Bahr & Krosshaug,

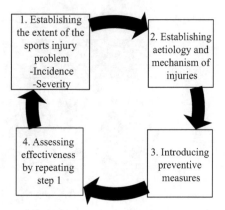

FIGURE 4.3   Sequence of Prevention Framework. Adapted from van Mechelen et al. (1992).

2005). The typical biomechanical methods use 2D and 3D movement analysis equipment, wearable sensors, electromyography, and other imaging techniques (Hamill & Knutzen, 2006). These methods often require expensive equipment with specialised operators – thus limiting large-scale use. Visual inspection or observational movement diagnosis is one alternative screening method which would reduce screening time and cost, while potentially preserving relatively high-risk assessment accuracy (Knudson, 2013; Petushek et al., 2015c). For example, a practitioner (i.e. coach, athletic trainer, physical therapist, etc.) could almost instantaneously assess injury risk by observing a task where movement patterns are similar to those that cause the injury (i.e. jump landing or cutting for anterior cruciate ligament (ACL) injury) (Nilstad et al., 2014). Identifying athletes or patients with abnormal/flawed or inefficient movement is a common task for many coaches (Knudson, 2000), sport judges, and sports medicine practitioners (i.e. physical therapist, athletic trainer, etc.) (Jensen et al., 2000). One of the main problems, however, is determining the accuracy and reliability of visual inspection for estimating injury risk.

To address this problem, Petushek and colleagues (Petushek et al., 2015a) designed an efficient five-item test to assess individual's capability of estimating injury risk based on visual inspection of a routine movement screening test – termed the Anterior Cruciate Ligament Injury-Risk-Estimation Quiz or ACL-IQ (Petushek et al., 2015a, 2015b, 2015c) (see Figure 4.4 for screenshot of video/example item).

Because this assessment involves perceptual-cognitive tasks, the ACL-IQ uses psychometric, verbal protocol, survey, and eye-tracking methods to develop and add validity evidence to this test. The test was developed by comparing the assessors visual rating to a gold-standard 3D biomechanical assessment (Hewett et al., 2005). This five-item assessment demonstrated sufficient test–retest reliability (r = .90), sensitivity to change, discriminant and convergent validity (Petushek et al., 2015a). Following test development, Petushek and colleagues (Petushek et al., 2015c) examined inter-professional/individual differences in risk estimation ability and found substantial differences warranting improvement. For example, individuals with routine contact (i.e. coaches) or who contribute to pre-participation/return-to-sport decisions (i.e. physicians) had poorer risk estimating ability than other sports medicine

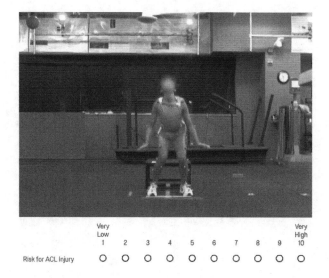

FIGURE 4.4   Screenshot of a video item from the Anterior Cruciate Ligament Injury Risk Estimation Quiz. (Adapted from Petushek et al., 2015a.)

professionals (i.e. athletic trainers, physical therapists) (see Figure 4.5: Petushek et al., 2015c). Thus, the next step was to develop decision support tools and training to address these performance differences.

To understand the perceptual-cognitive characteristics underlying risk estimating skill to inform the design of decision support and training, various HFE methods were used to converge on the essential visual cues. In one experiment, cue-utilization (Loveday et al., 2014) was assessed using survey methods. Three cues were found to be highly important for assessing injury risk: medial knee motion, landing stiffness, and landing symmetry (Petushek et al., 2015b). Network analysis was also used to explore open response data to the question: What information were you using when assessing ACL injury risk? The results of the network analysis revealed that medial knee motion was the cue that was mentioned most frequently similar to the cue-utilisation results. Finally, eye-tracking and retrospective verbal reporting of experts was undertaken to create concept maps of this risk estimating task (see Figure 4.6).

All of these methods seemed to converge on similar cues (Figure 4.7). From this information a fast-and-frugal decision tree (see Figure 4.7) (Martignon et al., 2003; Phillips et al., 2017) and a short animated video (see Figure 4.7) highlighting the effective cue/strategies for estimating ACL injury risk were developed. The video was tested in a randomised controlled trial using sport coaches and revealed improvements in ACL-IQ scores of six percentage points (Cohen's d of .45) from baseline (unpublished data).

The previous studies are an example of successfully applying HFE approaches to reducing sport injuries by targeting and improving efficiency in the second step of the widely-used sequence of prevention research framework – risk assessment. Holistically, the test development and cross-professional difference assessments provides evidence for using convenient, cost-effective, strategies (i.e. visual assessment)

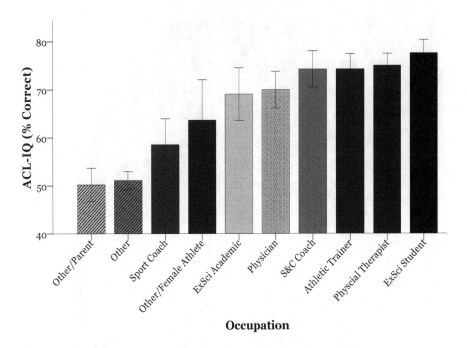

FIGURE 4.5 Anterior Cruciate Ligament Injury Risk Estimation Quiz (ACL-IQ) scores across professions. ExSci – Exercise Science, S&C – Strength and Conditioning. Data extracted from Petushek et al. (2015).

to assess injury risk. Screening or assessing injury risk using movement tasks such as the drop vertical jump, is not without limitation (Roald Bahr, 2016). For example, screening using this movement task does not appear to be accurate in elite athletes (Krosshaug et al., 2016; Mørtvedt et al., 2019). However, the cost of misclassification is typically low due to the type of preventive intervention introduced (e.g. exercise-based training), thus assessing risk may help improve uptake and adherence to effective risk reduction strategies – which is a major challenge in preventing injuries.

4.4 OPTIMISING UPTAKE AND ADHERENCE TO EVIDENCE-BASED PREVENTIVE STRATEGIES (STEP 3)

Injury prevention strategies in sport have been categorised into equipment, training, or regulations (McBain et al., 2012). In medical or health fields, the uptake of and adherence to evidence-based prevention strategies is low, and some have suggested it takes about 17 years to translate research evidence into practice (Morris et al., 2011). So how can we improve this time-lag from evidence to practice? HFE, system engineering, and implementation sciences are well positioned to help address these problems (Carayon et al., 2018; Donaldson & Finch, 2013; Gurses et al., 2012; Russ et al., 2013). The following section will highlight the application of HFE methods to help improve the use of effective preventive strategies.

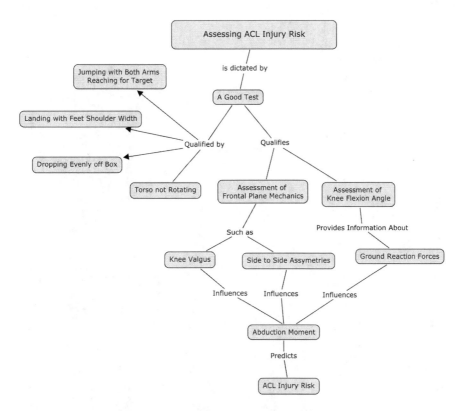

FIGURE 4.6 Example concept map from a retrospective verbal report of an expert (unpublished data).

Donaldson and colleagues (2016) built on the sequence of prevention framework to expand the third step – introducing preventative measures as this step seems to be the most complex and in need of further guidance (Donaldson et al., 2016). The initial step in this process requires the combination of synthesising the current research evidence and clinical experience to ensure the effective components, such as the specific physical exercises are selected for implementation. Typically, meta-analyses are used to quantitatively synthesise research evidence to create best practices. However, many meta-analyses do not explicitly outline the effective components of various interventions especially when they involve several complex components such as exercise-based injury prevention training programmes. One way guidelines for effective components of injury prevention programmes can be communicated are checklists.

Checklists, if designed and tested using HFE approaches such as usability testing, can effectively improve understanding and change practice (see examples such as Bergs et al., 2014; Weiser et al., 2010). Exercise-based injury prevention training programs have been shown to reduce injuries especially for ACL injuries in female athletes (Webster & Hewett, 2018). However, use of ACL injury prevention training is low (~13–20%) nationally and very low in rural areas (~4%) (Joy et al., 2013;

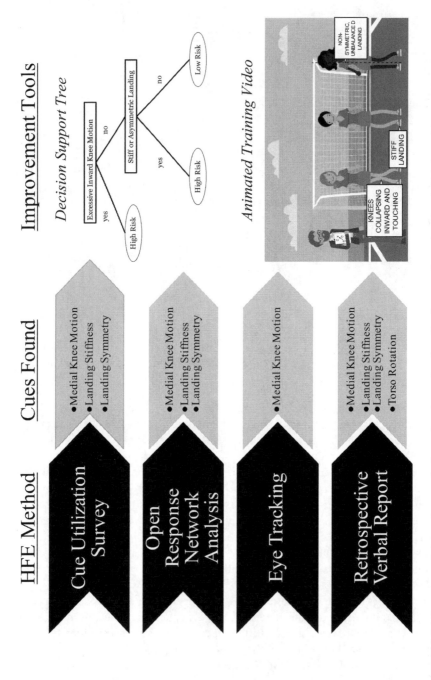

FIGURE 4.7 Summary of the various methods used to determine the effective cues for estimating ACL injury risk as well as the tools created from these convergent findings.

Norcross et al., 2016; Sugimoto et al., 2012). Even when injury prevention training is studied in controlled research environments, compliance is relatively low (e.g. ~50%) (Sugimoto et al., 2012, 2017). Compliance deficits can, to a significant extent, be attributed to the coach not the athlete. For example, when injury prevention training is introduced to teams in a research setting, only about 50% of coaches use the program (Sugimoto et al., 2012, 2017). However, within those teams that used the programmes, nearly 90% of athletes adhered to them. Thus, coach adherence to implementing injury prevention training is a key requirement for ACL injury prevention (Iversen & Friden, 2009; Pollard et al., 2006; Pryor et al., 2017). Coach knowledge and comprehension of evidence-based prevention strategies has been identified as the most important modifiable barrier to implementation (Donaldson et al., 2019; Iversen & Friden, 2009; McKay et al., 2014; Norcross et al., 2016; O'Brien & Finch, 2016; Orr et al., 2013). For example, only 50% of athletes, parents, and coaches believe knee injuries can be prevented and only 15% know about effective prevention strategies (Orr et al., 2013). Based on this information, and user-experience data, a best-practice checklist for reducing ACL injuries using meta-analytic and qualitative techniques was created (Petushek et al., 2019). The main goal was to design a checklist that included the minimum essential components to effectively reduce injuries. The final checklist can be seen in Figure 4.8. This checklist is not only a practical tool to facilitate effective component use, it can serve as a research tool to quantify current use of evidence-based prevention practices.

Does your program/training include?

	No	Yes
Lunges	0	1
Hamstring Exercises (e.g., Nordic Hamstring)	0	1
Heel/Calf Raises	0	1

Total number of *landing stabilization* exercises **per session** (e.g., drop landings, jump/hop and holds)

| 0 | 1 | 2 | 3 | 4 | 5 |

Number of exercises

What is the age of athletes you work with?

Middle/High School 1   College/Professional 0

How often do you perform the program?

Pre-Season Only 0   In-Season or Both Pre/In-Season 1

Has the person implementing the program (e.g., Coach) been trained or educated on ACL injury prevention programing (e.g., workshop, video/brochure)?

No 0   Yes 1

Total Score:____/11

Checklist Score	0	1	2	3	4	5	6	7	8	9	10	11
Interpretation	No Benefit			Small Benefit				Intermediate Benefit			Large Benefit	

FIGURE 4.8 Meta-analytic driven best practice checklist for reducing ACL injuries in female athletes. Reproduced from Petushek et al. (2018).

An often-neglected component of the injury prevention research process is the engagement of populations that will ultimately be required to implement the interventions (e.g. the athletes, coaches, sports administrators, sports/athletic trainers, and officials/referees/umpires). A key element to enhancing the uptake or implementation of any injury prevention strategy is to engage the end-users of the strategy in the whole process from needs assessment through to intervention development and implementation planning (Donaldson & Finch, 2012; Padua et al., 2014). This is important across the sports system (Donaldson et al., 2019; Finch & Donaldson, 2010) to maximise the 'fit' between the injury issue being addressed, the intervention developed and the implementation context. Two frameworks available to assist sports injury prevention researchers with intervention development and implementation planning are Bartholomew and colleagues' Intervention Mapping health promotion programme planning protocol (Donaldson & Poulos, 2014; Eldredge et al., 2016), and Donaldson and colleagues six-step process for developing evidence-informed and context specific sports injury prevention interventions (Donaldson et al., 2016). The two aforementioned frameworks utilise various HFE methods to elicit user knowledge and information such as Delphi consultation, concept mapping, prototyping, focus groups, and other stakeholder interview methods (Donaldson & Finch, 2012; Donaldson et al., 2015, 2019). The user-experience data is essential to design effective injury prevention initiatives, however, further research is needed to better understand how various users adapt to emerging knowledge and technology in order to sustain prevention efforts. In addition, efforts to explore the inter-relationships between the various system components and stakeholders and how this influences uptake and adherence to prevention efforts would further improve the design of sustainable prevention strategies (Hulme & Finch, 2015).

4.5 FUTURE WORK

The field of sport injury prevention could benefit from designing more efficient and usable tools for each step of the 'sequence of prevention' framework. In addition, systems-based approaches such as the Systems Engineering Initiative for Patient Safety could be applied to athlete safety to better understand and improve other important facets of athlete safety systems (Carayon et al., 2006; Hulme & Finch, 2015). An area not discussed here, but important, is injury surveillance and workload monitoring which has many HFE considerations. For example, usability analysis could be used for assessing surveillance system characteristics to optimise interaction and adherence. Furthermore, future work is needed to optimise the uptake of evidence-based interventions with a focus on adaptation to user preference and emerging technologies. Methods such as task and usability analyses and computational models (see Chapter 20) could aid in this area. Finally, the examples presented in this chapter could be applied to other burdensome injuries such as concussions (see Chapter 19) and ankle sprains (Comstock & Pierpoint, 2019).

4.6 CONCLUSION

Historically, sport injury prevention has borrowed methods from diverse fields such as epidemiology, psychology, exercise, health promotion, implementation science, and computer sciences as well as HFE. The focus on design within HFE has the potential to progress the field and make an impact on reducing injuries for all.

REFERENCES

Avraham, J. B., Bhandari, M., Frangos, S. G., Levine, D. A., Tunik, M. G., & DiMaggio, C. J. (2019). Epidemiology of paediatric trauma presenting to US emergency departments: 2006–2012. *Injury Prevention*, 25:2, 136–143. doi: 10.1136/injuryprev-2017-042435.

Bahr, R. (2016). Why screening tests to predict injury do not work – and probably never will...: A critical review. *British Journal of Sports Medicine*, 50:13, 776–780. doi: 10.1136/bjsports-2016-096256.

Bahr, R., & Krosshaug, T. (2005). Understanding injury mechanisms: A key component of preventing injuries in sport. *British Journal of Sports Medicine*, 39:6, 324–329. doi: 10.1136/bjsm.2005.018341.

Bergs, J., Hellings, J., Cleemput, I., Zurel, Ö., Troyer, V. D., Hiel, M. V., Demeere, J. -L., Claeys, D., & Vandijck, D. (2014). Systematic review and meta-analysis of the effect of the World Health Organization surgical safety checklist on postoperative complications. *British Journal of Surgery*, 101:3, 150–158. doi: 10.1002/bjs.9381.

Carayon, P., Hundt, A. S., Karsh, B. T., Gurses, A. P., Alvarado, C. J., Smith, M., & Brennan, P. F. (2006). Work system design for patient safety: The SEIPS model. *BMJ Quality and Safety*, 15:suppl 1, i50–i58.

Carayon, P., Wooldridge, A., Hose, B.-Z., Salwei, M., & Benneyan, J. (2018). Challenges and opportunities for improving patient safety through human factors and systems engineering. *Health Affairs*, 37:11, 1862–1869. doi: 10.1377/hlthaff.2018.0723.

Comstock, d., & Pierpoint, L. (2019). *National High School Sports-related Injury Surveillance Study: 2018–2019 School Year.* Presented by the Centre for Injury Research and Policy.

Donaldson, A., Callaghan, A., Bizzini, M., Jowett, A., Keyzer, P., & Nicholson, M. (2019). A concept mapping approach to identifying the barriers to implementing an evidence-based sports injury prevention programme. *Injury Prevention*, 25:4, 244–251.

Donaldson, A., Cook, J., Gabbe, B., Lloyd, D. G., Young, W., & Finch, C. F. (2015). Bridging the gap between content and context: Establishing expert consensus on the content of an exercise training program to prevent lower-limb injuries. *Clinical Journal of Sport Medicine* 25:3, 221–229. doi: 10.1097/JSM.000000000000012

Donaldson, A., & Finch, C. F. (2012). Planning for implementation and translation: Seek first to understand the end-users' perspectives. *British Journal of Sports Medicine*, 46:5, 306–307. doi: 10.1136/bjsports-2011-090461.

Donaldson, A., & Finch, C. F. (2013). Applying implementation science to sports injury prevention. *British Journal of Sports Medicine*, 47:473–475.

Donaldson, A., Lloyd, D. G., Gabbe, B. J., Cook, J., Young, W., White, P., & Finch, C. F. (2016). Scientific evidence is just the starting point: A generalizable process for developing sports injury prevention interventions. *Journal of Sport and Health Science*, 5:3, 334–341. doi: 10.1016/j.jshs.2016.08.003.

Donaldson, A., & Poulos, R. (2014). Planning the diffusion of a neck-injury prevention programme among community rugby union coaches. *British Journal of Sports Medicine*, 48:2, 151–159. doi: 10.1136/bjsports-2012-091551.

Eldredge, L. K. B., Markham, C. M., Ruiter, R. A. C., Fernández, M. E., Kok, G., & Parcel, G. S. (2016). *Planning Health Promotion Programs: An Intervention Mapping Approach.* John Wiley & Sons.

Finch, C. (2006). A new framework for research leading to sports injury prevention. *Journal of Science and Medicine in Sport*, 9:1–2, 3–9.

Finch, C. F., & Donaldson, A. (2010). A sports setting matrix for understanding the implementation context for community sport. *British Journal of Sports Medicine*, 44:13, 973–978. doi: 10.1136/bjsm.2008.056069.

Gurses, A. P., Ozok, A. A., & Pronovost, P. J. (2012). Time to accelerate integration of human factors and ergonomics in patient safety. *BMJ Quality and Safety*, 21:4, 347–351. doi: 10.1136/bmjqs-2011-000421.

Hamill, J., & Knutzen, K. M. (2006). *Biomechanical Basis of Human Movement.* Philadelphia, PA: Lippincott Williams & Wilkins.

Hewett, T. E., Myer, G. D., Ford, K. R., Heidt, R. S., Colosimo, A. J., McLean, S. G., van den Bogert, A. J., Paterno, M. V., & Succop, P. (2005). Biomechanical measures of neuromuscular control and valgus loading of the knee predict anterior cruciate ligament injury risk in female athletes: A prospective study. *The American Journal of Sports Medicine*, 33:4, 492–501. doi: 10.1177/0363546504269591.

Hootman, J. M., Dick, R., & Agel, J. (2007). Epidemiology of collegiate injuries for 15 sports: Summary and recommendations for injury prevention initiatives. *Journal of Athletic Training*, 42(2), 311.

Hulme, A., & Finch, C. F. (2015). From monocausality to systems thinking: A complementary and alternative conceptual approach for better understanding the development and prevention of sports injury. *Injury Epidemiology*, 2(1), 31. doi: 10.1186/s40621-015-0064-1.

Iversen, M. D., & Friden, C. (2009). Pilot study of female high school basketball players' anterior cruciate ligament injury knowledge, attitudes, and practices. *Scandinavian Journal of Medicine and Science in Sports*, 19:4, 595–602.

Jensen, G. M., Gwyer, J., Shepard, K. F., & Hack, L. M. (2000). Expert practice in physical therapy. *Physical Therapy*, 80:1, 28–43. doi: 10.1093/ptj/80.1.28.

Joy, E. A., Taylor, J. R., Novak, M. A., Chen, M., Fink, B. P., & Porucznik, C. A. (2013). Factors influencing the implementation of anterior cruciate ligament injury prevention strategies by girls soccer coaches. *The Journal of Strength and Conditioning Research*, 27:8, 2263–2269.

Knudson, D. (2000). What can professionals qualitatively analyze? *Journal of Physical Education, Recreation and Dance*, 71:2, 19–23. doi: 10.1080/07303084.2000.10605997.

Knudson, D. V. (2013). *Qualitative Diagnosis of Human Movement: Improving Performance in Sport and Exercise.* Champaign, IL: Human Kinetics.

Krosshaug, T., Steffen, K., Kristianslund, E., Nilstad, A., Mok, K.-M., Myklebust, G., Andersen, T. E., Holme, I., Engebretsen, L., & Bahr, R. (2016). The vertical drop jump is a poor screening test for ACL injuries in female elite soccer and handball players: A prospective cohort study of 710 athletes. *The American Journal of Sports Medicine*, 44:4, 874–883. doi: 10.1177/0363546515625048.

Loveday, T., Wiggins, M. W., & Searle, B. J. (2014). Cue utilization and broad indicators of workplace expertise. *Journal of Cognitive Engineering and Decision Making*, 8:1, 98–113. DOI: 10.1177/1555343413497019.

Martignon, L., Vitouch, O., Takezawa, M., & Forster, M. R. (2003). Naive and yet enlightened: From natural frequencies to fast and frugal decision trees. *Thinking: Psychological Perspective on Reasoning, Judgment, and Decision Making*, 189–211.

McBain, K., Shrier, I., Shultz, R., Meeuwisse, W. H., Klügl, M., Garza, D., & Matheson, G. O. (2012). Prevention of sport injury II: A systematic review of clinical science research. *The British Journal of Sports Medicine*, 46:3, 174–179.

McKay, C. D., Steffen, K., Romiti, M., Finch, C. F., & Emery, C. A. (2014). The effect of coach and player injury knowledge, attitudes and beliefs on adherence to the FIFA 11+ programme in female youth soccer. *The British Journal of Sports Medicine*, 48:17, 1281–1286.

Meeuwisse, W. H. (1994). *Assessing Causation in Sport Injury: A Multifactorial Model.* Philadelphia, PA: LWW.

Morris, Z. S., Wooding, S., & Grant, J. (2011). The answer is 17 years, what is the question: Understanding time lags in translational research. *Journal of the Royal Society of Medicine*, 104:12, 510–520.

Mørtvedt, A. I., Krosshaug, T., Bahr, R., & Petushek, E. (2019). I spy with my little eye … a knee about to go 'pop'? Can coaches and sports medicine professionals predict who is at greater risk of ACL rupture? *British Journal of Sports Medicine*. doi: 10.1136/bjsports-2019-100602.

Nilstad, A., Andersen, T. E., Kristianslund, E., Bahr, R., Myklebust, G., Steffen, K., & Krosshaug, T. (2014). Physiotherapists can identify female football players with high knee valgus angles during vertical drop jumps using real-time observational screening. *Journal of Orthopaedic and Sports Physical Therapy*, 44:5, 358–365.

Norcross, M. F., Johnson, S. T., Bovbjerg, V. E., Koester, M. C., & Hoffman, M. A. (2016). Factors influencing high school coaches' adoption of injury prevention programs. *Journal of Science and Medicine in Sport*, 19:4, 299–304.

O'Brien, J., & Finch, C. F. (2016). Injury prevention exercise programmes in professional youth soccer: Understanding the perceptions of programme deliverers. *BMJ Open Sport and Exercise Medicine*, 2:1, e000075.

Orr, B., Brown, C., Hemsing, J., McCormick, T., Pound, S., Otto, D., Emery, C. A., & Beaupre, L. A. (2013). Female soccer knee injury: Observed knowledge gaps in injury prevention among players/parents/coaches and current evidence (the KNOW study). *Scandinavian Journal of Medicine and Science in Sports*, 23:3, 271–280.

Padua, D. A., Frank, B., Donaldson, A., de la Motte, S., Cameron, K. L., Beutler, A. I., DiStefano, L. J., & Marshall, S. W. (2014). Seven steps for developing and implementing a preventive training program: Lessons learned from JUMP-ACL and beyond. *Clinics in Sports Medicine*, 33:4, 615–632. doi: 10.1016/j.csm.2014.06.012.

Petushek, E. J., Cokely, E. T., Ward, P., Durocher, J. J., Wallace, S. J., & Myer, G. D. (2015a). Injury risk estimation expertise: Assessing the ACL injury risk estimation quiz. *The American Journal of Sports Medicine*, 43:7, 1640–1647.

Petushek, E. J., Cokely, E. T., Ward, P., & Myer, G. D. (2015b). Injury risk estimation expertise: Cognitive-perceptual mechanisms of ACL-IQ. *Journal of Sport and Exercise Psychology*, 37:3, 291–304.

Petushek, E. J., Sugimoto, D., Stoolmiller, M., Smith, G., & Myer, G. D. (2019). Evidence-based best-practice guidelines for preventing anterior cruciate ligament injuries in young female athletes: A systematic review and meta-analysis. *The American Journal of Sports Medicine*, 47:7, 1744–1753.

Petushek, E. J., Ward, P., Cokely, E. T., & Myer, G. D. (2015c). Injury risk estimation expertise: Interdisciplinary differences in performance on the ACL injury risk estimation quiz. *Orthopaedic Journal of Sports Medicine*, 3:11, 232596711561479. *DOI*: 10.1177/2325967115614799.

Phillips, N. D., Neth, H., Woike, J. K., & Gaissmaier, W. (2017). FFTrees: A toolbox to create, visualize, and evaluate fast-and-frugal decision trees. *Judgment and Decision Making*, 12:4, 344–368.

Physical Activity Council. (2019). *2019 Physical Activity Council's Overview Report on U.S. Participation.*

Pollard, C. D., Sigward, S. M., Ota, S., Langford, K., & Powers, C. M. (2006). The influence of in-season injury prevention training on lower-extremity kinematics during landing in female soccer players. *Clinical Journal of Sport Medicine*, 16:3, 223–227.

Poulsen, E., Goncalves, G. H., Bricca, A., Roos, E. M., Thorlund, J. B., & Juhl, C. B. (2019). Knee osteoarthritis risk is increased 4–6 fold after knee injury – A systematic review and meta-analysis. *British Journal of Sports Medicine*, 53:23, 1454–1463. doi: 10.1136/bjsports-2018-100022.

Pryor, J. L., Root, H. J., Vandermark, L. W., Pryor, R. R., Martinez, J. C., Trojian, T. H., Denegar, C. R., & DiStefano, L. J. (2017). Coach-led preventive training program in youth soccer players improves movement technique. *Journal of Science and Medicine in Sport*, 20:9, 861–866.

Russ, A. L., Fairbanks, R. J., Karsh, B.-T., Militello, L. G., Saleem, J. J., & Wears, R. L. (2013). The science of human factors: Separating fact from fiction. *BMJ Quality and Safety*, 22:10, 802–808. doi: 10.1136/bmjqs-2012-001450.

Sugimoto, D., Mattacola, C. G., Bush, H. M., Thomas, S. M., Foss, K. D. B., Myer, G. D., & Hewett, T. E. (2017). Preventive neuromuscular training for young female athletes: Comparison of coach and athlete compliance rates. *Journal of Athletic Training*, 52:1, 58–64.

Sugimoto, D., Myer, G. D., Bush, H. M., Klugman, M. F., McKeon, J. M. M., & Hewett, T. E. (2012). Compliance with neuromuscular training and anterior cruciate ligament injury risk reduction in female athletes: A meta-analysis. *Journal of Athletic Training*, 47:6, 714–723.

van Mechelen, W., Hlobil, H., & Kemper, H. C. (1992). Incidence, severity, aetiology and prevention of sports injuries. *Sports Medicine*, 14:2, 82–99.

Verhagen, E. a. L. M., & van Mechelen, W. (2010). Sport for all, injury prevention for all. *British Journal of Sports Medicine*, 44:3, 158–158. doi: 10.1136/bjsm.2009.066316.

Webster, K. E., & Hewett, T. E. (2018). Meta-analysis of meta-analyses of anterior cruciate ligament injury reduction training programs. *Journal of Orthopaedic Research*, 36:10, 2696–2708. doi: 10.1002/jor.24043.

Weiser, T. G., Haynes, A. B., Lashoher, A., Dziekan, G., Boorman, D. J., Berry, W. R., & Gawande, A. A. (2010). Perspectives in quality: Designing the WHO Surgical Safety Checklist. *International Journal for Quality in Health Care*, 22:5, 365–370.

5 Physical Ergonomics of Distance Running Footwear

Laurent Malisoux and Daniel Theisen

CONTENTS

5.1 INTRODUCTION

Recreational distance running is one of the most popular sports in the world (Hulteen et al., 2017). It can be practised virtually anywhere and requires mainly a pair of 'good' running shoes. However, the term 'good' is ambiguous, and physical ergonomics of distance running footwear has always generated heated debates among all types of runners, from beginners to 'specialists' (Napier & Willy, 2018). Indeed, running shoes are thought to affect both performance through improved running economy, and injury risk via a modification of the repetitive external mechanical load applied to the musculoskeletal system. Therefore, this chapter will focus on running footwear and how shoes for distance runners can potentially enhance performance and protect the athlete from injury. This chapter will address 1) the key shoe features for performance in long-distance running, 2) the mechanisms underlying the occurrence of running-related injury, 3) the research on the influence of running shoes on injury risk, 4) future directions for running shoe development, and 5) practical applications in sports.

5.2   PERFORMANCE

One of the key questions regarding performance in distance running is whether a sub-two-hour marathon is possible (Hoogkamer et al., 2017). Although such a performance is affected by many aspects, the selection of appropriate footwear is often advocated as a critical aspect. While there is a lack of research investigating the effect of footwear on running performance directly (Fuller et al., 2015), several studies have investigated the effect of footwear on running economy, a surrogate measure of running performance. Running economy is determined from the oxygen demand at a given velocity of running and is a good predictor of distance running performance (Saunders et al., 2004). Thus, higher running economy indicates that less amount of oxygen is required at submaximal running speed, or that a higher running speed can be maintained with the same amount of oxygen utilisation.

5.2.1   SHOE MASS

Globally, the metabolic cost of running increases linearly with increasing shoe mass (Franz et al., 2012; Frederick, 1984; Fuller et al., 2015). At moderate running speed, the rate of oxygen consumption increases by approximately 1% for each 100 g of mass added to each shoe (Franz et al., 2012; Frederick, 1984). Other studies have confirmed the 1% rule for running velocities up to 3.5 m/s (Hoogkamer et al., 2017). A linear relationship predicted that there would be no difference in the metabolic cost between barefoot and shod running for footwear weighing less than 440 g per pair (Fuller et al., 2015). Extrapolation of this relationship to the hypothetical situation where shoe mass is zero indicates that footwear features other than shoe mass could have a small beneficial effect on running economy. Based on this finding, shoes were classified as light or heavy using 440 g as a cut-off value in a meta-analysis. The findings revealed that running in light shoes and barefoot running induced a lower metabolic cost compared with heavy shoes. However, there was no difference between running in light shoes and barefoot running (Fuller et al., 2015).

5.2.2   MINIMALIST SHOES AND BAREFOOT RUNNING

Given the strong linear relationship between shoe mass and metabolic cost during distance running, the reduction of shoe mass resulting from running barefoot or in minimalist shoes* may contribute to an enhanced running economy compared to shod running. This question was specifically addressed in a meta-analysis which found that both barefoot running and running in minimalist shoes are potentially more economical than shod running (Cheung & Ngai, 2015). Theoretically, this higher running economy may improve performance, but no direct effect on long-distance running performance has been demonstrated so far. Thus, this relationship remains speculative. Also, a high risk of bias was reported in the studies addressing the question, caused by methodological shortcomings.

When controlling for differences in shoe mass, foot strike and cadence, the use of minimalist shoes induced a lower metabolic cost compared to conventional shoes (Fuller et al., 2015; Perl et al., 2012). However, this observation might be explained by the previous experience of the participants regarding minimalist shoe use which implies that the findings may not apply to all populations of runners (Fuller et al., 2015; Perl et al., 2012).

5.2.3 Shoe Cushioning and Barefoot Running

Although barefoot running might enhance running economy by eliminating shoe mass (Fuller et al., 2015), it also requires additional muscular effort to cushion the impact of the foot with the ground. In cushioned shoes, this effect is achieved by the midsole material, but the additional mass increases the metabolic cost. Thus, shoe cushioning and shoe mass affect running economy in opposing directions. The 'cost of cushioning' hypothesis during barefoot running was first suggested by Frederick et al. (1983) and later confirmed in a study where slabs of shoe midsole material were attached to a treadmill to eliminate the confounding factor of shoe mass (Tung et al., 2014). As predicted, running barefoot on foam cushioning slabs required 1.83% less metabolic energy compared to running barefoot on the rigid treadmill belt. Additionally, running in cushioned shoes (230 g each; 10 mm of foam midsole) required a similar metabolic cost to running barefoot in the rigid condition. The theory was additionally supported by a study demonstrating that shod running had a lower metabolic cost than barefoot running for footwear conditions of equal mass (achieved through attaching small lead strips to each foot/shoe) (Franz et al., 2012). The latter observation suggests that running barefoot offers no metabolic advantage over running in lightweight, cushioned shoes.

5.2.4 Energy Return from Midsole Material

Many different shoe cushioning materials have been used by the running footwear industry, and these materials differ in many aspects such as stiffness, hardness, or energy restitution. Therefore, their impact on running economy might also vary. For

---

* The standardised definition of minimalist shoes developed by an international panel of experts is: 'Footwear providing minimal interference with the natural movement of the foot due to its high flexibility, low heel to toe drop, weight and stack height, and the absence of motion control and stability devices'. Esculier J. F., Dubois B., Dionne C. E., et al. A consensus definition and rating scale for minimalist shoes. *J Foot Ankle Res.* 2015; **8**:42.

example, the great percentage of energy returned from the Adidas boost midsole material has been found to provide ~1% improvement in running economy compared with conventional ethylene vinyl acetate (EVA) cushioning (Worobets et al., 2014). While such a small difference might still be of relevance for performance, the reproducibility of such small effects should be convincingly demonstrated.

5.2.5 LONGITUDINAL BENDING STIFFNESS

Running shoes with different levels of longitudinal bending stiffness (18, 38, and 45 N/mm, respectively) were compared (Roy & Stefanyshyn, 2006). Bending stiffness was increased with a carbon fibre plate through the full length of the midsole. Running economy was improved by 0.8% in the intermediate condition when compared to the other conditions, which suggests that there might be an optimal bending stiffness for running economy.

In conclusion, shoe technology provides some potential for facilitating a sub-two-hour marathon, as weight saving, midsole resilience, and increased midsole bending stiffness can provide a substantial improvement in running economy (Hoogkamer et al., 2017). Moreover, if footwear might contribute to performance enhancement via running economy improvement, it should be underlined that one of the key factors for high performance is the ability to practice and cumulate high training load. Consequently, any shoe design helping the athlete to tolerate the training load and to prevent injury must be considered as a key factor for performance enhancement.

5.3 THE AETIOLOGY OF RUNNING-RELATED INJURY AND THE ROLE OF RUNNING FOOTWEAR

Before addressing the role that running shoes might play in the development of running-related injury, it is important to first gain some insight into risk factors, causes and the mechanisms of injury aetiology.

5.3.1 MECHANISMS UNDERLYING RUNNING-RELATED INJURY

Most running-related injuries are 'overuse injuries', resulting from an imbalance between the repetitive loading of the musculoskeletal system and the tissue load capacity (Bertelsen et al., 2017). Ideally, minor tissue damage resulting from repetitive loading stimulates remodelling and adaptation, but chronic overload leads to tissue failure (i.e. tear, rupture, fracture) (Edwards, 2018). Thus, most running-related injuries have a complex multifactorial origin (Malisoux et al., 2015a; Meeuwisse et al., 2007).

A recent systematic review identified more than 70 studies that examined injury risk and protective factors for running-related injury in middle- and long-distance runners, such as footwear properties, training-related characteristics, anthropometric measurements, and demographics (Hulme et al., 2017). Although these factors can help identify a population of runners at higher or lower injury risk, some of these factors might not be causally related to running-related injury (Shrier & Platt, 2008). For example, an excessive fat mass might decrease the training load a runner can tolerate, but it will not be in itself the *direct cause* of an overuse injury, excessive running is. Therefore, an injury prevention measure targeting training load is much more likely to be effective than any weight-loss programme. Actually, training errors have been acknowledged as a major contributor to overuse injury in distance runners (Hreljac, 2004; Nielsen et al., 2012). A consensus statement recommended research should investigate the dose-response relationship between a sports participation-related factor and injury risk through prospective studies, as well as focus on the potential interaction with other risk factors (Soligard et al., 2016). In other words, the fundamental questions are 'how much running is too much?', and 'how can factors such as excessive fat mass, footwear and surface influence the dose-response relationship between running participation and injury risk?' (Bertelsen et al., 2017).

In line with these considerations, a conceptual framework for the complex, multifactorial aetiology of running injuries has been suggested (Bertelsen et al., 2017). The framework includes four parts: 1) the structure-specific capacity at the beginning of the session that can be influenced by factors such as running experience, previous injury, diet or age; 2) the structure-specific cumulative load resulting from the session, which depends on the number of strides, the magnitude of load applied per stride, and the distribution of load over tissue structures; 3) the reduction in the structure-specific capacity during the session which gradually decreases in response to repetitive loading (repetitive loading without recovery time); and 4) the balance between the cumulative load and the structure capacity which might lead to an injury if the latter is exceeded. This framework implies that a running-related injury does not occur because of footwear features, but in a situation where a runner increases their running participation so that, given the other risks factors (e.g. footwear features), the load capacity of a structure is exceeded. In conclusion, footwear does not *cause* injury, but can modify the global training load a runner can tolerate before sustaining an injury (Malisoux et al., 2015a).

5.3.2 Investigating the Role of Sports Footwear in Injury Development

The role of running shoes in the prevention of running-related injury is broadly discussed in the specialised press and media, on dedicated internet blogs, as well as amongst health professionals and coaches. However, a close look at the scientific literature reveals that the evidence for shoe prescription is lacking (Napier & Willy, 2018). Actually, the relationship between specific shoe features and injury is often unknown or controversial. To address such a complex problem, it is necessary to combine biomechanical analyses with long-term follow-ups of a large number of runners, a huge yet feasible challenge (Malisoux et al., 2017). To critically appraise the current evidence on the relationship between running shoes and injury, a framework was presented under the term 'Bermuda triangle', with reference to the lack of knowledge on the topic (Figure 5.1) (Theisen et al., 2016).

An association between footwear characteristics and injury is sometimes implicitly inferred based on studies on biomechanics. However, this inference is correct only if injury is the outcome investigated, which is hardly ever the case. Biomechanical

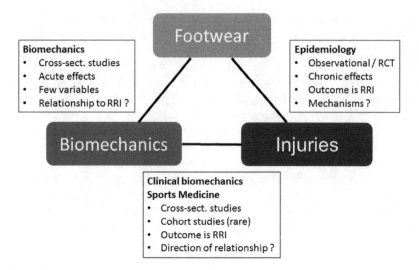

FIGURE 5.1 The 'Bermuda triangle framework', one way to look at and interpret the scientific literature on running-related injuries (RRI) and footwear (Reprinted from *Sport-Orthopadie – Sport Traumatologie – Sports Orthopaedics and Traumatology*, 32(2), Daniel Theisen, Laurent Malisoux, Paul Gette, Christian Nührenbörger, Axel Urhausen, 'Footwear and running-related injuries – Running on faith?', 169–176, Copyright (2016), with permission from Elsevier).

studies are mostly cross-sectional, involving data collection at a defined time in the laboratory, i.e. conditions poorly related to the real world. They allow analysing the acute effects of different shoe conditions on a few aspects of running biomechanics in healthy participants. One of the main limitations of this approach is that none of the differences observed between shoe conditions can be related to injury risk, since injury is not the investigated variable.

Another type of study, more closely related to the field of sports medicine, focuses on the association between biomechanics and injuries by comparing healthy and injured (or previously injured) runners (e.g. Mann et al., 2015). These so-called 'case-control' studies can establish a relationship between biomechanics and injury, but the direction of the association cannot be determined (the chicken-or-egg question).

A third approach is found in the form of epidemiological studies in which the main outcome of interest is running injury (Malisoux et al., 2016a; Malisoux et al, 2016b; Theisen et al., 2014). These studies generally involve several hundreds of participants who are followed-up prospectively or enrolled in a randomised trial. These designs make it possible to study the long-term effects of personal characteristics, training behaviour or a given shoe type on running injury. The level of evidence is higher with such study designs, but the risk factors identified cannot be mechanistically explained in the absence of biomechanical analyses.

Given the shortcomings of the above-described study types, it can be argued that a superior design would be to combine several methods. The ideal approach would be to monitor a large cohort of runners, analysing their running technique both in standard conditions as well as in their own habitual environment, and follow them up over a sufficiently long period (several months) regarding their exposure to running and their injuries encountered (Malisoux et al., 2017; Theisen et al., 2016).

5.4 THE EVIDENCE RELATING TO RUNNING FOOTWEAR AND INJURY

Some research suggests that modern running shoes have failed to decrease injury incidence (Davis, 2014; Knapik et al., 2015). Running shoes were introduced in Basic Combat Training in the US army in 1985. Joseph Knapik and co-workers 2015 compared the injury incidence before and after the introduction of running shoes in a meta-analysis and did not find any difference. This result is intriguing, since their reference was the classical military boots. In 2009, a research group searched the scientific literature for studies addressing the capacity of pronation control elevated cushioned heel running shoes to prevent injury (Richards et al., 2009). The authors did not find any original research on their effect on injury risk and concluded that the prescription of this shoe type to distance runners was not evidence-based at that time. Ten years later, experts are still calling for further high-quality randomised controlled trials, because evidence for shoe prescription with regards to injury prevention is still lacking (Napier & Willy, 2018). The aim of this section is to present the most relevant studies that investigated the impact of running footwear on injury risk. The research conducted on minimalist footwear will also be addressed, although it differs in many aspects from traditional cushioned shoes.

5.4.1 Once Upon a Time …

The first (retrospective) study that investigated an association between running shoes and injuries was published in 1984 (Pinshaw et al., 1984). The authors observed that amongst injured runners with shin splints, a disproportionately high number wore Adidas shoes while few used Nike shoes. Similarly, a disproportionately high number of runners with the iliotibial band syndrome wore New Balance shoes. The authors concluded that, with regards to optimal treatment, the first step is the prescription of the most appropriate running shoes. Although this seems to make sense, it is a highly speculative interpretation, since the study was not designed to support this conclusion! Twenty years later, Martin Schwellnus and George Stubbs (2006) investigated if this shoe prescription strategy is efficient. They compared the injury incidence between the runners who had been advised on running shoes following a clinical assessment with those who had only received general advice. No difference was found between the two groups suggesting that this strategy did not work.

In 1988, a Swiss team published another retrospective study on a large cohort of competitive runners (Marti et al., 1988). The week before a popular 16 km race (Grand-Prix of Bern), more than 4,000 participants filled out a questionnaire on training habits, injuries, and running shoes. The authors presented three main observations in their article. First, the runners who preferred one of the three most popular running shoes showed no difference in injury incidence when compared to those using other shoes. Second, the runners who had no preference had significantly fewer injuries. And third, expensive shoes were associated with a higher injury incidence. Although these results are interesting, the level of evidence is weak, given that this study was retrospective by design and poorly controlled for other important variables that could explain these observations.

In 2003, a first prospective study collected data on running shoes and injuries (Taunton et al., 2003). The authors analysed, amongst others, if shoe age was associated with injury risk. They did find an association, but the effect was in opposite directions in men and women. Therefore, no strong conclusion could be drawn.

5.4.2   THE MAIN SHOE FEATURES

Over the past five decades, various characteristics have been added to (and sometimes removed from) running footwear to influence biomechanics and indirectly prevent running injuries (Ramsey et al., 2019). According to a systematic review (Ramsey et al., 2019), footwear characteristics studied so far in relation to running injuries are heel-to-toe drop, midsole thickness, minimalist index (Esculier et al., 2015),* innersole thickness, mass, midsole hardness, stability elements, and shoe age/usage. Unravelling the contribution of each of these shoe features to running-related injury is extremely complex given that shoe models often differ in many aspects and footwear is usually classified in two or three categories (e.g. traditional/partial minimalist/full minimalist shoes, or highly cushioned/stability/motion control shoes). Additionally, shoe features are not consistently reported (Ramsey et al., 2019). However, a few studies have singled-out the effect of specific shoe features on the development of running-related injury (Theisen et al., 2016).

5.4.3   SHOE CUSHIONING

Shoe cushioning is the single most investigated characteristic, at least as regards the effect on running biomechanics. The shock absorption properties of footwear mainly result from the materials used in the sole (i.e. their type, density, structure and combination), as well as from the geometry of the shoe (i.e. the midsole thickness and the design of inserts). One of the most popular approaches has been to change the hardness of the shoe midsole (Baltich et al., 2015; Theisen et al., 2014). Only one study has investigated the association between shoe cushioning and injury risk so far. In 2012, two different pairs of shoes were randomly distributed to 247 recreational runners (Theisen et al., 2014). The running shoe models were exactly the same, except for midsole hardness. The participants had to report all their training sessions and injuries over five months in an electronic system called TIPPS (Training and Injury Prevention Platform for Sports (Malisoux et al., 2013)). Midsole hardness did not influence injury risk in this randomised control trial. Unfortunately, to date, no other study has confirmed or contradicted the conclusion of that randomised controlled trial.

---

* The minimalist index is a rating scale (%) used to determine the degree of minimalism of running shoes. It includes the following shoe features: weight, stack height, heel-to-toe drop, flexibility (longitudinal and torsional) as wll as motion control and stability technologies.

5.4.4 MOTION CONTROL SYSTEMS

It is a common belief that foot posture is linked to the risk of running-related injury to the lower extremities. Static foot posture is frequently assessed in clinical settings or, more recently, sports shops. The foot posture refers to different aspects such as the navicular drop, the calcaneal eversion, the longitudinal arch angle, or the footprint (Redmond et al., 2006). Briefly, supinated feet could be described as ankles leaning outward, combined with a high medial arch, while pronated feet appear flat, with ankles leaning inward. Foot posture has been presented as a key factor by the running shoe industry and has become the basis of a well-established shoe classification system. According to the 'running shoe shop theory', people with supinated feet should use highly cushioned shoes because of the greater stiffness of their foot structure. People with neutral feet should use stability shoes, and those with pronated feet should use motion control shoes (Richards et al., 2009).

The more fundamental question is whether foot posture is a risk factor for running-related injury. A prospective cohort study (DANORUN) with a one-year follow-up on more than 900 novice runners showed that foot pronation was not associated with injury risk (Nielsen et al., 2014). A second important question concerns the effectiveness of assigning specific running shoes based on the runner's foot posture to prevent running-related injury. This was tested in the US army (Knapik 2010a; Knapik et al., 2009; Knapik et al., 2010b; Knapik et al., 2014). In the experimental group, the new recruits received a pair of shoes 'adapted' to their plantar shape according to the 'shoe shop theory'. In the control group, all new recruits received stability shoes. The intervention was performed in three different services in the US army for a total of 7,000 new recruits. The results showed that there was no difference in injury risk between the two groups, lending no support for the shoe shop theory. There was even a trend towards higher injury risk in the experimental group amongst women in the Air Force (Knapik et al., 2010b). In another study, a similar approach was applied in female runners, and again, there was no evidence that the strategy works (Ryan et al., 2011).

Still, none of these previous investigations analysed if motion control systems are effective in reducing the risk of running-related injury. A randomised controlled trial was specifically designed to study this question, and to determine if this effect depended on the runner's foot posture (Malisoux et al., 2016). For example, motion control shoes could be appropriate for the runners with pronated feet, but harmful for those with neutral feet. More than 400 regular recreational runners were recruited for the study. Their foot posture was assessed according to the Foot Posture Index method and classified into five categories, according to normative values (Redmond et al., 2008). Then, based on their foot category, the participants were randomly allocated to one of the study groups: one group received a pair of standard neutral shoes exempt of any motion control technology, while the other received a pair of motion control shoes that had a dual-density midsole and an arch-supporting element placed on the medial midfoot part. The participants were then followed-up over a period of six months, during which 25% of the runners sustained at least one running-related injury. The primary analysis revealed that the rate at which injury occurred was lower in the group of runners who received the shoes equipped with the motion control system, regardless of foot type. When looking in detail at the different categories of runners classified according to their foot posture, the study revealed that motion control shoes were effective in lowering the risk of injury in runners with pronated feet. Equally interesting was that motion control shoes were not harmful for those with other foot types.

To date, this is the only study on motion control technology and injury risk. In addition, the results partly support the 'shoe shop theory' regarding footgear preference and foot posture. However, it is worth noting that the neutral shoes used in this study had no motion control technology at all, and that most models advertised today as 'neutral' or 'stability' shoes do have some features of that kind. This may then also explain the apparent contrast with the findings from the DANORUN study (see above) that used precisely such a neutral shoe (Supernova Glide 3, Adidas; this shoe includes medial arch support) for all their (beginner) runners.

5.4.5 SHOE DROP

One of the most popular shoe features scrutinised recently is the heel-to-toe drop, i.e. the difference in stack height between the heel and forefoot. Conventional cushioned running shoes are usually designed with a shoe drop of 10 to 12 mm, while, by definition, running barefoot represents a zero millimetre drop condition. Between these extreme shoe drops, shoes advertised as 'minimalist' come in a wide range of drops going from 0 mm to 8 mm. Some previous cohort studies investigated the association between minimalist shoes and injury risk (see below) (Goss & Gross, 2012; Ryan et al., 2014). However, this comparison implies that the shoe conditions differed in many other aspects than the heel-to-toe drop. Only one study investigated the specific effect of shoe drop on injury risk in standard cushioned running shoes (Malisoux et al., 2016). It was a randomised controlled trial with a six-month follow-up period in which three versions of the same shoe model that only differed in heel-to-toe drop were compared: 10, 6, and 0 mm, respectively. About 600 pairs of shoes were randomly distributed to the participants who reported all their training sessions as well as any injury into an electronic system. More than 550 participants were included in the analyses, and 25% of them reported at least one running-related injury. Overall, the injury risk was not influenced by shoe drop in the whole cohort. However, given that some studies have previously shown that running experience influences injury incidence, locations, and mechanisms (Kluitenberg et al.,2015; Malisoux 2015a; Videbaek et al., 2015), the relationship between shoe drop and injury risk might be influenced by running regularity over the previous year (short-term experience). Indeed, the stratified analysis showed that in the group of 'occasional' runners (having run weekly for less than 6 months over the 12 months prior to the study), the rate at which the injuries occurred was lower amongst those using low-drop shoe versions (drop 6 or 0 mm), whereas, in the group of 'regular' runners, the injury risk was higher in the group of participants who had received the low-drop versions. Based on this secondary analysis, it seems safe to recommend low-drop footwear for occasional or inexperienced runners. In contrast, regular runners having received low-drop shoes appeared to be at a higher risk compared with those using conventional shoes. Since the participants were required to use the study shoes for all their running sessions, it could be speculated that in the regular runners, the transition from their usual running shoes to the low-drop versions was not progressive enough and increased injury risk. The conclusion might be that the association between shoe drop and injury risk depends on the runners' profile.

5.4.6 Minimalist Shoes

A re-emergence of the minimal running shoe has been observed in the last decade (Davis, 2014). A retrospective study on US soldiers showed that those who declared running with minimalist shoes reported fewer injuries (Goss & Gross, 2012). Conversely, in a prospective study, the participants who received a minimalist (five fingers) or partial minimalist (Nike free) shoe were at a higher risk of injury (Ryan et al., 2014). In another study, bone marrow oedema was investigated in runners after a ten-week period of transitioning from traditional to very minimalist running shoes while maintaining the usual training schedule. Despite a lower training volume in the minimalist compared to the traditional shoe group, bone marrow oedema was more common in the former, i.e. those transitioning to minimalist footwear. The question remains whether the main reason for this higher risk of injury is the minimalist shoe itself or the transition to the minimalist shoe (Ridge et al., 2013). In a more recent trial with a six-month follow-up period, 61 habitual rearfoot strikers received either conventional or minimalist shoes. A total of 27 injuries was recorded. Surprisingly, shoe type was not associated with injury risk, but body mass was. Also, an interaction between shoe type and body mass was observed. In runners using minimalist shoes, sustaining an injury became increasingly more likely with increasing body mass above 71.4 kg, when compared with conventional shoes (Fuller et al., 2017). It should, however, be stressed that this study was underpowered and that the conclusions are weak. Unfortunately, no large randomised controlled trial has investigated the difference in injury risk between minimalist and conventional shoes after a transition period so far.

5.5   FUTURE DIRECTIONS

The evidence for the role of running footwear in injury risk is weak. Insufficient research has been carried out in that area with regards to the multi-billion-dollar industry represented by running footwear. For instance, the impact of shoe age and wear on injury risk has never been investigated. Therefore, the maximal distance for running shoes set at 800 to 1,000 km is simply popular belief, based on recommendations from the running shoe sector. Comfort has been suggested to be an important factor (Nigg et al., 2015), although it has never been seriously investigated either. Other footwear features such as toe box width, longitudinal bending stiffness or weight are systematically provided in the product specifications by manufacturers and have been the focus of their research and development services, yet limited research has focused on injury development in relation to those features (i.e. studies in which the outcome is running-related injury).

    Performance and protection against injury are considered the fundamental functions of running footwear, but the characteristics that meet these objectives might not be compatible. Extra lightweight shoes might be needed for optimal performance in long-distance running, while a greater level of cushioning (inducing additional weight) might be required for injury prevention during periods with high cumulative load. This issue can be circumvented to some degree as most recreational runners use several pairs of running shoes in parallel, depending on the surface (i.e. road, forest or mountain) or the purpose (i.e. training or competition) (Malisoux et al., 2015b). An observational study investigated if the regular alternation between different pairs of running shoes might play a role in injury prevention (Malisoux et al., 2015b). This five-month prospective study included some 250 recreational runners and showed a lower injury risk in those using different pairs of running shoes concomitantly *versus* those using a single pair. Of course, further research is needed to define good strategies for alternation in terms of frequency, shoe features involved, and context of utilisation. The transition period to a new pair of shoes was also suggested as a key feature to carefully control for in research, given that injury risk might be affected by the adaptation phase to some new shoe features (Napier & Willy, 2018).

Future studies should focus on prospective long-term follow-ups of runners under 'natural' training conditions. Performance and injury prevention should be the main outcomes of interest, while the metrics related to running technique and biomechanics, as provided by 'smart wearables', may provide insights into the underlying mechanisms. More research is required to investigate systematically individual or combined shoe features, involving large cohorts and runners with different profiles to allow for scientifically sound conclusions on matching shoe and runner profiles.

5.6   PRACTICAL APPLICATION IN SPORT

Shoe mass is so far the only feature that has been convincingly shown to influence running performance, as evaluated by running economy. A shoe mass of less than 220 g seems to be optimal, while greater reductions in shoe mass provide little additional advantage. However, it may be difficult to reconcile designing a shoe weighing no more than 220 g with implementing technology that could help in reducing running injury risk. Despite recently emerging evidence on the relationship between some shoe characteristics and running injuries, it is still too early to formulate precise prescriptions regarding the choice of running shoe features. Whether it will be possible at all one day to make scientifically sound prescriptions to prevent injury remains to be answered, despite the heavy opposite lobbying of the running shoe manufacturing sector. In short, it is possible that the role of running shoe technology for injury prevention has been largely overrated. An exception to this may be the very specific situation of a radical change of footwear use, for example, when a regular runner decides to switch from maximalist to minimalist running. A runner should plan any transition to a new, unusual footwear condition very progressively, over a period of several months and up to a year. Along the same line, it is probably a good idea to alternate between running shoe pairs to avoid systematic mechanical overload and allow progressive transitioning to new shoes.

Some arguments for shoe prescriptions can be identified, although the evidence is preliminary. It seems that a minimum of motion control in cushioned shoes, as provided in most 'classic' models, is of relevance to running injury risk, especially for runners with highly pronated feet. In addition, it seems safe to recommend low-drop footwear for occasional or inexperienced runners. The effectiveness of cushioning of running shoes is yet to be demonstrated.

Apart from these preliminary conclusions, it seems that some basic 'old rules' still have their validity, such as the subjective feeling of comfort when choosing a pair of running shoes, transitioning progressively and carefully into a new pair and listening to your body when training. However, the most important aspect of injury prevention might eventually be athlete education, to allow every runner to develop their own, optimal self-management strategies. The authors hope that this chapter represents a valuable, albeit very modest, contribution to this objective.

REFERENCES

Baltich J, Maurer C, Nigg BM. Increased vertical impact forces and altered running mechanics with softer midsole shoes. *PLoS One*. 2015;**10**:e0125196.

Bertelsen ML, Hulme A, Petersen J, et al. A framework for the etiology of running-related injuries. *Scand J Med Sci Sports*. 2017;**27**:1170–80.

Cheung RT, Ngai SP. Effects of footwear on running economy in distance runners: A meta-analytical review. *J Sci Med Sport*. 2015.

Davis IS. The re-emergence of the minimal running shoe. *J Orthop Sports Phys Ther*. 2014;**44**:775–84.

Edwards WB. Modeling overuse injuries in sport as a mechanical fatigue phenomenon. *Exerc Sport Sci Rev*. 2018;**46**:224–31.

Esculier JF, Dubois B, Dionne CE, et al. A consensus definition and rating scale for minimalist shoes. *J Foot Ankle Res*. 2015;**8**:42.

Franz JR, Wierzbinski CM, Kram R. Metabolic cost of running barefoot versus shod: is lighter better? *Med Sci Sports Exerc*. 2012;**44**:1519–25.

Frederick EC, Clarke TE, Larsen JL, et al. The effects of shoe cushioning on the oxygen demands of running. In: Nigg, B.M., and Kerr, B.A. (eds). *Biomechanical Aspects of Sport Shoes and Playing Surfaces*. University of Calgary, Calgary, Canada, 1983.

Frederick EC. Physiological and ergonomics factors in running shoe design. *Appl Ergon*. 1984;**15**:281–7.

Fuller JT, Bellenger CR, Thewlis D, et al. The effect of footwear on running performance and running economy in distance runners. Sports Med. 2015;**45**:411–22.

Fuller JT, Thewlis D, Buckley JD, et al. Body mass and weekly training distance influence the pain and injuries experienced by runners using minimalist shoes: A randomized controlled trial. *Am J Sports Med*. 2017;**45**:1162–70.

Goss DL, Gross MT. Relationships among self-reported shoe type, footstrike pattern, and injury incidence. *US Army Med Dep J*. 2012:25–30.

Hoogkamer W, Kram R, Arellano CJ. How biomechanical improvements in running economy could break the 2-hour Marathon barrier. Sports Med. 2017;**47**:1739–50.

Hreljac A. Impact and overuse injuries in runners. *Med Sci Sports Exerc*. 2004;**36**:845–9.

Hulme A, Nielsen RO, Timpka T, et al. Risk and protective factors for middle- and long-distance running-related injury. Sports Med. 2016.

Hulteen RM, Smith JJ, Morgan PJ, et al. Global participation in sport and leisure-time physical activities: A systematic review and meta-analysis. *Prev Med*. 2017;**95**:14–25.

Kluitenberg B, van Middelkoop M, Diercks R, et al. What are the differences in injury proportions between different populations of runners? A systematic review and meta-analysis. Sports Med. 2015;**45**:1143–61.

Knapik JJ, Brosch LC, Venuto M, et al. Effect on injuries of assigning shoes based on foot shape in air force basic training. *Am J Prev Med.* 2010;**38** Supplement:S197–211.

Knapik JJ, Jones BH, Steelman RA. Physical training in boots and running shoes: a historical comparison of injury incidence in basic combat training. *Mil Med.* 2015;**180**:321–8.

Knapik JJ, Swedler DI, Grier TL, et al. Injury reduction effectiveness of selecting running shoes based on plantar shape. *J Strength Cond Res.* 2009;**23**:685–97.

Knapik JJ, Trone DW, Swedler DI, et al. Injury reduction effectiveness of assigning running shoes based on plantar shape in Marine Corps basic training. *Am J Sports Med.* 2010;**38**:1759–67.

Knapik JJ, Trone DW, Tchandja J, et al. Injury-reduction effectiveness of prescribing running shoes on the basis of foot arch height: summary of military investigations. *J Orthop Sports Phys Ther.* 2014;**44**:805–12.

Malisoux L, Chambon N, Delattre N, et al. Injury risk in runners using standard or motion control shoes: a randomised controlled trial with participant and assessor blinding. *Br J Sports Med.* 2016;**50**:481–7.

Malisoux L, Chambon N, Urhausen A, et al. Influence of the heel-to-toe drop of standard cushioned running shoes on injury risk in leisure-time runners: A randomized controlled trial with 6-month follow-up. *Am J Sports Med.* 2016;**44**:2933–40.

Malisoux L, Delattre N, Urhausen A, et al. Shoe cushioning, body mass and running biomechanics as risk factors for running injury: a study protocol for a randomised controlled trial. *BMJ Open.* 2017;**7**:e017379.

Malisoux L, Frisch A, Urhausen A, et al. Monitoring of sport participation and injury risk in young athletes. *J Sci Med Sport.* 2013;**16**:504–8.

Malisoux L, Nielsen RO, Urhausen A, et al. A step towards understanding the mechanisms of running-related injuries. *J Sci Med Sport.* 2015;**18**:523–8.

Malisoux L, Ramesh J, Mann R, et al. Can parallel use of different running shoes decrease running-related injury risk? *Scand J Med Sci Sports.* 2015;**25**:110–5.

Mann R, Malisoux L, Nuhrenborger C, et al. Association of previous injury and speed with running style and stride-to-stride fluctuations. *Scand J Med Sci Sports.* 2015;**25**:e638–45.

Marti B, Vader JP, Minder CE, et al. On the epidemiology of running injuries. The 1984 Bern Grand-Prix study. *Am J Sports Med.* 1988;**16**:285–94.

Meeuwisse WH, Tyreman H, Hagel B, et al. A dynamic model of etiology in sport injury: the recursive nature of risk and causation. *Clin J Sport Med.* 2007;**17**:215–9.

Napier C, Willy RW. Logical fallacies in the running shoe debate: let the evidence guide prescription. *Br J Sports Med.* 2018;**52**:1552–3.

Nielsen RO, Buist I, Parner ET, et al. Foot pronation is not associated with increased injury risk in novice runners wearing a neutral shoe: a 1-year prospective cohort study. *Br J Sports Med.* 2014;**48**:440–7.

Nielsen RO, Buist I, Sorensen H, et al. Training errors and running related injuries: a systematic review. *Int J Sports Phys Ther.* 2012;**7**:58–75.

Nigg BM, Baltich J, Hoerzer S, et al. Running shoes and running injuries: mythbusting and a proposal for two new paradigms: 'preferred movement path' and 'comfort filter'. *Br J Sports Med.* 2015;**49**:1290–4.

Perl DP, Daoud AI, Lieberman DE. Effects of footwear and strike type on running economy. *Med Sci Sports Exerc.* 2012.

Pinshaw R, Atlas V, Noakes TD. The nature and response to therapy of 196 consecutive injuries seen at a runners' clinic. *S Afr Med J.* 1984;**65**:291–8.

Ramsey CA, Lamb P, Kaur M, et al. How are running shoes assessed? A systematic review of characteristics and measurement tools used to describe running footwear. *J Sports Sci.* 2019:1–13.

Redmond AC, Crane YZ, Menz HB. Normative values for the Foot Posture Index. *J Foot Ankle Res.* 2008;**1**:6.

Redmond AC, Crosbie J, Ouvrier RA. Development and validation of a novel rating system for scoring standing foot posture: the foot posture index. *Clin Biomech (Bristol, Avon)*. 2006;**21**:89–98.

Richards CE, Magin PJ, Callister R. Is your prescription of distance running shoes evidence-based? *Br J Sports Med*. 2009;**43**:159–62.

Ridge ST, Johnson AW, Mitchell UH, et al. Foot bone marrow edema after a 10-wk transition to minimalist running shoes. *Med Sci Sports Exerc*. 2013;**45**:1363–8.

Roy JP, Stefanyshyn DJ. Shoe midsole longitudinal bending stiffness and running economy, joint energy, and EMG. *Med Sci Sports Exerc*. 2006;**38**:562–9.

Ryan M, Elashi M, Newsham-West R, et al. Examining injury risk and pain perception in runners using minimalist footwear. *Br J Sports Med*. 2014;**48**:1257–62.

Ryan MB, Valiant GA, McDonald K, et al. The effect of three different levels of footwear stability on pain outcomes in women runners: a randomised control trial. *Br J Sports Med*. 2011;**45**:715–21.

Saunders PU, Pyne DB, Telford RD, et al. Factors affecting running economy in trained distance runners. *Sports Med*. 2004;**34**:465–85.

Schwellnus MP, Stubbs G. Does running shoe prescription alter the risk of developing a running injury?. *International SportMed Journal*. 2006;**7**:138–53.

Shrier I, Platt RW. Reducing bias through directed acyclic graphs. *BMC Med Res Methodol*. 2008;**8**:70.

Soligard T, Schwellnus M, Alonso JM, et al. How much is too much? (Part 1) International Olympic Committee consensus statement on load in sport and risk of injury. *Br J Sports Med*. 2016;**50**:1030–41.

Taunton JE, Ryan MB, Clement DB, et al. A prospective study of running injuries: the Vancouver Sun Run "In Training" clinics. *Br J Sports Med*. 2003;**37**:239–44.

Theisen D, Malisoux L, Genin J, et al. Influence of midsole hardness of standard cushioned shoes on running-related injury risk. *Br J Sports Med*. 2014;**48**:371–6.

Theisen D, Malisoux L, Gette P, et al. Footwear and running-related injuries—Running on faith? *Sports Orthopaedics and Traumatology Sport-Orthopädie - Sport-Traumatologie*. 2016;**32**:169–76.

Tung KD, Franz JR, Kram R. A test of the metabolic cost of cushioning hypothesis during unshod and shod running. *Med Sci Sports Exerc*. 2014;**46**:324–9.

Videbaek S, Bueno AM, Nielsen RO, et al. Incidence of running-related injuries per 1000 h of running in different types of runners: A systematic review and meta-analysis. Sports Med. 2015;**45**:1017–26.

Worobets JT, Wannop JW, Tomaras E, et al. Softer and more resilient running shoe cushioning properties enhance running economy. *Footwear Sci*. 2014;**6**:147–53.

6 The Evolution of Steering Wheel Design in Motorsport

James W. H. Brown, Neville A. Stanton, and Kirsten M. A. Revell

CONTENTS

6.1 INTRODUCTION

The level of technology in modern racing cars, coupled with the requirement to improve performance wherever possible has led to steering wheels becoming highly complex interfaces. The primary task of the racing driver and its physical demands, in combination with the secondary task of interacting with these complex interfaces has led to both driving and interface-based errors. Initial examination of four recent F1 steering wheels revealed considerable differences in design philosophies indicating that teams may not be fully aware of the potential advantages of applying HFE methods. There may therefore be scope for modifications that could improve both safety and performance. This chapter aims to examine how HFE methods can be applied within motorsport, starting with an example of a modern wheel. The drivers' primary and secondary tasks are explored, together with their environment and associated performance effects. The rationales for re-design are discussed, and relevant aspects of usability identified. Some of the key usability aspects focus on the interface being effective, operable, error resistant, memorable, and efficient. These, and additional aspects define the associated HFE methods that would be most appropriate for use in analyses. Appropriate HFE methods and their applications are considered. An initial analysis of four existing wheels is described that examines efficiency and difficulty of usage, highlighting differing design philosophies within each team. A final example of two less complex steering wheels is provided that further highlights the wide variation in current designs.

6.2   A BRIEF HISTORY OF STEERING WHEEL DEVELOPMENT IN MOTORSPORT

Motorsport began in France in the 1890s; vehicles were primitive and very slow by modern standards due to the concept of the motor car still being in its infancy. The sport of racing automobiles against each other quickly became popular, however, and was soon embraced internationally. Throughout the years, engine and chassis design improved, increasing power outputs and cornering capabilities. In 1950 the Fédération Internationale de l'Automobile (FIA) was formed to regulate what is considered by many to be the highest echelon of motorsport, Formula One. Steering wheel designs in the 1950s remained simple, the wheels' sole tasks being lateral control and the feedback through the wheel rim of the car's dynamic state. Cockpits were large and relatively spacious with drivers sitting in upright positions similar to those of modern road cars. The remaining primary controls consisted of a gearstick and foot-operated throttle, brake, and clutch. The steering wheels were also of large diameter, providing the drivers with the leverage necessary to control the cars with the required accuracy and finesse. The few displays that existed generally consisted of basic analogue dials on the dashboard for RPM, oil pressure, fuel level, and water temperature, additional controls were sparse and either dashboard or floor mounted.

The following decade, however, saw radical conceptual changes; engines became mid-mounted, and coupled with new understandings of aerodynamics and vehicle dynamics, cockpits became much more closely enveloping in order to reduce frontal area and lower the centre of gravity. As a result, the radius of steering wheels was reduced. These more radical changes represented the start of the drivers' environment being influenced in order to improve vehicle performance. The use of aerodynamic wings, a now fundamental part of F1, was introduced in the late 1960s, and resulted in higher cornering speeds. The downforce increased tyre grip and inflicted higher lateral and longitudinal loadings on the drivers. It was also in the 1960s that teams started to place some controls on the steering wheel. Initially this took the form of a single toggle switch, mounted unilaterally with a hand-painted label. The 1970s saw the introduction of slick tyres, generating even higher lateral and longitudinal forces, and in the latter part of the decade, turbos were fitted to the vehicles. The turbos generally featured adjustable boost via a knob fitted in the cockpit, usually on the dashboard. This was one of the first instances of a control being provided to the driver to enable direct adjustment of vehicle performance. This boost control allowed drivers to select higher settings for additional power, with the associated risks of higher fuel consumption and engine wear.

The 1980s saw the number of secondary controls and instruments began to increase significantly, although they stayed primarily on the dashboard in most cases. Technological advances resulted in cars fitted with sequential gearboxes that removed the requirement for a clutch pedal and allowed gears to be changed using paddles mounted on the back of the steering wheel. Active suspension systems were also fitted that had various settings that could be configured via cockpit controls. Liquid Crystal Displays (LCD) were introduced to replace analogue dials, allowing a wealth of additional data to be displayed to the drivers. In the 1990s for a brief few years, traction control and Anti-Lock Braking systems (ABS) were sanctioned by the FIA. Both of these driver aids featured scope for adjustability with associated performance benefits, and additional controls were sited within the cockpit to provide those advantages.

The end of the 1990s and early 2000s signalled a significant shift in both placement and numbers of secondary controls. The championship winning F1 car in 1997 featured four steering wheel-based controls, this increased to 9 in 1998, 18 in 1999, and 20 in 2001. Since then, the numbers of steering wheel-based controls have varied between approximately 14 and 25. Wheels also fundamentally changed in terms of shape, from the traditional round design, they became rectangular, partially to allow the driver an improved view of the track over the top of the wheel that would otherwise be obstructed due to the very low seating position. New forms of controls were introduced, pushbuttons and toggles were initially employed; however, around 2000, rotary-based encoders started to appear, followed by thumb wheel-based encoders around 2010. It should be noted that on modern steering wheels, the number of controls does not necessarily correlate with the number of available functions. The functionality of some individual controls can be adjusted based on mode settings, the modes being selected via a specific rotary encoder usually placed near the centre of the wheel. In response to the higher number of controls, steering wheel labelling, and colour coding has become prevalent. The championship winning car in 2007 featured nine different colours and 27 separate labels. Also, the FIA mandated LCD driver display system can display up to 100 unique pages of information. Safety regulations have also introduced a range of factors that affect the design of steering wheels; drivers are required to wear helmets, fireproof race suits, and gloves. Despite the gloves being specifically designed to enable maximum tactile throughput, controls need to be designed to account for reduced sensation.

6.3 CONTEMPORARY STEERING WHEEL DESIGNS IN MOTORSPORT

The main rationale for the complexity of modern steering wheels is considered to be the benefits that can be gained by optimising the cars' settings and maximising performance. Over half of the controls on a modern F1 car are dedicated to performance optimisation. The illustration in Figure 6.1 shows the steering wheel used by Nico Rosberg at the 2015 Suzuka Grand Prix. Three types of control are visible, 12 pushbuttons, three rotary encoders along the bottom, and three thumbwheel encoders mounted around each thumb aperture. In the centre is the LCD with Light Emitting

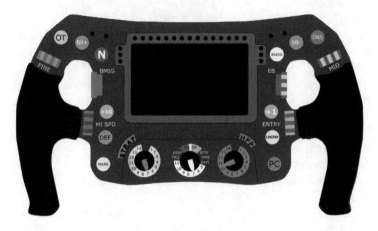

FIGURE 6.1   2015 Mercedes Benz F1 steering wheel.

TABLE 6.1

Controls Featuring on the Front of Nico Rosberg's 2015 Mercedes Benz Wheel

Control Label	Function	Improves Performance
OT	Overtake, used to deploy additional power for a short time.	Yes
DRS	Drag Reduction System. Adjusts rear wing to reduce drag.	Yes
BB–	Move brake balance rearwards to alter handling.	Yes
BB+	Move brake balance forward to alter handling.	Yes
N	Select neutral.	No
Radio	Talk to engineers in the pitlane.	No
+10	Cycle through settings in groups of ten.	Some settings
+1	Cycle through settings individually.	Some settings
DEF	Selects default modes.	No
Limiter	Activates pit lane speed limiter.	No
Mark	Places a reference marker in the car's data log.	No
PC	Pit Confirm. Acknowledge information from the pitlane.	No
Fine	Likely to be a fine adjustment for torque or differential.	Yes
Mid	Differential adjustment to alter handling mid-corner.	Yes
BMIG	Adjust brake migration to change handling when braking.	Yes
EB	Brake balance adjustment.	Yes
HI SPD	Differential adjustment to alter high-speed handling.	Yes
Entry	Differential adjustment to alter handling on corner entry.	Yes
Strat	Strategy, multiple settings for different track scenarios.	Yes
Tyre/Bite	Multiple nested setting for optimising handling.	Yes
HPP	High Performance Powertrain, power optimisations.	Yes

Diodes (LEDs) that indicate engine Revolutions Per Minute (RPM) across the top. The extensive use of colour and labelling is apparent, and some drivers also like sequential instructions to be printed and displayed on the wheel to ensure complex procedures are correctly carried out.

The controls and associated functions on the Mercedes wheel are listed in Table 6.1.

6.4 THE FORMULA ONE DRIVING TASK

6.4.1 PRIMARY DRIVING TASK

The main goal of the racing driver can be simply described as to win races. However, this constitutes a set of specialist and interlinked complex tasks that must be carried out effectively. The primary task, that of driving the vehicle, was described by Henderson (1968) as being a combination of a pursuit task and a compensatory task. 'Racing lines' are employed by drivers to maximise speed through corners. This enables them to be negotiated on the largest radius possible, thus enabling the highest speed for the given maximum attainable lateral acceleration, limited by available tyre adhesion. This is the fundamental basis of the tracking task that drivers carry out, and is defined by the equation below (Frere, 1963) where a represents lateral acceleration in m/s^2, v is velocity in m/s, m is mass in kg, and r is radius in m.

$$a = \frac{mv^2}{r}$$

Whilst this equation defines the ideal scenario, drivers will adjust their racing lines dependent upon track conditions, traffic, and the nature of the following track sequence, in order to maximise performance.

Figure 6.2 illustrates an idealised 90° right-hand turn. The inside line indicates a driver following what might be perceived as the 'shortest route'; however, the small radius results in a lower maximum corner speed. Employing the racing line as shown by the alternate line, allows a larger radius and consequently a higher cornering speed.

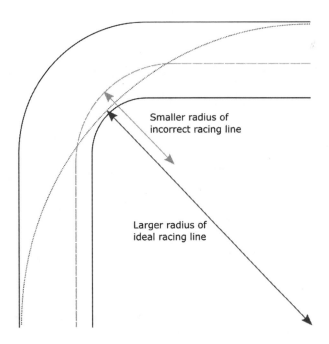

FIGURE 6.2 An idealised racing line (left) compared to an incorrect racing line (right).

In addition to the previously mentioned pursuit task, drivers carry out a compensatory task that involves managing the vehicle's dynamics and maintaining control at the limit of adhesion. Tyres can only provide a limited amount of grip; they do so both laterally and longitudinally. The limit of adhesion is often illustrated as a friction circle (Holbert et al., 1982), which represents lateral and longitudinal forces on a vehicle.

Figure 6.3 shows an example friction circle; it is shown as a non-symmetrical ellipse, as most vehicles can generate higher longitudinal forces when braking compared to accelerating. The example drivers' start at point A, whilst braking for a left-hand corner; they then turn into the corner (B), generating a lateral acceleration to the right; they then exit the corner and accelerate (C). The driver with the dashed line is initially not braking as hard as possible and is then releasing the brakes too much prior to turning in. This results in them wasting the available grip as they enter the corner, and in turn they then accelerate too late, again, wasting the available grip. The driver with the dotted line transitions from brake to steering and back to throttle in such a way that the majority of the available grip is exploited, resulting in considerably faster cornering and resultant lap time.

Drivers will generally try to use as much of the available grip as possible, resulting in the highest levels of longitudinal and/or lateral acceleration. When they demand too much from the tyres and exceed the limit of adhesion, they will start to slide, resulting in one of three vehicle states: understeer, oversteer, or neutrality. These states, whilst occasionally intentional and beneficial, are usually unwanted and a result of the driver over-driving the car or misjudging the available grip level. When these states occur, they require correction, often almost instantaneously. Indeed, drivers are usually able to sense the dynamic state of the car via equilibrioception

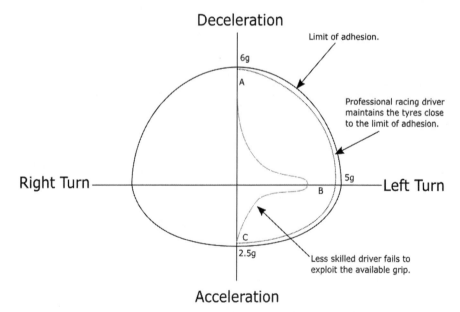

FIGURE 6.3 A friction circle illustrating the limit of adhesion and how skilled drivers exploit it.

and react pre-emptively. Depending upon the setup of the vehicle, track conditions and driving technique, these handling states can occur at relatively high frequencies. Corrections, particularly to oversteer require a high level of precision, as overcorrections can result in exacerbation of the state and ultimately loss of control. It is this balancing of control, as the driver attempts to maintain the car at the limit of adhesion that constitutes the compensatory task as defined by Henderson (1968).

6.4.2 SECONDARY TASK DEMANDS

The requirement for performance optimisation of many of the systems featured on modern vehicles presents drivers with a considerable secondary task. Aspects such as brake bias and differential settings provide a means to optimise the vehicle's handling characteristics. It can be beneficial to adjust these sometimes on a per-corner basis. In addition to these controls, there exist additional controls for adjusting aspects such as fuelling, energy recovery, and settings profiles that are selected based upon the current tyre compound. A substantial load is placed on the driver when attempting to maintain awareness of the car's current status and to adjust settings as necessary. Drivers will usually adjust chassis balance-related settings based on their own considerations and feeling of the car; however, team engineers will also speak to the drivers via radio to provide them with requests for settings based on strategy or feedback from in-car systems that they monitor via telemetry. The two-way verbal communication with engineers by radio represents an additional load on the driver. Despite the teams having an awareness of the distractive nature of talking to drivers, they can still occasionally be heard on broadcasts being told to stop talking by the driver who is trying to concentrate. In 2016, the FIA introduced article 20.1 into the regulations stating that, 'The driver must drive the car alone and unaided' (FIA, 2017). Whilst this reduced a significant amount of the distraction associated with radio communications, it also meant that the driver was required to be fully aware of the multitude of functions built into their interface. In the event of the driver forgetting a function or needing assistance to select a correct setting, as long as there was no safety implication, the team engineers were not allowed to provide any help. This rule was widely criticised by drivers (PlanetF1, 2016), as the wheels had been designed to a level of complexity with the understanding that engineers could provide guidance as necessary. This additional driver requirement was ultimately considered to be too demanding, and the rule was revoked.

Drag Reduction System (DRS) is employed prior to long straights for maximum benefit. Handling adjustments (BB+/−, Entry/MID) are made prior to the corner for which they are required. The radio is likely to be used at the start of a straight to minimise distraction. Strategy changes such as selecting a wet weather setting are rare and would be made on a straight as necessary. The overtake button (OT) would usually be employed on a straight to enable the passing of another vehicle. (Figure 6.4.)

6.5   ENVIRONMENTAL DEMANDS

The drivers' environment has a large influence on their task difficulty. In addition to the cramped conditions within the cockpit, and the regulatory requirement to

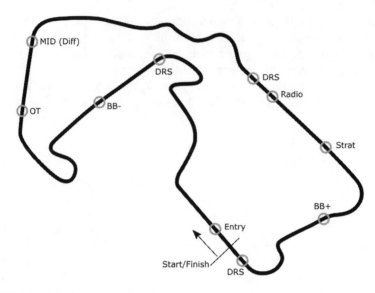

FIGURE 6.4 Points at which specific controls might be activated during practice on the Silverstone GP track.

wear fireproof clothing and a full-face crash helmet, drivers are strapped in tightly with multi-point harnesses. Their seating position is as low as possible and reclined with high cockpit sides to provide lateral protection. All of these features mean that there are multiple constraints in terms of movement and visibility. Bertrand et al. (1983) identified five distinct stresses to which drivers are exposed: g-forces, temperature, emotion, vibration, and muscular effort. Despite being strapped into their cars, lateral and longitudinal forces of up to 6 g require drivers to have very strong neck muscles to maintain their head position under braking and cornering. The performance requirements and FIA mandated regulations also require the cars to run with a very low ride height with associated stiff suspension settings. This means that when the cars traverse even slightly bumpy track surfaces the undulations are transmitted straight to the driver through the carbon tub. The clothing requirements, in combination with the physical exertions and enclosed cockpit and proximity to the cars' engine, often makes the drivers' environment very hot. As a result of the emotional stress, competition, and danger, drivers exhibit high levels of stress hormones and adrenaline when racing (Watkins, 2006). In addition to the muscular effort required to maintain head position, drivers are required to exert high levels of forces on the brake pedal and steering system (Pruett, 2012). These inputs take place at high frequencies throughout the race, potentially over durations of up to two hours in F1, and even longer in endurance racing.

6.6   RATIONALE FOR RE-DESIGN

Multiple rationales exist for improving the usability of multi-function steering wheels in motorsport, the main being the resultant improvement in safety. Distraction appears to be one of the disadvantages of a complex steering wheel design that requires multiple interactions at relatively high frequencies. Distracted drivers may not be paying the requisite attention to their driving task, and this can result in mis-judgements, car control errors, and a reduction in situation awareness. All of these can lead to a loss of control of the vehicle, and/or a collision with a competitor's car or trackside object. The maximum speeds in the higher echelons of motorsport within which complex steering wheels are commonly found can be in excess of 200 mph in F1 or Le Mans Prototype (LMP) categories. The margin for error is small, but the risks in terms of kinetic energies are high and collisions can be catastrophic. Much is invested in improving safety in other areas of the car; therefore, the driver's user interface should receive the same attention. Expense is another rationale for interface improvements, with both driving errors and interface errors potentially costly to teams that run the cars. In the event of an accident caused by an interface issue, repair costs are accrued. Racing cars are often engineered using expensive materials such as carbon fibre and titanium, meaning that accident damage is rarely insignificant even for minor incidents.

The essence of motorsport is competition, teams and drivers compete to win races and ultimately championships. Employing wheels that allow a high level of perfor-mance optimisation may, under optimal conditions, allow the driver to set the fast-est lap time by, at most, a few tenths of a second on some circuits. However, when circumstances are less than ideal, and a driver's cognitive resources become more taxed, fewer resources are available to make wheel-based optimisations (Wickens, 1981). This is when distraction starts to become an issue, and either driving errors or interface-based errors begin to occur. There are multiple circumstances in which this can happen; for example, if weather conditions change then more cognitive effort may be required to maintain control in slippery conditions. A driver might be rac-ing closely with a competitor and having to expend more effort in keeping track of their spatial position whilst defending or attacking. Even if the driver's task remains at the same level of difficulty, there are other influences that may affect their task performance, fatigue can be an issue in long races, and dehydration is known to have a cognitively detrimental effect (Lieberman, 2007). Literature suggests that almost all of the major stresses which racing drivers undergo have some negative cognitive effects associated with them (Brown et al., 2018a). The design intent to maximise drivers' ability to optimise the cars' settings in terms of performance may therefore not necessarily be appropriate. A driver who might be 0.2 s faster per lap for 20 laps by making multiple adjustments would find themselves 4 s ahead, however, the complexity of their secondary task may raise the probability of a driving error, particularly at later stages of a race when fatigue is setting in. The smallest of errors in motorsport can easily cost between 0.5 s to 1 s, therefore only four small errors of perhaps missing a braking or turn-in point are required to lose all the gained advan-tage, in addition to the associated cognitive load and frustration.

There are multiple documented incidents that have occurred in motorsport that can be traced back to steering wheel control design issues. One of the most recent involved a GT team at the Bathurst 12-hour endurance race in Australia (Van Leeuwen, 2019). The drivers had expressed a desire for the engine stop button to be moved to the front panel of the steering wheel. This was carried out by engineers and it was placed on the upper right portion of the wheel face, labelled, and highlighted in a bright orange colour. During the race, one of the drivers was rejoining the track after a pit-stop when, instead of pressing the pit lane speed limiter button, they pressed the stop button, and the car rolled to a halt. Multiple seconds were lost restarting the car and getting back up to speed. This same mistake occurred a second time later in the race. Had the car not lost time due to these mistakes, the team may have finished in a higher position. On examination of the car's steering wheel, the newly placed orange engine stop button was in close proximity to the red pit lane speed limiter button, both situated in the top right quadrant of the steering wheel.

In Formula One, a strategy mode rotary switch on the wrong setting at the start of the 2016 Barcelona GP ultimately resulted in both the Mercedes Benz teams cars retiring despite starting in first and second places (Collins, 2016). The incorrect setting caused the lead car to reduce the amount of deployed power three corners into the race, ultimately resulting in a collision between teammates' vehicles. On more than one occasion, drivers have complained that their workload associated with using the controls on their steering wheels is too high. This has resulted in losing control through simply driving off the track due to distraction (Galloway, 2014), and braking too late in the pit lane and colliding with the pit crew because they were concentrating on wheel settings (Autoweek, 2017).

6.7 HUMAN FACTORS AND ERGONOMICS METHODS IN WHEEL DESIGN AND EVALUATION

A range of aspects of usability apply to motorsport, and by studying documented cases of issues relating to interface usage, the relevant aspects reveal themselves and their level of influence. Issues generally fall into two categories: drivers carrying out erroneous secondary (steering wheel-based) control operations, or primary task (driving) errors as a result of distraction by secondary control operations. Through the identification of usability criteria that map to these categories within the context of motorsport, a set of motorsport specific usability criteria can be established, allowing the selection of appropriate human factors methods for analysis. Harvey and Stanton (2013) identified a set of core aspects of usability for In Vehicle Information Systems (IVIS); these provided a basis for usability aspect selection. For example, the interface should be efficient (Nielson, 1993), and operations should be quick as drivers ideally need to keep their hands firmly on the wheel. Current designs typically place multiple buttons and thumbwheels in an arc around where the drivers' thumbs can reach whilst their hands remain on the steering wheel. The rationale is to ensure that drivers can remain in control of the car with both hands whilst simultaneously retaining the ability to access secondary control functionality. Less frequently used controls are generally clustered in the middle of the wheel and require a hand to be removed from the wheel to operate them. Controls should be error resistant (McGrenere & Ho, 2000), reducing the potential for mistakes, as well as the seriousness of the consequences. They should be effective (Shackel, 1991), enabling even complex functions to be carried out simply to reduce the potential for error and cognitive load. The interface should be highly operable (Bevan, 2001), as this may improve control efficiency and reduce effort. A memorable interface is also important (Nielson, 1993), as drivers who forget functionality or control placements may be severely distracted whilst trying to read labels on their wheel. The fact that current cars feature a high number of colours and labels is suggestive that drivers may already be experiencing issues related to memorability. Due to space constraints, safety issues, and uncontrollable confounding variables in real cars, initial usability testing should ideally be carried out using a high-fidelity motorsport simulator.

Through the application of a motorsport specific usability toolset of appropriate HFE methods, existing and new wheel designs may be assessed for a range of usability aspects and potential improvements. Through iterative application of the toolset, it may be possible to increase driver performance and reduce errors, thus improving safety and competitiveness. The toolset includes interface analysis methods, such as Fitts' Law (Fitts, 1954), and Link Analysis, and Layout Analysis (Stanton et al., 2013). Through applying the methods to wheel interfaces insight can be gained into the efficiency of control layouts and their difficulty of usage. Error identification can be carried out at an early stage using methods such as the Human Error And Reduction Technique (HEART) or the Human Error Template (HET) (Stanton et al., 2013). Workload is an important aspect for analysis, and this can be assessed through subjective measures such as the NASA TLX (Hart & Staveland, 1988) or DALI (Pauzie, 2008), as well as physiological measures such as Heart Rate Variability (HRV) (Brookhuis & de Waard, 2010) and pupillometry (Marshall, 2002). Performance measures for both the driving task and interface tasks, such as lap times and operation completion times, would provide some insight into interactions between primary and secondary tasks. For example, a driver setting fast lap times may be expected to make more mistakes when asked to activate elements of their interface. Quantifying performance may provide some evidence of how interface designs are affecting drivers' cognitive load.

The analysis of existing wheel designs provides data on aspects that are beneficial or disadvantageous. Examination of current literature on steering wheel-based controls provides design guidance and innovative control methods that should be tested where applicable. For example, the memorability and operability of interfaces may be improved through laterally coding control placements with respect to function (Xiong & Proctor, 2015), and grouping controls based on function, applying the principle of objective primacy (Gkikas, 2011). In terms of innovative control methods, the use of integrated switches to combine controls (Murata, Yamada, & Moriwaka, 2009) could improve efficiency and reduce complexity. Touch screens are not commonly seen on steering wheels; however, they have a faster throughput than currently used controls such as rotary encoders. (Thomas, 2018; Stanton et al., 2013b). This might improve safety and reduce driving errors by minimising the amount of time drivers do not have both hands placed optimally on the steering wheel.

The existence of different control types featuring on wheels over the years reflects the required demands or currently available technology. The most recent type being thumbwheel-based rotary encoders that require sufficient torque to operate that they cannot be moved accidentally, and intended movements result in definitive haptically-communicated clicks. There may, however, be other types of control that should be explored, such as those used in Hands On Throttle And Stick (HOTAS)-based control systems that feature an environment similar to that found in racing cars. Point-of-view or 'Hat' style joysticks could provide some benefits such as lowering subjective workload due to resistance to error in high vibration environments (Thomas, 2018). Their operation is well understood, they provide good feedback, and they are well proven on control sticks in military aviation, which is a comparable domain in terms of physical environment. Touchscreens specifically designed to be wheel-mounted could be of some benefit, it would be possible for the interface to adjust dynamically based on demands. For example, in complex parts of a circuit, or during specific race phases, it could present a simpler interface with larger buttons, hiding those that would not normally be utilised.

Once a steering wheel design has been refined in simulator-based experiments, it should then be assessed in a real car ideally under conditions similar to those found when racing. Whilst the design should have taken into account the demands of a real car, this final step will identify the need for any further improvements.

6.7.1 A Case Study Examining Efficiency and Difficulty of Usage

An analysis of four teams' steering wheel designs from 2017 (Brown et al., 2018b) was carried out using Fitts' Law (MacKenzie, 1992) and link analysis (Stanton et al., 2013) in order to assess the efficiency and difficulty of usage of the layouts. The positions and dimensions of controls common to all wheels were calculated based on the known dimensions of the LCD screen that is standardised across teams, control function, name, and type were also noted. A set of 155 common control interactions were derived from in-car footage that included processes such as preparing the car at the start, initial laps and a pit-stop. These interactions comprised lap number, distance from the start, the control activated, and race stage. The control data and interaction data were coded into comma separated value (CSV) files. A software application was developed in C# to load these CSV files, this generated a graphic for each wheel illustrating the link analysis output, and calculated bi-lateral values for Index of Difficulty (ID) for each interaction as well as traversal distances between controls and rest positions. This revealed some considerable variations between wheel designs in terms of both traversal distances to controls, and the side of the wheel on which specific controls are placed.

All four wheels analysed were used by right-handed drivers. Two teams appeared to favour the drivers' right hand for the majority of a common set of controls, one team favoured the driver's left hand and one team split secondary tasks evenly across both hands. This is suggestive that there is no consensus on which hand should ideally operate secondary controls. If the drivers' non-dominant hand is employed for secondary tasks, they may take longer to complete due to being less dextrous. However, the driver's dominant and therefore stronger and more dextrous hand remains on the wheel and a more precise primary task performance is therefore likely. The advantage of the inverse scenario is a faster and potentially less error-prone secondary task performance, albeit with compromised precision of primary task performance. Traversal distances when activating controls also showed considerable variations between teams. For a given set of 155 control operations carried out in sequence, including returning hands to the driving position, one wheel required a total traversal distance of 7 m, whereas another team's design required 13.5 m. This again may indicate specific design philosophies, by placing controls closer together, they can be activated more efficiently; however, there is potentially a greater chance of activating an adjacent control in error. Controls vary in terms of the criticality of their functions, and in their frequency of usage. There is a balance to be found when considering where controls should be sited taking these factors into account. Critical controls such as the pit lane speed limiter should be able to be activated in very short durations, as activating too late would risk a potential time penalty. However, other controls that see frequent use should also be sited close to the driver's hands, as combined high traversal times will result in driver's hands not being in full contact with the steering wheel for significant durations, which may detrimentally affect primary performance.

Examination of wheels from four F1 teams indicated that they share some commonality in critical control positioning. For example, the button to select neutral is generally placed top left and easily accessible by the drivers' thumbs; however, one team placed it low on the left side. The pit lane speed limiter button is also sited usually on the top right, again with one team placing it low on the right side. Frequently used controls such as the overtake buttons were placed in the upper easily accessible quadrants by three teams, with one placing it on the back of the wheel. The radio button is placed top left on three wheels, and top right on another. This is suggestive that frequency and criticality are likely accounted for, but perhaps not optimised.

There is also the aspect of individual design input, as typically in Formula 1, drivers are allowed to customise their wheels to an extent. Wheel designs are in a state of flux even during racing seasons, with changes implemented between races based on the track characteristics and driver requests. Whilst iterative development is advantageous, it raises the issue of memorability, introducing the risk of a driver forgetting that a control has moved.

Multiple variations in steering wheel design are illustrated in Figure 6.5 from two road car derived racing cars. The Mercedes AMG GT employs mainly icons or acronyms placed on the buttons themselves, whereas the Porsche Carrera Cup car features more explicitly expressed functions, using whole words placed adjacent to the controls. It is interesting to note the guard around the pitlane speed limiter (PIT) button on the Mercedes to increase error resistance. Such a guard is not present on

FIGURE 6.5 The steering wheel controls on a Mercedes Benz AMG GT (left) and 2019 Porsche Carrera Cup (right). These wheels only feature pushbuttons; however, it highlights not only the variation in fundamental steering wheel shape, but also positioning of controls, use of colour and word/acronyms/icons used to denote function.

the Porsche (Pit Speed) perhaps due to deliberate separation from other controls with the same rationale.

Ideally, racing teams would potentially benefit from either integrating human factors methods into the roles that undertake any aspect of human/vehicle interaction, or creating a small dedicated human factors sub-team to assess and improve systems as necessary. The competitive nature of motorsport where teams vie for the slightest advantage would likely see a propagation of human factors methods and practitioners within the sport once the advantages are both understood and proven.

6.8   CONCLUSIONS

The introduction of steering wheel-based controls in motorsport was made primarily to provide the drivers with the ability to further improve the performance of their vehicles by adjusting systems whilst they are on track. Whilst these adjustments do provide definite performance advantages, there is a balance that needs to be investigated between the gains that can be made versus the potential for distraction and error due to this secondary task. By identifying the most appropriate aspects of usability within the context of motorsport and applying human factors methods to each, the design of steering wheel-based controls may be improved. The demand of the drivers' primary task coupled with the unique physical demands of motorsport must additionally be assessed and considered.

Human factors methods can also be applied to a wide range of additional areas of motorsport. These can aid improvements in safety as well as competitiveness. Salmon and Neville (2014) highlighted the need to take a holistic approach when examining events with a view to improving safety. They considered the fatal accident involving Jules Bianchi where he left the track during a caution period and collided with a recovery vehicle. They describe the propensity for many to blame an individual element of the system that can lead to additional contributing factors being overlooked. Salmon and Neville (2014) advise examining events such as these as interrelated systems that may require amendments to multiple elements. The complex and time-critical sequence of events that takes place during pitstops is a good example of human factors being well addressed in Formula One. Indeed, they have been honed to the point that other domains have learned from the methods employed. Catchpole et al. (2007) examined F1 pitstop techniques and aviation models to improve the quality and safety of patient handovers between operating theatres and intensive care units. Ultimately, there appears to be a great opportunity for the application of human factors methods in many aspects of motorsport. In an industry that invests considerable amounts in the optimisation of other areas of engineering, this could bring valuable benefits.

REFERENCES

Autoweek (2017). Formula One: Kubica loses second with pit-stop error during Belgian Grand Prix [online]. Available at: http://autoweek.com/article/formula-one/formula-one-kubica-loses-second-pit-stop-error-during-belgian-grand-prix, accessed on 31 March 2017.

Bertrand, C., Keromes, A., Lemeunier, B. F., Meistelmann, C., Prieur, C., & Richalet, J. P. (1983). Physiologie des Sports Mécaniques. In: 1st International Congress of Sport Automobile, Marseilles.

Bevan, N. (2001). International standards for HCI and usability. *International Journal of Human-Computer Studies*, 55:4, 533–552.

Brookhuis, K. A., & de Waard, D. (2010). Monitoring drivers' mental workload in driving simulators using physiological measures. *Accident Analysis and Prevention*, 42:3, 898–903.

Brown, J. W. H., Stanton, N. A., & Revell, K. M. (2018a). A review of the physical, psychological and psychophysiological effects of motorsport on drivers and their potential influences on cockpit interface design. In: International Conference on Applied Human Factors and Ergonomics (pp. 514–522). Cham: Springer.

Brown, J. W. H., Stanton, N. A., & Revell, K. M. (2018b). Software analysis of racing drivers' interfaces using link analysis and Fitts' law. In: *Proceedings of Ergonomics and Human Factors Conference*.

Catchpole, K. R., De Leval, M. R., McEwan, A., Pigott, N., Elliott, M. J., McQuillan, A., Macdonald, C., & Goldman, A. J. (2007). Patient handover from surgery to intensive care: Using Formula 1 pit-stop and aviation models to improve safety and quality. *Pediatric Anesthesia*, 17:5, 470–478.

Collins, S. (2016). Gravel trap: How Rosberg's switch error caused first lap crash – Racecar Engineering [online]. *Racecar Engineering*. Available at: http://www.racecar-engineering.com/blogs/gravel-trap-did-rosbergs-switch-error-cause-first-lap-crash/, accessed on 27 March 2017.

FIA. (2017). Federation Internationale de l'Automobile [online]. Available at: http://www.fia.com/regulation/category/110, accessed on 21 March 2017.

Fitts, P. M. (1954). The information capacity of the human motor system in controlling the amplitude of movement. *Journal of Experimental Psychology*, 47:6, 381.

Frere, P. (1963). *Competition Driving*. London, UK: Batsford Ltd.

Galloway, J. (2014). Pastor Maldonado takes blame for mistakes after crash follows earlier spin in China [online]. *Sky Sports*. Available at: http://www.skysports.com/f1/news/24231/9271393/pastor-maldonado-takes-blame-for-mistakes-after-crash-follows-earlier-spin-in-china, accessed on 31 March 2017.

Gkikas, N. (2011). Formula 1 steering wheels: A story of ergonomics. *Ergonomics in Design*, 19:3, 30–34.

Hart, S. G., & Staveland, L. E. (1988). Development of NASA-TLX (Task Load Index): Results of empirical and theoretical research. In: *Advances in Psychology* (Vol. 52, pp. 139–183). North-Holland: Elsevier.

Harvey, C., & Stanton, N. A. (2013). *Usability Evaluation for In-Vehicle Systems*. Boca Raton, FL: CRC Press.

Henderson, M. (1968). *Motor Racing in Safety. The Human Factors*. London, UK: Stephens.

Holbert, A., Holbert, B., & Bochroch, A. (1982). *Driving to Win*. Tucson, AZ: Aztex Corporation.

Lieberman, H. R. (2007). Hydration and cognition: a critical review and recommendations for future research. *Journal of the American College of Nutrition*, 26(sup5), 555S–561S.

MacKenzie, I. S. (1992). Fitts' law as a research and design tool in human-computer interaction. *Human-Computer Interaction*, 7(1), 91–139.

Marshall, S. P. (2002). The index of cognitive activity: Measuring cognitive workload. In: *Proceedings of the IEEE 7th conference on Human Factors and Power Plants* (p. 7). IEEE.

McGrenere, J., & Ho, W. (2000). Affordances: Clarifying and evolving a concept. In: *Graphics Interface* (Vol. 2000, pp. 179–186).

Murata, A., Yamada, K., & Moriwaka, M. (2009). Design method of cockpit module in consideration of switch type, location of switch and display information for older drivers. In: *Proceedings of the 5th International Workshop on Computational Intelligence and Applications* (Vol. 2009, No. 1, pp. 258–263). IEEE SMC Hiroshima Chapter.

Nielsen, J. (1993). Iterative user-interface design. *Computer*, 26(11), 32–41.

Pauzié, A. (2008). A method to assess the driver mental workload: The driving activity load index (DALI). *IET Intelligent Transport Systems*, 2(4), 315–322.

Planet F1. (2016). Alonso takes Hamilton's side on radio ban. *PlanetF1.com* [online]. Available at: https://www.planetf1.com/uncategorized/alonso-takes-hamiltons-side-on-radio-ban/, accessed on 28 December 2019.

Pruett, M. (2012). Dario Franchitti: So you think driving an Indy car is easy? Try steering – Part 2 [online]. *Road & Track*. Available at: http://www.roadandtrack.com/motorsports/news/a9143/dario-franchitti-so-you-think-driving-an-indy-car-is-easy-try-steering-part-2-37917/, accessed on 15 April 2017.

Salmon, P., & Neville, T. (2014). Jules Bianchi and sharing the responsibility for catastrophe. [online]. *The Conversation.com*. Available at: https://theconversation.com/jules-bianchi-and-sharing-the-responsibility-for-catastrophe-32937, accessed on 28 December 2019.

Shackel, B. (1991). Usability-context, framework, definition, design and evaluation. *Human Factors for Informatics Usability*, 21–37.

Stanton, N. A., Harvey, C., Plant, K. L., & Bolton, L. (2013b). To twist, roll, stroke or poke? A study of input devices for menu navigation in the cockpit. *Ergonomics*, 56:4, 590–611.

Stanton, N., Salmon, P. M., & Rafferty, L. A. (2013). *Human Factors Methods: a Practical Guide for Engineering and Design*. London, UK: Ashgate Publishing, Ltd.

Thomas, P. R. (2018). Performance, characteristics, and error rates of cursor control devices for aircraft cockpit interaction. *International Journal of Human-Computer Studies*, 109, 41–53.

Van Leeuwen, A. (2019). Bathurst 12 Hour: Bentley's Soucek takes blame for 'amateur' errors [online]. *Autosport.com*. Available at: https://www.autosport.com/gt/news/141305/soucek-takes-blame-for-costly-amateur-errors, accessed on 28 December 2019.

Watkins, E. S. (2006). The physiology and pathology of formula one Grand Prix motor racing. *Clinical Neurosurgery*, 53(14), 145–152.

Wickens, C. D. (1981). Processing resources inattention, dual task performance and workload assessment. Technical Report EPL-81-3/ONR-81-3, Champaign, IL: University of Illinois.

Xiong, A., & Proctor, R. W. (2015). Referential coding of steering-wheel button presses in a simulated driving cockpit. *Journal of Experimental Psychology: Applied*, 21:4, 418.

7 Current Methods for Optimising Sports Wheelchairs at an Individual Level

David S. Haydon and Ross A. Pinder

CONTENTS

7.1 INTRODUCTION

Wheelchair sport has received increased levels of published research in recent years (Goosey-Tolfrey et al., 2018; Mason et al., 2018), with a particular focus on Wheelchair Rugby (WCR) (Haydon et al., 2018a), Wheelchair Basketball (WCB) (de Witte et al., 2018; van der Slikke et al., 2018), and Wheelchair Tennis (WCT) (de Groot et al., 2017). Performance in these wheelchair court sports can primarily be considered in three main areas: (i) the athlete; (ii) the wheelchair; and (iii) the athlete–wheelchair interaction (Mason et al., 2013). The athlete component considers the physical and psychological capabilities of the individual, including the sport-specific skills, and physical activity limitation resulting from their impairment (i.e. physical capabilities). Replicating match demands (both physiological and skill-based) in training through small-sided games has been a particular focus of recent work in order to further develop these skills (Mason et al., 2018). The wheelchair aspect focuses on the mechanical design, where improvements in factors such as rolling resistance and (in some cases) reduced mass can improve performance. The athlete-wheelchair interaction relies on the wheelchair being designed based on a detailed understanding and consideration of the individual's impairment and the sport-specific skills required to maximise performance (Mason et al., 2013) – effectively the application of physical ergonomics. This results in wheelchair configurations varying substantially due to the performance goals and needs, impairment type and severity (Haydon et al., 2016). The variance in these factors leads to the need for individual assessments to optimise a wheelchair configuration and performance.

This chapter focuses on the emergence of research exploring the physical ergonomics of athlete–wheelchair interaction and implications for performance in elite para-sport. While sports such as wheelchair racing place a similar importance on the athlete–wheelchair interaction as court sports, they have substantially different performance goals and configurations; the desire for top-end speed and long propulsive strokes in wheelchair racing leads to a kneeling position (Lewis, 2019), whereas a need for ball-handling, agility, and stability in WCR leads to a more conventional seating position (Mason et al., 2010; van der Slikke et al., 2016b). Whilst learnings may have some transfer to wheelchair racing contexts, in this chapter we focus primarily on research in WCR and other court sports, with an aim to overview wheelchair design parameters, emerging technologies and testing approaches. Finally, we provide the outline of a case-study example from WCR and provide insights and recommendations for practitioners aiming to optimise individual wheelchair ergonomics in elite sports.

7.2   DESIGN PARAMETERS

Para-sports, particularly team sports, use classification systems with the intention of minimising impairment effects on the competition outcome (Altmann et al., 2015). In WCR, common impairment types include limb deficiencies and impaired muscle power, with each individual assigned a classification score based on the sport-specific activity limitation caused by their impairment (Vanlandewijck et al., 2011). Each team is then restricted to four athletes, and a total of eight points on-court at a time. Individual classification scores range from 0.5 to 3.5 points, where a lower score indicates greater activity limitations (Altmann et al., 2013). These are based on a range of assessments for strength, range of motion, and coordination of the trunk, arm, and hand (Molik et al., 2008). Therefore, the same classification score can be assigned to individuals with substantially different impairments. Players then attempt to carry a ball over a line in order to score (typically offensive players, 2.0–3.5 classification scores), with opposition players trying to physically prevent them from achieving this by blocking with their wheelchair (i.e. defensive players, 0.5–1.5 point scores) (Molik et al., 2008). In order to escape or execute blocks, quick accelerations and changes of directions are required and are therefore seen as key performance factors in WCR (Mason et al., 2010). The different impairments and on-court roles result in various propulsion approaches across individuals, with consideration of this crucial to optimising wheelchair configuration (Haydon et al., 2018a). WCB implements a similar classification system, although individual and total point scores differ, as does the exact classification method (International Wheelchair Basketball Federation, 2014). WCT utilises a different approach with two divisions: 'Open' and 'Quad'. The open division allows for players with impairments substantially impacting function of one or both legs but not upper body function. The Quad division allows for players with impairments that limit the ability to manoeuvre around the court, grip the racquet, and perform shots including overhead serves, forehands, and backhands (International Tennis Federation, 2019).

The athlete-wheelchair interaction can generally be considered in two main sections of the wheelchair design: (i) the seat position (see Figure 7.1); and (ii) the set-up of the main wheels. These two components are related, with seat position influencing the mass distribution and access to the wheel (for a given configuration). Whilst other configuration parameters have an impact on performance (e.g. footplate position, strapping), it is beyond the scope of this chapter to overview all factors.

7.2.1 SEAT POSITION

Seat position considers the height of the seat from the ground (Figure 7.1B), the horizontal position of the seat relative to the wheel axle (often referred to as the fore-aft position or seat depth – Figure 7.1A), and the seat angle (Figure 7.1– C). All of these factors impact performance factors including acceleration, agility, stability, and ball-handling – all crucial to on-court performance in WCR and WCB (Mason et al., 2010, 2013). The major difficulty in optimising these parameters is the trade-off effect they often have on each performance factor.

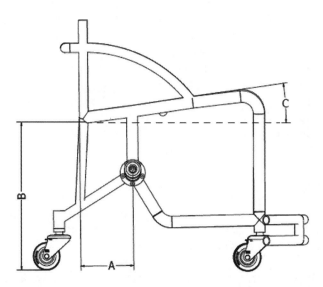

FIGURE 7.1 Seat position parameters include: A – seat depth; B – seat height; and C – seat angle.

An increase in seat height has been reported by players to allow for improved ball-handling and a better view of the court for WCR and WCB players (Mason et al., 2010). WCB and WCT typically have higher seat positions in comparison with WCR to allow for improved shooting and extra serve height, respectively (Laferrier et al., 2012). However, this comes with reduced access to the wheel/pushrim as well as decreasing the stability of the wheelchair as the centre of mass height is increased. For players with greater trunk function (where *function* refers to strength, range of motion, and co-ordination), they are able to compensate for these reductions by flexing their trunk or manipulating their mass distribution – hence the increased seat heights for high-point players (e.g. 3.0–3.5 for WCR) compared with mid- (2.0–2.5 points) and low-point (0.5–1.5 points) players (Haydon et al., 2016). Low-point players – who often have a reduced level of trunk function – are unable to do this as effectively, and therefore rely on a wheelchair configuration that affords more stability. Recommendations for daily propulsion suggest that the elbow should be at an angle of 100–130 degrees when the hand is positioned at top dead centre (TDC) of the wheel for best physiological performance (namely VO_2 and mechanical efficiency) (van der Woude et al., 2009). These results do not transfer directly to wheelchair sports due to the varying requirements, although research has recommended inclusion of normalised parameters based on individual's anthropometrics in future work (Haydon et al., 2016; Mason et al., 2013). While research is now emerging in elite sports (van der Slikke et al., 2018), much more is required with a focus on factors affecting propulsion in both sub-maximal and maximal acceleration in wheelchair propulsion.

Fore-aft position (or seat depth – see Figure 7.1A) has a major influence on the manoeuvrability of the wheelchair. Rather than affecting the vertical position of the centre of mass, it alters the centre of mass horizontally in relation to the wheel axle. An increased fore-and-aft position (seat further behind the wheel axle) is therefore suggested to result in a more 'reactive' wheelchair (i.e. faster turning speeds) (Mason et al., 2013). This may have benefits for agility, but it also makes the wheelchair harder to control. Furthermore, considerations are required regarding the individual athlete and their propulsion technique. The fore-and-aft position will also affect where the players can contact the wheel/pushrim, and hence the range in which the can apply force (Mason et al., 2013). This is particularly important to consider for low-point players in WCR, who often rely on biceps brachii function to generate motion; an increased fore-aft position is therefore adopted as it maximises the region in which they are able to 'pull' the wheel around (Haydon et al., 2018a), utilising their biceps.

To aid in stability of the wheelchair, the seat angle (the angle of the seat above the horizontal – Figures 7.1 to 7.2) can be increased; this places the athlete in a more reclined position, but has been shown to limit trunk function (including when testing with able-bodied participants) and therefore results in decreased acceleration in the first three seconds of propulsion from standstill (Vanlandewijck et al., 2011), a critical component in most wheelchair sports, but in particular WCR. High-point players (i.e. those that typically have greater trunk function) tend to have flatter seat angles compared with low-point players (Haydon et al., 2016) to allow for a greater trunk contribution in propulsion, but also to allow for more mobility in ball-handling

FIGURE 7.2 A comparison of a new WCT design with optimised frame structure and customised seating interface (left) in comparison with the player's previous wheelchair.

activities. Similar emphasis on seat angle has been reported in WCB (Mason et al., 2010).

7.2.2 THE WHEELS

The diameter of the wheel and camber angle (angle of the wheel from the vertical) also influence performance factors for wheelchair sports. The diameter of the wheel is often based on an athlete's preference, with wheel sizes in court sports typically ranging from 24–28 inches (Haydon et al., 2016; Laferrier et al., 2012). Larger wheel sizes have been reported by athletes to be more difficult to accelerate from standstill (Mason et al., 2010) due to an increase in the moment of inertia, potentially resulting in low-point players preferring smaller wheels compared with high-point players (Haydon et al., 2016). However, larger wheels have also been reported to allow athletes to achieve higher top end speeds. Our personal observations suggest that WCT players tend to use larger wheels, sacrificing acceleration for increased height for serving/racquet skills. Further work is required to quantify any effects of wheel diameter on initial acceleration, particularly for players with impairments that have a larger impact on activity limitation (i.e. low-point players in WCR). Compared with daily wheelchairs (which have little to no camber), sports chairs have a much greater degree of wheel camber; in WCR this can be up to 24 degrees (Laferrier et al., 2012). An increase in camber angle results in a wider wheel base, which can have multiple benefits, including an increase in stability due to the increase support base, improving the ability of players to perform blocks, improving ease of turning, and improved hand protection (Faupin et al., 2004; Laferrier et al., 2012; Mason et al., 2012a, 2013). However, too much camber can increase rolling resistance (due to increased contact between the tyre and ground), affecting acceleration and velocity. Following on-court testing for WCB and WCT players, Mason et al. (2012a) recommended a camber angle of 18 degrees for new players. This was due to an increase in linear sprint performance compared with 24 degrees, and improved manoeuvrability performance compared with 15 degrees. However, it was acknowledged that this is only a recommendation as it is unlikely to result in improved performance for all players. Further work is required in this area to develop the understanding of camber angle impacts across various sports and individuals, and harnessing experiential knowledge in elite populations is an important starting point (Greenwood et al., 2012; Haydon et al., 2016; Paulson & Goosey-Tolfrey, 2017).

Although camber angle has an effect on the contact area between the tyre and ground, this is potentially more affected by tyre pressure. Decreasing pressures results in more contact with the ground and more grip, but greater rolling resistance. Conversely, increasing tyre pressures has the opposite effects and can result in reduced grip (Mason et al., 2010). Currently, players select their own tyres and tyre pressures from personal feel, with these varying substantially; recent work with WCR players has shown that pressures vary from 80–200 psi (Haydon et al., 2019). To our knowledge, there has been no work investigating the interaction of camber angle and tyre pressure on ground contact area and subsequent accelerations, velocities, and rotations; there is clearly scope for substantial improvement in our understanding in wheelchair sports performance.

Existing studies highlighted throughout this section have largely investigated single parameters, and whilst this can provide detailed information on that single parameter (i.e. athlete response, impact on performance factors), it involves a time-consuming process to do this for the range of parameters that affect wheelchair performance. Furthermore, researchers and practitioners may not fully capture the potential impact on performance due to the complex interactions between key chair parameters. Recent work has investigated the potential of testing and modelling approaches that are able to reduce the amount of athlete testing required whilst maintaining the ability to assess the effects of individual parameters (further detail provided in *Design parameters*) (Haydon et al., 2019; Usma-Alvarez et al., 2014). Many of these emerging approaches need further work, however, initial results suggest they will begin to provide more comprehensive approaches for assessing the effect of manipulating multiple parameters in an efficient manner.

7.3 EMERGING TECHNOLOGIES

7.3.1 EQUIPMENT DESIGN

Development of equipment and monitoring techniques in para-sport has received increased attention over the past ten years. This has included handcycle designs (e.g. Alex Zanardi for the London 2012 Paralympic Games), and partnerships between BMW and the UK WCB team as well as Honda and the Japanese team for racing wheelchairs. These racing designs in particular are focused on frames constructed from carbon fibre, allowing for substantial reductions in mass as well as increasing the stiffness of the frame, which allows for increased power transfer from the athlete to the wheelchair (Laferrier et al., 2012). An additional feature of these designs is the individually moulded seat that conforms to the individual's body. These are produced using 3D scanning of the athlete's lower body, with the mould then developed to fit this geometry. The premise of a customised seating interface that conforms to the athlete's body provides benefits for improved power transfer, improved wheelchair control, and a reduced frontal area (and therefore usually more aerodynamic overall as the size of the seat can be minimised) (Lewis, 2019). The interface can also be adjusted to ensure that the athlete is positioned to promote symmetrical motion between left and right sides; this has benefits in consistent power production, steering, and is thought to reduce injury risk (Goosey-Tolfrey & Campbell, 1998; Lewis, 2019). However, due to the nature of this work (i.e. elite level, expensive, individualised), there is currently limited information or research exploring the specific effects on performance factors. Three-dimensional assessments of the individual propulsion approach can potentially identify injury risk through any excessive movements and ranges of motion, while the seat itself can reduce the likelihood of pressure sores (a common issue for wheelchair athletes (Mutsuzaki et al., 2014)) by minimising high pressure regions.

Whilst aerodynamic drag may not be a major consideration for wheelchair court sports due to the frequent changes in direction and upright seating (cf. Lewis, 2019), the improved power transfer and control provided by a customised seating interface are of interest. These approaches have been implemented in WCT, where the need for wheelchair control and stability is paramount when completing shots. For a Paralympic gold medallist and multiple Grand Slam champion, this resulted in a completely new seating interface, as well as an altered frame structure (see Figure 7.2). Performance improvements included improved speed, ability to control the wheelchair, and maintain stability throughout sport-specific skills.

For WCR and WCB, there are currently limited examples of moulded seating interfaces; potentially related to the team aspect of the sport. In terms of overall wheelchair design, the frames for WCR and WCB are primarily made from aluminium. Personal discussions with players and coaches have resulted in the suggestion that heavier wheelchairs improve the ability to hold or avoid attempted blocks easier due to the increased momentum. While there is currently limited published work in this area (see van der Slikke et al., 2018 for an exception), there are likely multiple contributing factors, including: work is often completed by national organisations and seen as a competitive advantage; final designs are highly individualised, with the process of that potentially of more interest; and current restrictions in the peer-review processes may not value the impact of individual case-study designs, and hence it is more difficult to publish this work (Haydon et al., 2018b; Pinder et al., 2015).

7.3.2 Sensor Technology and Testing Approaches

Improvements in sensor capabilities have also transferred to the analysis of wheelchair sports. This has been particularly evident in wheelchair court sports, where inertial measurement units (IMUs) are beginning to be used to monitor and quantify on-court performance factors (van der Slikke et al., 2015a, 2015b). This is an important advancement, as less invasive testing approaches allow for a greater degree of fidelity in testing, ensuring that measures are representative of specific performance contexts (Pinder et al., 2011). Previous testing of wheelchair design parameters have often relied on the use of ergometers or other testing approaches which fail to replicate on-court propulsion (Mason et al., 2014; Stephens & Engsberg, 2010), and there has been a growing call for alternative options (Haydon et al., 2018b). Initially, an understanding of the performance factors to be monitored is crucial in determining the testing requirements. Test selection and design should reflect the performance goals; the question may be best answered through controlled, laboratory testing approaches. Each testing method has benefits and limitations; however, it is recommended that tests that are more representative of on-court performance (when controlled appropriately) have the ability to provide better translation to on-court results (Haydon et al., 2018b).

For wheelchair court sports, there is often an emphasis on acceleration from standstill, agility, and ball-handling for on-court performance (Mason et al., 2013; van der Slikke et al., 2016b). Testing should therefore, at a minimum, consider these factors. In team environments, this may include standardised tests such as full-court sprints and Illinois agility tests. However, the specific design of tests should consider previous research showing how simple changes to test design can have significant effects on performance and propulsion measures (Haydon et al., 2018b). In this work, players completed 5-metre sprints under two conditions: (i) an acceleration from standstill initiated in their own time (a typical approach in testing) and (ii) and acceleration after completing a catch and pass to a teammate whilst facing away from the intended sprint direction. The second condition required the player to pass and then immediately turn 180 degrees (replicating a frequent skill in WCR) and complete the 5-metre sprint. Results revealed significant differences in both time to complete the sprint, and peak accelerations across the first three strokes as a result of changes to propulsion measures such as contact and release angles. In addition, group analysis of testing conditions masked changes to individual performance and propulsion. These results suggest that emphasis should be placed on ensuring test design is able to represent the desired performance factor under conditions relevant to match performance, with analysis performed at an individual level. Recent work by de Witte et al. (2018) has also developed a wheelchair mobility performance (WMP) measure to quantify performance in WCB. This involves completing 15 activities including sprinting, turning, stop–starts, and skills with standardised rest periods. Analysis of this method confirmed the construct validity and reliability of this approach to quantify overall performance, with applications in assessing wheelchair configurations or monitoring performance progression of an individual. Other simple examples could include completing testing in WCT with a racquet (de Groot et al., 2017), which can immediately enhance the transfer of results and fidelity of testing to on-court performance, but much more focus in research is required here. Until a suitable level of control is reached in representative test designs, it is recommended that a combination of representative and controlled test designs are incorporated into testing protocols to ensure there is an ability to monitor the effects of configuration on both physical and sport-specific skill performance (Haydon et al., 2018b). Specific test designs can be developed, although it is suggested there is input from players and coaches and clear understanding on their purpose (see below case-study).

To support more detailed on-court testing and analysis, IMUs have been utilised, typically placed on each wheel and the frame. By utilising accelerometer and gyroscope measurements, a measure of each wheel's rotation and frame orientation can be obtained. After applying a correction for the camber angle (Pansiot et al., 2011), this then allows for tracking of the frame movement. To ensure accurate tracking, van der Slikke et al. (2011) also recommends applying a wheel skid correction factor. Using this approach for variables such as linear speed, rotation centres, and rotational speed, intra-class correlation coefficients (ICC) were >0.9 when compared with the gold standard of motion capture. However, when higher intensity exercise was performed including collisions and skidding, the system lost accuracy (van der Slikke et al., 2015a). Similar work has been completed by Shepherd et al. (2016), where implementing a Madgwick filter as part of an Attitude Heading Reference System (AHRS) algorithm was used for tracking motion; however, this approach was tested at low velocities outside of normal match conditions. Implementing these systems for on-court monitoring has benefits over radio-frequency based indoor tracking systems in terms of frequency (IMUs at ~500 Hz compared with 8–16 Hz) (Rhodes et al., 2014) and image-based processing (time-consuming analysis conducted post-event (Barris & Button, 2008)). Implementation of these approaches allows for tracking analysis and monitoring of overall performance factors (linear speed, rotational velocities, etc.) (van der Slikke et al., 2016b), within tests (i.e. within the weave section during an Illinois agility test) (Haydon et al., 2019). There is also the potential for improved monitoring of propulsion (number and frequency of strokes (van der Slikke et al., 2016a), contact and release times (Haydon et al., 2019)), particularly if practitioners are able to synchronise IMU data with video. These capabilities improve the ability to assess the impact of wheelchair parameters on performance in contexts representative of competition. As sensors and processing algorithms continue to develop, opportunities to monitor on-court performance will improve allowing for more representative analysis of performance.

7.4 AN APPLIED APPROACH IN ELITE WHEELCHAIR RUGBY

Following the review provided so far in this chapter, we now demonstrate how this information has been considered and implemented in the pursuit of optimising performance in an elite WCR squad by the authors (see Haydon (2018) for more details), as well as presenting some broader research and practical implications from a range of works. We provide some insights into the process completed and provide a range of practical considerations and recommendations for sports scientists, coaches and other specialists (Haydon et al., 2019).

7.4.1 UNDERSTANDING PERFORMANCE NEEDS

A range of considerations in testing approach, monitoring, and analysis are required to effectively assess wheelchair performance, particularly when altering configurations. This includes understanding the performance factors, current context, and coach and individual player's goals. For example, we purposefully began by exploring current wheelchair configurations and player perceptions of parameter effects on performance goals (to see this approach in detail, see Haydon et al., 2016), and increased our understanding of the various propulsion approaches that are implemented across a squad at an individual level (Haydon et al., 2018a). It is strongly recommended that any seating intervention attempts to utilise the anecdotal knowledge and fully harness the experiential knowledge from coaches and players (Paulson & Goosey-Tolfrey, 2017), which can guide testing approaches, emphasise important performance factors for each player, and guide the development of practical suggestions based on an player's level of activity limitation (Haydon et al., 2019). Establishing simple perceptual scales can be effective, and should also be considered to gauge athlete and coach insights during the testing phase (see below).

7.4.2 Design Parameters

Due to the large number of potential configuration parameters, all of which have wide range of potential settings, testing all potential configurations of a wheelchair is difficult (and undesirable!). For elite players, it would be preferable to identify individual parameters that could potentially be optimised for an individual; this parameter could then be tested in isolation as highlighted in previous research (Mason et al., 2009, 2012a, 2012b). In some cases, this can be achieved in the player's wheelchair – e.g. for increasing seat height, the thickness of the seat cushion can be increased (van der Slikke et al., 2018). In other cases (wheel diameter) adjusting a single parameter would require an alternate wheelchair with all parameter settings adjusted accordingly. This can then be a difficult process requiring substantial time and cost commitments and no guarantee of improved performance.

To address these issues, practitioners could consider the use of a sport-specific adjustable wheelchair. This allows for an increase in the number of parameters that can be adjusted across the wheelchair, including seat position parameters. Although considering all configuration parameters is impractical, this method provides opportunities to investigate a number of parameters throughout a single testing protocol. However, with this comes additional difficulties – how do you select then assess the wide range of potential parameters and settings in a practical manner?

A potential solution for this involves utilising orthogonal design testing approaches (Burton et al., 2010; Mori & Tsai, 2011; Usma-Alvarez et al., 2014). Such approaches systematically vary design parameters (e.g. seat height, seat depth, seat angle) at various levels (e.g. current, 10% increase, 10% decrease) to reduce the number of trials required when determining the effect of specific parameter level. A reduction in the number of trials has benefits for minimising any effects of player fatigue whilst shortening time requirements – both crucial considerations when working in elite environments. For wheelchair sports, there is the additional benefit of limiting the number of 'transitions' in and out of the wheelchair to allow for configuration adjustments; an important practical consideration, particularly for low-point WCR players (see International Wheelchair Rugby Federation (2015) for more detail on player capabilities). Using this approach, testing of four parameters (seat height, seat depth, seat angle, tyre pressure) at three levels (current, increase, decrease) results in a reduction of the number of trials from 81 for full-factorial testing (i.e. testing each possible combination) to nine following an orthogonal design

(Haydon et al., 2019). A limitation of using this approach is that it assumes that each parameter can be adjusted in isolation and have limited to no confounding effects on performance goals (Mori & Tsai, 2011). Therefore, in some cases, this approach may not be appropriate.

In this case, the seat position parameters were selected to focus on the change in position relative to the wheel and the subsequent effect on performance. Then, to reduce the complexity of wheelchair adjustment, the design was limited to a constant wheel size and camber angle. Tyre pressure was also investigated due to the lack of published research on tyre pressures in wheelchair court sports. From discussions with coaches and players, it was expected chosen parameters would have the largest influences on performance factors including acceleration, agility, and ball-handling. Practitioners should consider existing evidence and experience to consider the priority areas for adjustment, and make suitable compromises to provide appropriate control within their design.

7.4.3 TEST SELECTION

For testing, it is important to get a balance between fidelity and experimental control (Haydon et al., 2018b). For example, our testing protocol included a linear sprint (5 m test), an agility test (Illinois agility test), and a skill test. The skill test was developed in conjunction with the coach to focus on the ability to pass, receive, and change direction with and without the ball. The skill test was specifically designed to provide insights into how the athlete could control the chair while rotating and ball-handling, and supplement existing test protocols. To complete this, players followed a figure eight type path whilst completing passes against a wall (meaning there is no reliance on any external influence). The combination of these tests provides insight into the key performance factors of WCR with very short time commitment (sprint test – ~3 s, agility test – ~30 s, skill test – ~15 s). This provided benefits over options such as the WMP test (de Witte et al., 2018; van der Slikke et al., 2018) or Beck Battery (Yilla & Sherril, 1998); however, practitioners should decide testing conditions based on performance factors, ability to translate results to on-court, and time requirements.

7.4.4 MONITORING

Aligning with test selection, ensuring a clear and practical method of monitoring performance is crucial to assessing impact of wheelchair configuration. Monitoring methods depend on how performance is being assessed: is it purely by time, where comparison of testing times in various configurations is sufficient? Although this is likely the clearest identifier of improved performance, in most cases it is suggested that additional information is important and beneficial. Measures such as peak accelerations in a sprint (Haydon et al., 2018a) or rotational velocities in a turn (van der Slikke et al., 2016b) can provide more detail on how or why the performance time occurred. To achieve these, IMU set-ups as discussed in Section 7.3.2 are recommended – provided that IMUs and corresponding expertise in analysis are available.

Furthermore, and particularly for linear sprints, this data can be synchronised with video to allow for detailed assessment of propulsion. Once synchronised, the acceleration data can be used to investigate contact and release angles – providing more detail on the effect a configuration has on performance (see Haydon et al. (2019) for a full method/example). This approach can also provide detail on *where* the peak accelerations are occurring within a stroke – providing a clearer understanding of the player's strongest regions and how chair design can further enhance this. This information again adds to the understanding of why measures such as testing times may be changing, providing opportunities for the practitioner to maximise on or address these issues.

7.4.5 Analysis

Once testing has been completed, analysis of the results is then required. Depending on the sport and requirements, the focus of the analysis is likely to vary. In wheelchair court sports, the focus is typically on acceleration from standstill, agility, and ball-handling. For wheelchair racing, the analysis would primarily focus on top end speed, propulsion efficiency, and potentially identifying any asymmetries during the propulsion stroke. In the case of wheelchair court sports, and in our case WCR, more detailed performance goals have been developed in conjunction with the improvements in monitoring methods. These include the peak accelerations and rotational velocities mentioned above, as well as propulsion features (contact and release position on the wheel) (Haydon et al., 2019).

Due to the limited number of trials and need for individual analysis, it is often difficult to perform statistical analysis on results. Alternate approaches to determine the impact of specific parameter levels is therefore required. For the WMP, an overall time to complete all the tests is used to provide an indication of overall performance – a reduced time shows improved performance (de Witte et al., 2018; van der Slikke et al., 2018). Other approaches, including ranking and normalising times of specific tests, can also provide an overall view of the configuration on performance. In our case, we still used simple statistical methods with Bonferroni corrections to identify any changes in performance; achieving significance was difficult due to the small sample sizes involved in intra-athlete comparisons but provided some initial insights on the potential magnitude of changes in performance factors.

Visually assessing the impact of parameter settings should complement and reduce the reliance on any statistical analysis, as well as provide benefits in communicating these results with coaches and players. A simple method to view the impact of various testing conditions on a range of performance factors is through the use of radar plots. Radar plots have been used to present comparisons between players of various classification groups in WCB (see exemplar data in Figure 7.3a) (van der Slikke et al., 2016b), and have potential for use in assessing wheelchair configurations or tracking player progression. Similar approaches were implemented to aid the statistical approach for WCR, with radar plots aiding analysis and presentation of performance times and mobility measures (Figure 7.3b) (Haydon et al., 2019). The visual presentations promoted discussions with coaches and the athlete surrounding the most important performance factors for the player,

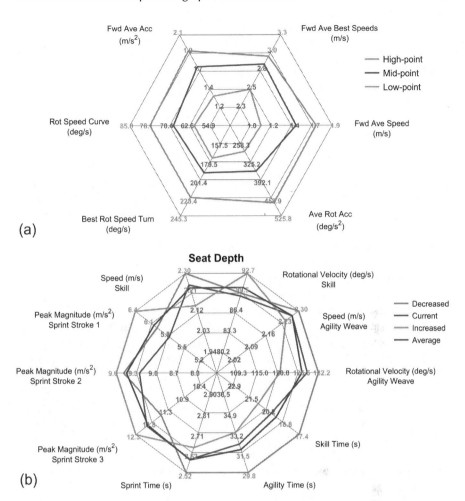

FIGURE 7.3 Example radar plots for: (a) comparing classification groups across wheel-chair mobility performance variables using exemplar data (adapted from van der Slikke et al. (2016b)); and (b) comparing effects of changing seat depth settings on performance times and mobility measures.

and the parameters that were having the largest effect on these. This approach resulted in similar or improved performance for all players involved in testing. Some preferred the new configuration immediately, whilst for those who didn't (they preferred their current configuration), the reasoning given was due to comfort (this requires further understanding of player perceptions). These results strongly suggest this method was effective, particularly as it improved performance for elite international players across a broad range of the WCR classification scale (for more detail on this approach and results see Haydon et al., 2019). Here, we reiterate the need for more case-study approaches and for literature to embrace these approaches due their ability to produce important and interesting findings, particularly in para-sport.

7.5   PRACTICAL IMPLICATIONS

The above sections provide an overview of the current approaches in analysing wheelchair configurations, particularly for WCR. In order to apply this to wheelchair sports, it is suggested that practitioners consider:

- Representative test designs. Depending on the sport and intended performance factors, test designs should be selected to ensure that results transfer to on-court performance.
- Individual case-study approaches. Due to the various impairments and severities involved in para-sports, individual case-studies can provide greater insights and performance improvement. It is suggested that case-study approaches should be embraced further through the peer-review process.
- Coach and player perceptions. In many cases, coaches and players will have extensive experience in the sport. Capitalising on this knowledge for performance factors, representative testing, and interpretation of results will substantially add to the design of interventions.
- The most important design factors. Wheelchair set-up has numerous configuration parameters where testing all is often impractical; it is suggested that testing should focus on the crucial design factors for largest impact on performance. Required compromises and the most important factors can be determined from previous research, pilot testing, and discussions with coaches and players.
- Assessing multiple parameters. Whilst isolated parameter testing provides detailed insight, testing multiple parameters enables practitioners to provide more feedback regarding the configuration from similar amounts of testing. This not only reduces the time demands from testing, but also allows improved wheelchair configurations to be achieved in a shorter period of time.
- Effective methods for data presentation. If utilising approaches where limited statistical analysis is possible, there needs to be clear ways to view and assess data. This can be through visual means (such as the radar plots above), or alternate numerical methods; practitioners should always consider the audiences needs and experiences.

7.6   CONCLUSION

The optimisation of sports wheelchair design is a highly individualised problem, dependent on the specific sport, associated performance goals, and player constraints (e.g. impairment type/ severity). It is clear that the application of physical ergonomics research including appropriate considerations for wheelchair design are necessary to optimise sports performance and allow for athletic adaptation, depending on the context. Due to the improvements of equipment and technology, it is now possible to achieve customised seating interfaces that allow for improved power transfer and control in sports such as wheelchair racing and WCT, and improved monitoring methods through the use of IMUs and associated algorithms as well as radio-frequency based tracking systems. These advancements increase the potential for detailed assessments of configurations to occur, provided appropriate care is taken with test design and analysis. Emphasis should be placed on ensuring that simple test designs are able to reflect the desired and individualised goals for optimisation (e.g. acceleration or agility time), as well as exploring more detailed measurements (e.g. peak accelerations, propulsion kinematics) in order to assess why any changes in performance occur. Crucially, optimisation approaches in larger squads should be viewed as a series of individual interventions – each individual will have unique physical constraints, varying on-court demands and propulsion approaches, which all result in differences in an optimised wheelchair design. To maximise potential translation of interventions to improve on-court performance, researchers should strive to balance experimental control with more practical approaches that allow for better replication of the performance context of interest.

REFERENCES

Altmann, V., Groen, B., van Limbeek, J., Vanlandewijck, Y., & Keijsers, N. (2013). Reliability of the revised wheelchair rugby trunk impairment classification system. *Spinal Cord*, 51, 913–918.

Altmann, V. C., Hart, A. L., Vanlandewijck, Y. C., van Limbeek, J., & van Hooff, M. L. (2015). The impact of trunk impairment on performance of wheelchair activities with a focus on wheelchair court sports: a systematic review. *Sports Medicine – Open*, 2:1. doi:10.1186/s40798-015-0013-0.

Barris, S., & Button, C. (2008). A review of vision-based motion analysis in sport. Sports Medicine, 38:12, 1025–1043. doi:10.2165/00007256-200838120-00006.

Burton, M., Subic, A., Mazur, M., & Leary, M. (2010). Systematic design customization of sport wheelchairs using the taguchi method. *Procedia Engineering*, 2:2, 2659–2665. doi:10.1016/j.proeng.2010.04.048.

de Groot, S., Bos, F., Koopman, J., Hoekstra, A. E., & Vegter, R. J. K. (2017). Effect of holding a racket on propulsion technique of wheelchair tennis players. *Scandinavian Journal of Medicine and Science in Sports*, 27:9, 918–924.

de Witte, A. M. H., Hoozemans, M. J. M., Berger, M. A. M., van der Slikke, R. M. A., van der Woude, L. H. V., & Veeger, D. (2018). Development, construct validity and test-retest reliability of a field-based wheelchair mobility performance test for wheelchair basketball. *Journal of Sports Sciences*, 36:1, 23–32. doi:10.1080/02640414.2016.1276613.

Faupin, A., Campillo, P., Weissland, T., Gorce, P., & Thevenon, A. (2004). The effects of rear-wheel camber on the mechanical parameters produced during the wheelchair sprinting of handibasketball athletes. *Journal of Rehabilitation Research and Development*, 41: 3B, 421–428.

Goosey-Tolfrey, V. L., & Campbell, I. G. (1998). Symmetry of the elbow kinematics during racing wheelchair propulsion. *Ergonomics*, 41:12, 1810–1820. doi:10.1080/001401398185983.

Goosey-Tolfrey, V. L., Vegter, R. J. K., Mason, B. S., Paulson, T. A. W., Lenton, J. P., van der Scheer, J. W., & van der Woude, L. H. V. (2018). Sprint performance and propulsion asymmetries on an ergometer in trained high- and low-point wheelchair rugby players. *Scandinavian Journal of Medicine and Science in Sports*. doi:10.1111/sms.13056.

Greenwood, D., Davids, K., & Renshaw, I. (2012). How elite coaches' experiential knowledge might enhance empirical research on sport performance. *International Journal of Sports Science and Coaching*, 7:2, 411–422. doi:10.1260/1747-9541.7.2.411.

Haydon, D. S. (2018). *Optimisation of the Rugby Wheelchair for Performance* (Doctor of Philosophy). Adelaide: University of Adelaide.

Haydon, D. S., Pinder, R. A., Grimshaw, P. N., & Robertson, W. S. P. (2016). Elite wheelchair rugby: a quantitative analysis of chair configuration in Australia. *Sports Engineering*, 19:3, 177–184. doi:10.1007/s12283-016-0203-0.

Haydon, D. S., Pinder, R. A., Grimshaw, P. N., & Robertson, W. S. P. (2018a). Overground propulsion kinematics and acceleration in elite wheelchair rugby. *International Journal of Sports Physiology and Performance*, 13:2, 156–162.

Haydon, D. S., Pinder, R. A., Grimshaw, P. N., & Robertson, W. S. P. (2018b). Test design and individual analysis in wheelchair rugby. *Journal of Science and Medicine in Sport*, 21:12, 1262–1267. doi:10.1016/j.jsams.2018.04.001.

Haydon, D. S., Pinder, R. A., Grimshaw, P. N., & Robertson, W. S. P. (2019). Wheelchair rugby chair configurations: An individual, robust design approach. *Sports Biomechanics*, 1–16. doi:10.1080/14763141.2019.1649451.

International Tennis Federation. (2019). *Provisional Wheelchair Tennis Classification Rules*. London: ITF.

International Wheelchair Basketball Federation. (2014). *Official Player Classification Manual*. Canada: IWBF.

International Wheelchair Rugby Federation. (2015). *International Wheelchair Rugby Federation Classification Manual*. 3rd edition. Retrieved from http://www.iwrf.com/resources/iwrf_docs/IWRF_Classification_Manual_3rd_Edition_rev-2015_%28English%29.pdf.

Laferrier, J., Rice, I., Pearlman, J., Sporner, M. L., Cooper, R., Liu, T., & Cooper, R. A. (2012). Technology to improve sports performance in wheelchair sports. *Sports Technology*, 5:1–2, 4–19. doi:10.1080/19346182.2012.663531.

Lewis, A. R. (2019). *Performance Benefits of Customised Seating Interfaces for Elite Wheelchair Racing Athletes* (Doctor of Philosophy). Adelaide: University of Adelaide.

Mason, B., Lenton, J., Leicht, C., & Goosey-Tolfrey, V. (2014). A physiological and biomechanical comparison of over-ground, treadmill and ergometer wheelchair propulsion. *Journal of Sports Sciences*, 32:1, 78–91. doi:10.1080/02640414.2013.807350.

Mason, B., Porcellato, L., van der Woude, L. H., & Goosey-Tolfrey, V. L. (2010). A qualitative examination of wheelchair configuration for optimal mobility performance in wheelchair sports: a pilot study. *Journal of Rehabilitation Medicine*, 42:2, 141–149. doi:10.2340/16501977-0490.

Mason, B., Van der Slikke, R. M., Hutchison, M. J., Berger, M. A., & Goosey-Tolfrey, V. (2018). The effect of small-sided game formats on physical and technical performance in wheelchair basketball. *International Journal of Sports Physiology and Performance*, 13:7, 891–896. doi:10.1123/ijspp.2017-0500.

Mason, B., van der Woude, L., Tolfrey, K., & Goosey-Tolfrey, V. (2012a). The effects of rear-wheel camber on maximal effort mobility performance in wheelchair athletes. *International Journal of Sports Medicine*, 33:3, 199–204. doi:10.1055/s-0031-1295443.

Mason, B., van der Woude, L. H., & Goosey-Tolfrey, V. L. (2013). The ergonomics of wheelchair configuration for optimal performance in the wheelchair court sports. Sports Medicine, 43:1, 23–38. doi:10.1007/s40279-012-0005-x.

Mason, B., Van Der Woude, L. H., Tolfrey, K., Lenton, J. P., & Goosey-Tolfrey, V. L. (2012b). Effects of wheel and hand-rim size on submaximal propulsion in wheelchair athletes. *Medicine and Science in Sports and Exercise*, 44:1, 126–134. doi:10.1249/MSS.0b013e31822a2df0.

Mason, B. S., van der Woude, L. H., & Goosey-Tolfrey, V. L. (2009). Influence of glove type on mobility performance for wheelchair rugby players. *American Journal of Physical Medicine and Rehabilitation*, 88:7, 559–570. doi:10.1097/PHM.0b013e3181aa41c5.

Molik, B., Lubelska, E., Kosmol, A., Bogdan, M., Yilla, A. B., & Hyla, E. (2008). An Examination of the international wheelchair rugby federation classification system utilizing parameters of offensive game efficiency. *Adapted Physical Activity Quarterly*, 25, 335–351. doi:10.1123/apaq.25.4.335.

Mori, T., & Tsai, S.-C. (2011). *Taguchi Methods: Benefits, Impacts, Mathematics, Statistics, and Applications*. New York: ASME.

Mutsuzaki, H., Tachibana, K., Shimizu, Y., Hotta, K., Fukaya, T., Karasawa, M., Ikeda E., & Wadano, Y. (2014). Factors associated with deep tissue injury in male wheelchair basketball players of a Japanese national team. *Asia-Pacific Journal of Sports Medicine, Arthroscopy, Rehabilitation and Technology*, 1:2, 72–76. doi:10.1016/j.asmart.2014.01.002.

Pansiot, J., Zhang, Z., Lo, B., & Yang, G. Z. (2011). WISDOM: Wheelchair inertial sensors for displacement and orientation monitoring. *Measurement Science and Technology*, 22:10, 105801. doi:10.1088/0957-0233/22/10/105801.

Paulson, T. A. W., & Goosey-Tolfrey, V. (2017). Current perspectives on profiling and enhancing wheelchair court sport performance. *International Journal of Sports Physiology and Performance*, 12:3, 275–286. doi:10.1123/ijspp.2016-0231.

Pinder, R. A., Davids, K., Renshaw, I., & Araújo, D. (2011). Representative learning design and functionality of research and practice in sport. *Journal of Sport and Exercise Psychology*, 33:1, 146–155. doi:10.1123/jsep.33.1.146.

Pinder, R. A., Headrick, J., & Oudejans, R. R. D. (2015). Issues and challenges in developing representative tasks in sport. In: J. Baker & D. Farrow (eds.), *The Routledge Handbook of Sports Expertise* (pp. 269–281). London: Routledge.

Rhodes, J., Mason, B., Perrat, B., Smith, M., & Goosey-Tolfrey, V. (2014). The validity and reliability of a novel indoor player tracking system for use within wheelchair court sports. *Journal of Sports Sciences*, 32:17, 1639–1647. doi:10.1080/02640414.2014.910608.

Shepherd, J., Wada, T., Rowlands, D., & James, D. (2016). A novel AHRS inertial sensor-based algorithm for wheelchair propulsion performance analysis. *Algorithms*, 9:3, 55. doi:10.3390/a9030055.

Stephens, C. L., & Engsberg, J. R. (2010). Comparison of overground and treadmill propulsion patterns of manual wheelchair users with tetraplegia. *Disability Rehabilitation: Assistive Technology*, 5:6, 420–427. doi:10.3109/17483101003793420

Usma-Alvarez, C. C., Fuss, F. K., & Subic, A. (2014). User-centered design customization of rugby wheelchairs based on the Taguchi method. *Journal of Mechanical Design*, 136:4, 041001. doi:10.1115/1.4026029.

van der Slikke, R., Berger, M., Bregman, D., & Veeger, D. (2016a). Push characteristics in wheelchair court sport sprinting. *Procedia Engineering*, 147, 730–734. doi:10.1016/j.proeng.2016.06.265.

van der Slikke, R. M., Berger, M. A., Bregman, D. J., Lagerberg, A. H., & Veeger, H. E. (2015a). Opportunities for measuring wheelchair kinematics in match settings; reliability of a three inertial sensor configuration. *Journal of Biomechanics*, 48:12, 3398–3405. doi:10.1016/j.jbiomech.2015.06.001.

van der Slikke, R. M. A., Berger, M. A. M., Bregman, D. J. J., & Veeger, H. E. J. (2015b). Wheel skid correction is a prerequisite to reliably measure wheelchair sports kinematics based on inertial sensors. *Procedia Engineering*, 112, 207–212. doi:10.1016/j.proeng.2015.07.201.

van der Slikke, R. M. A., Berger, M. A. M., Bregman, D. J. J., & Veeger, H. E. J. (2016b). From big data to rich data: The key features of athlete wheelchair mobility performance. *Journal of Biomechanics*, 49:14, 3340–3346. doi:10.1016/j.jbiomech.2016.08.022.

van der Slikke, R. M. A., de Witte, A. M. H., Berger, M. A. M., Bregman, D. J. J., & Veeger, D. J. H. E. J. (2018). Wheelchair mobility performance enhancement by changing wheelchair properties; what is the effect of grip, seat height and mass? *International Journal of Sports Physiology and Performance*, 13:8, 1050–1058. doi:10.1123/ijspp.2017-0641.

van der Woude, L. H., Bouw, A., van Wegen, J., van As, H., Veeger, D., & de Groot, S. (2009). Seat height: Effects on submaximal hand rim wheelchair performance during spinal cord injury rehabilitation. *Journal of Rehabilitation Medicine*, 41:3, 143–149. doi:10.2340/16501977-0296.

Vanlandewijck, Y. C., & Thompson, W. R. (2011). *Handbook of Sports Medicine and Science: the Paralympic Athlete*. John Wiley & Sons.

Vanlandewijck, Y. C., Verellen, J., & Tweedy, S. (2011). Towards evidence-based classification in wheelchair sports: impact of seating position on wheelchair acceleration. *Journal of Sports Sciences*, 29:10, 1089–1096. doi:10.1080/02640414.2011.576694.

Yilla, A. B., & Sherril, C. (1998). Validating the beck battery of quad rugby skill tests. *Adapted Physical Activity Quarterly*, 15, 155–167.

8 Optimising Athlete Performance in Race Vehicle Systems Exposed to Mechanical Shock and Vibration

Neil J. Mansfield

CONTENTS

8.1 INTRODUCTION

There is an adage that the first bicycle race occurred shortly after the second bicycle was manufactured. Whilst this statement may or may not be strictly true, there is a strong desire for humans to compete, or to watch others competing. The first official cycling races are thought to have occurred in the 1870s (Ritchie, 1999). When competition involves the racing of vehicles, the winner is the competitor able to maximise the speed/route most optimally. There is an advantage to being able to accelerate and decelerate quickly, generating forces in the direction of travel, and there is an advantage to being able to turn corners quicker, generating forces not in the direction of travel. During linear motion, interaction with the environment will generate vibration (Ahlin & Granlund, 2002). In race vehicles the athlete's skill is key to maintaining control and situation awareness during these dynamic stresses.

This chapter will consider how the dynamic environment affects the human body in sport where athletes are in control of a race vehicle. It will consider how good human factors and good design can optimise the performance and well-being of athletes.

8.2   THE DYNAMIC ENVIRONMENT

People are exposed to shock and vibration through all surfaces in contact with a moving vehicle. For driven vehicles this will typically occur through the feet, seat and hands. The source of the vibration naturally depends on the task being undertaken – skiers, surfers, and skaters will primarily be loaded through the feet, those riding horses will also be exposed through the pelvis, and those riding bicycles will also experience input at the hands. Luge and skeleton sliders lie prone or supine on a sliding surface and therefore inputs to the body include the chest and back.

At any one moment, the body's response to vibration depends on the magnitude, waveform, direction, and point of application (Mansfield, 2004). How the vibration affects the body also depends on the physical and situational condition of the exposed person. Understanding of the biomechanical response of the body enables systems to be designed to minimise the effects of the vibration on the body.

Biomechanical responses of the standing person to dynamic loading is different to that for a sitting person, and the response is affected by posture. For example, the transmission of shock and vibration to the head has been shown to have a resonance at about 5–7 Hz in the vertical direction and in the fore-and-aft direction (Figure 8.1; Paddan & Griffin, 1993). This resonance corresponds to pitching of the head. However there is very little relative movement of the head when the body is exposed to lateral movement. The presence of a backrest increases the transmission to the head, but this also has the effect of distributing the dynamic load across the spine, thereby reducing the dynamic compression in the lumbar region (Guo et al., 2016).

Whilst helmets are desirable or mandated for many sports where head injury is a risk, the additional mass on the head changes the dynamics of the head–neck system. Increasing helmet mass increases the static loading on the neck and increases the

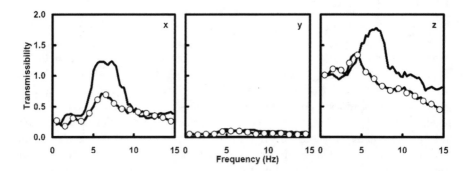

FIGURE 8.1   Median transfer functions between vertical vibration at the seat and translational vibration at the head for subjects seated with a backrest (----) and without a backrest (--0--) in the fore-and-aft (x), lateral (y), and vertical (z) direction. Data from Paddan and Griffin (1993); from Mansfield (2004).

rotational moment of inertia (Mathys & Ferguson, 2012). This results in a change in biomechanical response (Derouin & Fischer, 2019) meaning the resonance of the head occurs at a lower frequency, and therefore the angular motion increases for the same acceleration input. Lightweight helmets that meet impact requirements are superior than heavier alternatives as they reduce the loading on the neck. Even light-weight sports helmets can affect the head–neck dynamcs; for example, vibration tests of participants lying prone on a skeleton sled, with head cantilevered showed increased motion when wearing a helmet (Figure 8.2).

Low-frequency motion can induce motion sickness for susceptible people. Motion sickness occurs when there is a sensory mismatch between motion cues. Primary cues are vision, vestibular (balance), tactile, and knowledge of control (Mansfield, 2004, Figure 8.3). If these four channels are in agreement with what is expected, then motion sickness can occur. For example, sickness is common for rally car co-drivers (navigators) who have a requirement to read maps and course notes, rather than to attend to the details of the road profile such as rises and falls or corners (Perrin et al., 2013). This attention away from the road has also been associated with the higher prevalence of neck pain in co-drivers rather than drivers (Mansfield & Marshall, 2001). The fact that sickness was less common in competition in comparison to reconnaissance illustrates the strong psychological and motivational component to motion sickness, as observed by Darwin (1796) who described how feelings of sea sickness disappeared when passengers on a boat were in fear for their lives. Sea sickness affects sailors of small boats, affecting performance and causing distraction. Turner and Griffin (1995) reported daily incidence of feelings of sickness, which peaked at over 40% for sailors competing in a round-the-world yacht race, but also showed that incidence reduced as each stage of the race progressed, demonstrating habituation.

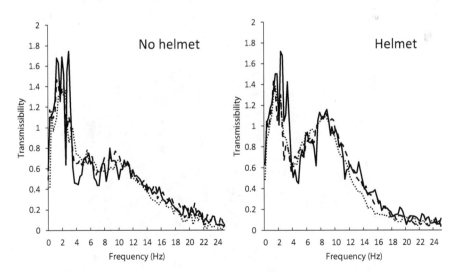

FIGURE 8.2 Transmission of vertical vibration from skeleton sled to head for prone slider with and without helmet. Data from three repeats measured in the laboratory.

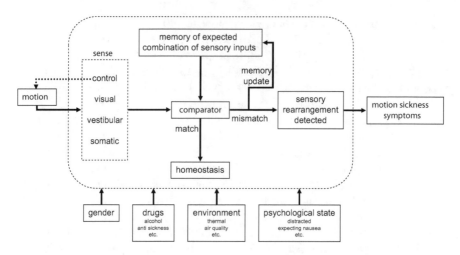

FIGURE 8.3 Model of motion sickness including sensory rearrangement and habituation (Mansfield, 2004).

8.3 HUMAN SHOCK AND VIBRATION MEASUREMENT AND ASSESSMENT

The measurement and assessment of human shock and vibration has become easier with very low-cost data loggers and accelerometers, but there remain fundamental difficulties with interpretation and analysis of the data (Figure 8.4). In many cases low cost systems can be used to make accurate measurements of vibration exposure but they require expert understanding of digital signal processing in order to make valid conclusions from the data (Mansfield et al., 2016). For measurements of vibration at the seat or saddle, multi-axial accelerometers are mounted in standardised seat pads on which the driver sits and holds in place with body weight. For large saddles as found on horses or motorcycles, the standard circular accelerometer mount can be adapted to a saddle-specific design (Figure 8.5).

FIGURE 8.4 Typical equipment for measuring shock and vibration on a seat or saddle. This example shows equipment for measuring acceleration on a horse saddle.

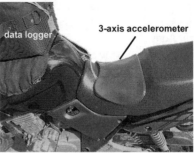

FIGURE 8.5   Accelerometer mounting in a Ferrari F355 Challenge Series race car and on a motorcycle. Accelerometers are embedded within flexible mounts that are compressed on to the seat by body weight.

Standard methods of assessing human response to shock and vibration involve measuring vibration at the input point (floor, seat, hands, for example) and applying digital filters to the data. For steady-state vibration frequency weightings can be applied. These weightings emphasise some frequencies of motion more than others, corresponding to models of the human response. The weightings are defined in International Standards ISO 8041, ISO 2631-1, and ISO 5349. Different weightings are used depending on point of application, the posture of the person, and the direction of the vibration. For example, if there is vertical motion at the seat, a weighting of W_k is used that has a peak at 5–10 Hz; vibration at the hands uses the W_h weighting.

Weighted acceleration signals can be used to give an indication of the acute vibration exposure at that moment in the environment, or can be combined with exposure times to produce a measurement of dose, and to allow for comparison of exposures of different durations. Vibration dose is usually measured in terms of A(8) where N is the number of different exposures being considered, a_{wn} is the frequency-weighted root mean square (r.m.s.) acceleration for exposure n and t_n is the duration of exposure n (ISO 2631-1 incorporating 2010 amendment, International Organization for Standardization, 1997):

$$A(8) = \sqrt{\frac{1}{8} \sum_{n=1}^{n=N} a_{wn}^2 t_n}$$

This method is not suitable for assessment of signals where they are dominated by shocks. For shock signals, the method defined in ISO2631-5 (International Organization for Standardization, 2018) can be used. This uses a model with a sixth power exponent, meaning that shocks are emphasised. It also only considers the highest peaks in the signal by summing those peaks to a dose value D, where A_i is the ith highest peak in the measured signal:

$$D = 1.07 \left[\sum_i A_i^6 \right]^{\frac{1}{6}}$$

It is important to consider the applicability of standardised methods to sports applications. ISO2631-1 has been written with the general public as an assumed user group. Whilst the standard works well for public transport settings, criteria within it may not be applicable where the primary purpose of a task is to maximise speed and therefore comfort is not an important consideration. Furthermore, the original experimentation on which the standards are based were the general public and not focused on athletes. Nevertheless, in most cases it is desirable to reduce the measured values of shock and vibration exposure in order to improve the physical environment in which the competitor races.

8.4 OPTIMISING THE DYNAMIC ENVIRONMENT THROUGH SYSTEM DESIGN

Optimisation of the dynamic environment can be achieved through optimisation of sources of dynamic loading, optimisation of transmission paths, and optimisation of the biomechanical response of the affected athlete. Figure 8.6 illustrates some key sources of dynamic loading that can induce vibration in the athlete–vehicle system. It shows those elements that can influence the transmission of shock and vibration through the system and some considerations for biomechanical influences.

8.4.1 OPTIMISATION OF SOURCES OF DYNAMIC LOADING

Aerodynamic effects can be used to improve performance through slip-streaming/drafting. Many sports such as cycling, motorsport, and sliding events rely on the best possible aerodynamics to give the competitive advantage. The athlete–vehicle system can be optimised for the ideal sporting conditions but when out of those design parameters aerodynamic loading causes buffeting and can make the system difficult to control or for speed to be lost. The high-profile 2019 crash of cyclist Chris Froome,

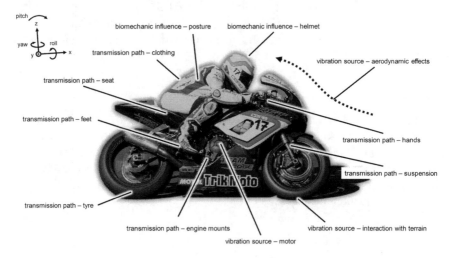

FIGURE 8.6 Sources of shock and vibration in vehicle-athlete systems, with transmission paths and factors that influence the human response to the dynamic environment. Photo: Chris Knowles.

the Tour de France race winner, was attributed to an unexpected gust of wind forcing him to lose control on a training ride. His cycle was optimised for headwind (i.e. relative movement against air in the direction of travel), but not for lateral air movement. Designs that aerodynamically optimise under perfect conditions should also be optimised to minimise adverse effects in poor weather, or when affected by aerodynamic loading from others on the course. Athletes should be trained how to recognise and respond to aerodynamic buffeting. This can be achieved in the wind tunnel or in training settings.

A second source of loading is interaction with the terrain. Purpose-built sports facilities may aim to minimise the dynamic loading and to be as smooth as possible, allowing for maximised speed. For many dynamically challenging sports, the terrain and how to handle it in varying conditions is a key part of the athletic experience (Kirkwood et al., 2019). Minimising the loading from the terrain can be achieved through racing line selection, speed optimisation, and driving/handling skills. It has been shown that skill level is an important factor for vibration exposure in alpine skiing (Tarabini et al., 2015). This could mean simply avoiding obstacles and choosing the smoothest path, but this may not be the fastest option and therefore unlikely to be chosen by competitors. For marine sports the sea state will affect the speed at which the craft can travel. Direction of travel relative to wave direction and boat speed can both be used to mitigate the loading due to the waves, but still the peak accelerations that boat crew are exposed to can be excessive as the boat hits subsequent wave troughs after freefall, or crests if planing (Rantaharju et al., 2015; Figure 8.7).

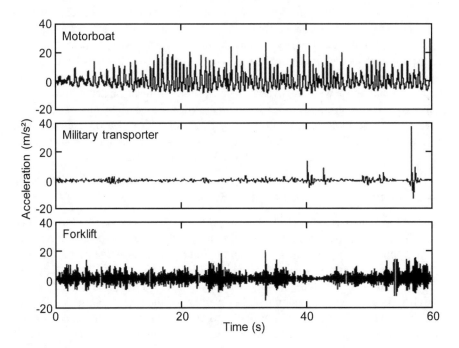

FIGURE 8.7 Example data measured in a planing motorboat, military transporter and industrial machine. (Data from Rantaharju et al..)

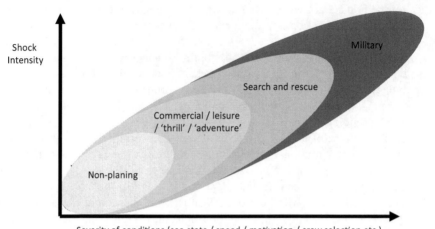

Severity of conditions (sea state / speed / motivation / crew selection etc.)

FIGURE 8.8 Illustration of how shock intensity increases for high-speed marine craft. Shocks increase with sea state and speed, which are influenced by the crew and motivation.

The intensity of shocks tends to also be a function of the user group and task, such that military crews are exposed to the greatest shocks, and non-planing craft the smallest (Figure 8.8).

The final source of dynamic loading is the motor and other internal mechanical devices. Internal combustion engines in competition vehicles do not normally need to comply with regulation that is required for road-going vehicles, and vehicle refinement is not a commercial priority for manufacturers. At its most extreme, niche spectator sports such as tractor pulling or drag racing can involve machines with multiple engines, although the race time and hence exposure time can be very short. Part of the attraction to those who are involved in such sports is the noise and vibration. The challenge of maintaining the attractive characterisics of an activity can be achieved. For example, the motorcycle brand Harley-Davidson retained a distinctive sound and tactile feel whilst reducing noise levels through using research on perception at contact points (e.g. Guest et al., 2002; Morioka & Griffin, 2008).

8.4.2 OPTIMISATION OF TRANSMISSION PATHS

Many of the transmission paths from the source of dynamic loading to the human can be tuned to improve the athlete's environment, but improving the athlete's physical environment does not necessarily result in improvements in race performance.

Engine mounts ensure that the motor remains securely in place and aligned to the drivertrain whilst applying rapidly changing loads. High frequency vibration can be well-isolated by engine mounts. A key element of the design optimisation is to tune the dynamic properties at low RPM where resonances occur in the engine-mount system (Rivin, 1985). The combination of low engine speed and the mount resonances can result in intermittent high-level vibrations as the machine accelerates through each gear.

Transmission of vibration from the environment through the race vehicle will include the contact surface and possibly suspension elements. Sliding sports such as skeleton prohibit suspension elements and restrict the design to a relatively small number of parts (International Bobsleigh and Skeleton Federation, 2018). For ground-based sports, most use tyres that have a primary purpose of providing grip and traction rather than improving rider/driver comfort. Where grip can substantially change between events or weather conditions, tyre pressures can be a key consideration, changing the dynamics of the system. For example, in cyclocross, tyres are often run at pressures as low as 20 PSI (1.3 bar) in order to maximise grip in slippery conditions (Forrester et al., 2018).

Suspension systems fitted to some types of sports vehicle are there to improve handling and ultimately speed. For example, downhill mountain bike suspension systems are designed around a best compromise of performance, durability, and commercial factors (Ulrich & Eppinger, 2015). Suspension systems aim to absorb energy and/or to change the characteristics of shocks from a very short high-acceleration exposure to a longer duration exposure but with a lower peak acceleration. Similar principals apply in seating systems that can be used as shock and vibration isolators. These can include in-built suspension systems designed to isolate the driver from shock and vibration (e.g. power boats).

The final stage of the transmission of dynamic loading to the athlete is the clothing. Compliant materials can be fitted to isolate the body from high frequency vibration, but is ineffective at isolating low-frequency motion. Padded cycling gloves and handlebar tape have been shown to be effective at reducing the shocks experienced in the hands for road cyclists (Drouet et al., 2018). Anti-vibration gloves as used in industry do not attenuate vibration below about 20 Hz, and many gloves are ineffective up to 100 Hz (Dong et al., 2009). Whether they are suitable in a sports context depends on the vibration to be isolated; to determine this, vibration measurement may be required and matching of the glove materials.

8.4.3   Optimisation of Biomechanic Responses

The posture of the athlete is a key element to optimisation of the biomechanic response. Jockeys can be observed using their legs as a suspension system and this principal is used across all 'rider' sports. The use of knee flexion is exploited in marine settings, as well in forestry, where 'jockey seats' are used to isolate from small magnitude motion, but the occupant can semi-stand with bent knees if they know that a shock is imminent (Mansfield et al., 2002).

Mountain bikers are trained to use their posture to isolate their torso mass and to allow the limbs to move with motion (Figure 8.9). This involves pushing the body weight down and rearwards. To facilitate this posture, many will fit a 'dropper' seat post, that lowers the saddle temporarily to give freer body movement. The saddle returns to the standard riding position through a control on the handlebars.

8.4.4   Optimisation of Performance

Optimising sporting performance involves a best compromise of multiple factors. How teams and athletes choose to make this compromise can be an attractive element

FIGURE 8.9 Use of body position to maintain upper body and head stability for elite mountain biker riding down a flight of steps on a cross-country race cycle with front suspension. In A and C, the rear and front tyres compress respectively due to shock loading. The fork cycles from extension (A–B) through to compression (C–D) mitigating some of the shock loading from the front wheel. Knees and arms pump to isolate the head, that maintains stability through the section of the training course.

to those competing and following the sport. Technical improvements can be constrained or short-lived by governing bodies. Regulations aim to maintain sporting competition whilst ensuring an appropriate level of safety and fairness. For example, after a trend for extreme lightweighting of bicycles through drilling and filing down components, structural integrity and safety were put at risk. In response to this, the regulatory body, UCI, introduced a minimum weight limit of 6.8 kg for road bikes in 2000. This weight limit can now be achieved through improvements in materials and design without compromise on safety such that professional teams need to add additional weight to some bikes to make them legal.

Performance optimisation in controlled condtions might not be replicated in competition. For example the most aerodynamic design of equipment in the wind tunnel might be too unstable to be controllable in real world conditions. Therefore human performance must be considered during the design process. Where power is derived from muscle a balance must be struck between comfort, power output, and aerodynamic efficiency (Faulkner & Jobling, 2019). An athlete unable to maintain the fastest posture for the duration of a race will compromise their performance as they fatigue. Furthermore, rapid recovery post-training or competition allows the athlete to return to the next bout of training more rapidly, such as was possible through optimisation of saddle design for elite track cyclists (Burt, 2014).

Optimisation of performance as a whole must be considered in a holistic manner as any one 'marginal gain' in one area could induce a 'marginal loss' in another area.

8.5 SUMMARY

Athletes in race vehicle systems are exposed to shock and vibration that puts dynamic demands on the body. The response of the body depends on the frequency, direction, and waveform of the motion. It is also affected by the point of application and the posture of the body. Whilst ISO standards can be helpful in evaluating shock and vibration in dynamic sports, the standards are general-purpose and not specifically designed for such applications. If standards are used for shock and vibration measurement and evaluation, careful interpretation is needed in order to ensure that misleading conclusions are avoided.

There are opportunities for optimisation of the athlete–vehicle system in order to maximise sporting performance, comfort, and health. These include control of mechanical loading at source through design of the environment in which the vehicle is travelling, although this is often deliberately designed to challenge the athlete. The vehicle itself can be optimised although the degrees of freedom in the optimisation is usually constrained by rules and regulations from the organising body, and the demands of maximising the competitive performance. Contact surfaces, suspension systems and athlete contact points can be optimised to isolate them from some adverse loading. The biomechanics of the body can be used to minimise the effects of shock and vibration through coaching to optimise posture.

There is often a balance to be struck to maximise the race performance and the human factors simultaneously. Well-written regulations can help in enforcing a base level of safety and well-being for the athlete exposed to physically demanding sports environments.

REFERENCES

Ahlin, K., & Granlund, N. J. (2002). Relating road roughness and vehicle speeds to human whole body vibration and exposure limits. *International Journal of Pavement Engineering*, 3:4, 207–216.

Burt, P. (2014). *Bike Fit: optimise Your Bike Position for High Performance and Injury Avoidance*. London: Bloomsbury Publishing.

Darwin, E. (1796). *Zoonomia; or, The Laws of Organic Life*.

Derouin, A. J., & Fischer, S. L. (2019). Validation of a three-dimensional visual target acquisition system for evaluating the performance effects of head supported mass. *Applied Ergonomics*, 76, 48–56.

Dong, R. G., McDowell, T. W., Welcome, D. E., Warren, C., Wu, J. Z., & Rakheja, S. (2009). Analysis of anti-vibration gloves mechanism and evaluation methods. *Journal of Sound and Vibration*, 321:1–2), 435–453.

Drouet, J. M., Covill, D., & Duarte, W. (2018). On the exposure of hands to vibration in road cycling: an assessment of the effect of gloves and handlebar tape. In: Multidisciplinary Digital Publishing Institute Proceedings (Vol. 2, No. 6, p. 213).

Faulkner, S. H., & Jobling, P. (2019). The impact of hip angle and time trial position on heat production during cycling. Proceedings of International Conference on Environmental Ergonomics, Amsterdam, p. 20.

Forrester, A. I. J., Adams, M., & Lamb, R. (2018). *The Cyclocross Bible*. Southampton, UK: How to Ride a Bike Limited.

Guest, S., Catmur, C., Lloyd, D., & Spence, C. (2002). Audiotactile interactions in roughness perception. *Experimental Brain Research*, 146:2, 161–171.

Guo, L. X., Dong, R. C., & Zhang, M. (2016). Effect of lumbar support on seating comfort predicted by a whole human body-seat model. *International Journal of Industrial Ergonomics*, 53, 319–327.

International Bobsleigh and Skeleton Federation. (2018). *International Skeleton Rules*.

International Organization for Standardization. (1990). *Human Response to Vibration – Measuring Instrumentation*. ISO 8041. Geneva: International Organization for Standardization.

International Organization for Standardization. (2001). *Mechanical Vibration – Measurement and Evaluation of Human Exposure to Hand Transmitted Vibration – Part 1: General Guidelines*. ISO 5349-1. Geneva: International Organization for Standardization.

International Organization for Standardization. 1997. *Mechanical Shock and Vibration – Evaluation of Human Exposure to Whole-Body Vibration – Part 1: General Requirements.* ISO 2631-1. Geneva: International Organization for Standardization.

International Organization for Standardization. (2018). *Mechanical Shock and Vibration – Evaluation of Human Exposure to Whole-Body Vibration – Part 5: Method for Evaluation of Vibration Containing Multiple Shocks, International Standard.* ISO 2631-5.

Kirkwood, L. A., Taylor, M. D., Ingram, L. A., Malone, E., & Florida-James, G. D. (2019). Elite mountain bike enduro competition: a study of rider hand-arm vibration exposure. *Journal of Science and Cycling,* 8:1, 18.

Mansfield, N. J., & Marshall, J. M. (2001). Symptoms of musculoskeletal disorders in stage rally drivers and co-drivers. *British Journal of Sports Medicine,* 35:5, 314–320.

Mansfield, N. J. (2004). *Human Response to Vibration.* Boca Raton, FL: CRC Press.

Mansfield, N. J., Holmlund, P., Lundstrom, R., Nordfjell, T., & Staal-Wasterlund, D. (2002). Vibration exposure in a forestry machine fitted with a saddle type suspension seat. *International Journal of Vehicle Design,* 30:3, 223–237.

Mansfield, N. J., Huang, Y., & Thirulogasingam, T. (2016). Does ultra low cost = ultra low quality? Dynamic performance of budget data logging accelerometers. Paper presented at UK Conference on Human Response to Vibration held at Institute of Naval Medicine.

Mathys, R., & Ferguson, S. J. (2012). Simulation of the effects of different pilot helmets on neck loading during air combat. *Journal of Biomechanics,* 45:14, 2362–2367.

Morioka, M., & Griffin, M. J. (2008). Absolute thresholds for the perception of fore-and-aft, lateral, and vertical vibration at the hand, the seat, and the foot. *Journal of Sound and Vibration,* 314:1–2, 357–370.

Paddan, G. S., & Griffin, M. J. (1993). The transmission of translational floor vibration to the heads of standing subjects. *Journal of Sound and Vibration,* 160:3, 503–521.

Perrin, P., Lion, A., Bosser, G., Gauchard, G., & Meistelman, C. (2013). Motion sickness in rally car co-drivers. *Aviation, Space, and Environmental Medicine,* 84:5, 473–477.

Rantaharju, T., Mansfield, N. J., Ala-Hiiro, J. M., & Gunston, T. P. (2015). Predicting the health risks related to whole-body shock and vibration: a comparison of alternative assessment methods for high-acceleration events in vehicles. *Ergonomics,* 58:7, 1071–1087.

Ritchie, A. (1999). The origins of bicycle racing in England: Technology, entertainment, sponsorship and advertising in the early history of the sport. *Journal of Sport History,* 26:3, 489–520.

Rivin, E. (1985). Passive engine mounts – Some directions for further development. *SAE Transactions,* 94, 582–591.

Tarabini, M., Saggin, B., & Scaccabarozzi, D. (2015). Whole-body vibration exposure in sport: four relevant cases. *Ergonomics,* 58:7, 1143–1150.

Turner, M., & Griffin, M. J. (1995). Motion sickness incidence during a round-the-world yacht race. *Aviation Space and Environmental Medicine,* 66:9, 849–856.

Ulrich, K. T., & Eppinger, S. D. (2015). *Product Design and Development.* 6th edition. McGraw-Hill Higher Education.

Section III

Cognitive HFE Applications

9 Decision-Making in Sport
Looking at and beyond the Recognition-Primed Decision Model

Anne-Claire Macquet

CONTENTS

9.1 INTRODUCTION

In sports, athletes, coaches, and referees (ACRs) have to make efficient decisions in a short time frame. In dynamic sports, they have to rapidly determine what is happening, where and when (Macquet, 2009). Information may be available rapidly or conversely may not be available rapidly enough, even when such information seems important for decision-making. This situation's uncertainty and time constraints involve a conflict: ACRs have to act promptly even though they may lack the information to rapidly understand a given situation. Given the time constraints, they may have to settle for partial and sufficient situation understanding in order to still have the time to carry out an action (Amalberti, 2001; Klein, 2008). Where this is the case, they can anticipate situation evolution to assist in decision-making.

Nevertheless, they risk implementing a decision that is not adapted to the actual situation. Alternatively, they can wait for the situation to evolve before making a decision, and in doing so, take the risk of implementing a decision too late. In addition to dealing with time and uncertainty, ACRs also have to manage stress and emotions. It has been suggested that stress and emotions can lead to a distortion of situation understanding (Macquet, 2016).

Most competitive situations in sports resemble dynamic situations studied using the Naturalistic Decision-Making (NDM) approach. Dynamic situations present the following key contextual factors: (a) time pressure and ill-structured problems; (b) uncertain and dynamic environments; (c) competitive goals; (d) decision loops (more than one decision); (e) many players and high stakes; and (f) organisational goals and norms (Zsambok, 1997). NDM focuses on experts' decision-making when experts are dealing with complex, uncertain and dynamic environments as opposed to novices' decision-making in laboratory settings (e.g. Klein, 2008).

In addition to the complexity of decisions, sports situations require experts to implement decisions using their motor skills. The level of these skills varies over the competition, depending on fatigue and emotion (Macquet, 2016). Athletes need to make decisions they are able to undertake efficiently, both physically and mentally.

Using NDM, decision-making models have been developed in dynamic and natural environments (e.g. Hoc & Amalberti, 2007; Klein, 1997; Rasmussen, 1983) to provide an alternative to the classical models of decision-making developed in static environments (e.g. Tversky & Kanheman, 1974). NDM models showed that experts did not often compare possible options; rather, they often took the first option that came to mind (e.g. Klein, 2008). Klein, Calderwood, and Clinto-Cirocco (1986) and Klein (1997) developed the Recognition Primed-Decision (RPD) model to explain how experts use their experience to make good decisions in natural settings. The RPD model postulates that experts rely on intuition to make their decisions. They use their experience to quickly draw a picture of the situation and make rapid decisions without having to compare options (Klein, 1997).

This chapter aims to provide an overview of studies using the RPD model in sports and to then revisit the RPD model in order to account for anticipation and planned decisions. More specifically, it aims to (a) explain the RPD model; (b) analyse the studies using the RPD model and their limitations; and (c) present the revisited RPD model.

9.2   THE RPD MODEL

The RPD model is based on three assertions. First, experts use their experience to decide to implement a possible course of action similar to the first one they consider. Second, because they use their experience and pattern-matching ability, time pressure does not affect performance. They recognise the typicality of a situation and generate a typical course of action. For Klein (2009, 2015), pattern-matching involves intuition. Intuition is used when decision-makers rely on a repertoire of patterns built up from experience to make decisions. It differs from automaticity trained under controlled situations (Klein, 2015). Experience enables experts to make decisions in challenging situations with a sufficient level of success. Third, that decision-makers can generate options without comparing different options. Decision-making is an intuitive process that enables experts to know what course of action will work. Decision-makers distinguish the information needed to make a satisfactory decision (Klein, 2009).

The RPD model presents three levels: (a) simple match; (b) diagnose the situation; and (c) evaluate course of action (Figure 9.1). The first refers to rapid situation recognition and the implementation of the typical action corresponding to the typical situation. For example, a fencer extended his/her elbow to hit the opponent's arm; the opponent implemented a circular parry to prevent the sword from reaching his/ her arm and riposted by touching the fencer's thigh. The second is related to the difficulty in recognising the situation rapidly, because information is not yet available or the situation has suddenly changed. Understanding is consequently hampered by inconsistencies and anomalies. The expert needs more time to diagnose the situation in order to recognise it as typical; once he/she recognises it, he/she adapts the typical action to the current situation and implements it. For example, a volleyball player decided to move back to defend a powerful shot at the back of the court; suddenly, the ball was deflected by the block and fell into the middle of the court. The defender moved forwards to defend. The third level refers to an assessment of the workability of a course of action. If the expert simulates that it could work, he/she implements it, if not, he/she changes it (Klein, 1997; Klein et al., 1986, see Figure 9.1). For example, a basketball player aimed to pass the ball to a teammate who was marked. She/he simulated that the opponent might catch the ball, consequently she/he passed the ball to another teammate who was free of any defenders.

The RPD model is based on two processes: situation recognition and mental simulation. Situation recognition enables the expert to assess the situation typicality using salient features that experience has shown useful and determine whether the situation is familiar/unfamiliar or atypical. It is based on the construction of a frame (i.e. mental model) of the actual situation, which is compared with a similar frame associated with a previous situation stored in memory from experience (Klein, 1998, 2003, 2009). Frames can also refer to knowledge about a job or life, for example, for fire-fighters, it might include the way a fire develops; for sports team players, how to coordinate players from the playbook. Such models enable understanding of how things and people work.

Recognition is achieved using four by-products: (a) relevant cues (e.g. the setter is setting the ball to the centre hitter); (b) expectancies (e.g. the blocker is expecting the hitter to attack from the centre of the pitch); (c) plausible goals (e.g., block the ball at the middle of the net in volleyball); and (d) typical action to be implemented (e.g. block the centre player's hit). These by-products are used to assess the situation and make sense of it. Bounded rationality (Simon, 1996) suggests that the decision-maker cannot connect all the data and has to prioritise certain information to reach a conclusion and complete a course of action in time. Recognition involves framing. Framing consists of fitting data into a frame and fitting the frame around data (Klein, Moon, & Hoffman, 2006a, 2006b). The frame is then compared to patterns

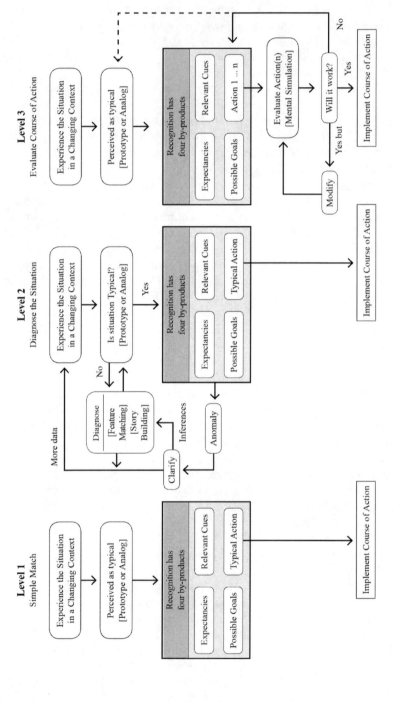

FIGURE 9.1 Recognition-primed decision model (Klein, 1997).

contained in memory to identify whether it matches a typical situation associated with a typical action. Framing relates to the 'simple match' level of the RPD model.

The frame is dynamic: it can be changed when inconsistencies are noticed and new information is received. It is then enriched by such information (i.e. reframing). Reframing concerns the 'diagnose the situation' level of the RPD model. Framing is related to Piaget's (1954) concept of assimilation, and reframing Piaget's (1954) concept of accommodation. Framing and reframing are central to the process of sense-making. Sense-making is the deliberate effort to understand situations (Klein et al., 2007), in order to direct control of a situation or crisis (Weick, 1995).

Mental simulation provides the means to assess the situation in time-pressure settings without having to compare options (Klein, 1998) and to use analytical decision-making and problem-solving strategies (Klein & Crandall, 1995). For example, the coach noticed that an athlete was being distracted by the speaker's voice and background noise. He/she mentally simulated that if he/she told the athlete to ignore the noise, the athlete would have paid even more attention to the noise. He/she did not have to investigate the possible options to prevent the athlete from being influenced by the noise. He/she just told the athlete to focus on his/her technique. Decision-makers use tacit knowledge based on experience to perceive affordances and make sense of a situation. Mental simulation is related to Piaget's (1954) phase of formal operations and ability to plan and anticipate possible situation development from a specific decision, using experts' experience of similar situations.

According to the RPD model, decision-making appears to be more intuitive than deliberative (Klein, 1997). Situation recognition or pattern-matching is intuitive, relating to Kahneman's (2011) system 1 (i.e. intuitive decision-making). System 1 refers to thinking fast and functioning in an autonomous way; it is effortless. For example, a boxer saw his/her opponent extended his/her elbow to hit the boxer face and the boxer rapidly moved his/her head to avoid the hit. Diagnosis and mental simulation are deliberate and analytical; they relate to Kahneman's system 2 (i.e. deliberative decision-making). System 2 relates to thinking more slowly, using more analytical and logical thinking; it requires effort. System 2 is slower than system 1 and is used when the individual faces a new problem he/she cannot solve. For example, a boxer thought about the possible ways to drive his/her opponent onto the ropes in order to restrict space and options. Recognition strategies are adaptive (Klein, 1997). They could be considered to be a source of power (Klein, 1997, 2003).

The RPD model provides a theoretical and practical model for understanding decision-making in sports. It accounts for unhurried decisions such as those taken in coaching (e.g. Abraham & Collins, 2015), as well as rapid decisions such as those made by athletes in fast sports (e.g. Macquet, 2009).

9.3 AN OVERVIEW OF STUDIES USING THE RPD MODEL IN SPORTS AND DISCUSSION

The RPD model has been used to study decision-making in sports for some ten years (Kermarrec, 2015). Electronic databases (PubMed, ScienceDirect, PsycINFO, Google scholar) were searched using combinations of the keywords *recognition-primed decision model, sports*. Studies were considered for inclusion in this overview if: (a) they were peer-reviewed; (b) they presented data on the by-products of the recognition process and/or levels of the RPD model; (c) participants were athletes, referees, or coaches; and (e) they were experts or there was a comparison between experts and novices. This overview presents nine studies (see Table 9.1). It consists of four parts: (a) study aims; (b) method; and (c) results and limitations.

9.3.1 STUDY AIMS

The studies exhibited three aims. The first was to identify how decisions were made with regard to the RPD model (all studies, see Table 9.1). The second was concerned with eliciting the information taken into account to make decisions (Abraham & Collins, 2015; Bossard, De Keukelaere, Cormier, Pasco, & Kermarrec, 2010; Kermarrec & Bossard, 2014; Macquet, 2009; Macquet & Lacouchie, 2015, 2017; Macquet & Skalej, 2015; Milazzo, Farrow, Ruffault, & Fournier, 2015). The third related to identification of the decisions made to manage time in a training and academic centre over the season (Macquet & Skalej, 2015; see Table 9.1).

9.3.2 METHOD

Studies involved expert athletes, coaches, and referees (see Table 9.1). They used different methods appropriate to a first-person approach in natural settings. Semi-directive interviews were conducted using video sequences (Abraham & Collins, 2015; Bossard et al., 2010; Kermarrec & Bossard, 2014; Macquet, 2009; Macquet & Skalej, 2015; Mulligan, McCracken, & Hodge, 2012), planning documents (Macquet & Skalej, 2015), and the critical decision-method (Macquet & Lacouchie, 2015), or conducted after the event the participant had just completed (Milazzo et al., 2015). Such interviews provide in-depth information that accounts for meaningful information in natural settings (Smith et al., 2017). They allow for a richer representation of the theoretical construct (Peräkylä & Ruusuviori, 2011; Sparkes, 2009). However, they are time-consuming, which can present a problem when access to elite athletes is very limited in time. To avoid direct intervention in the task being undertaken, Neville and Salmon (2017) analysed sequences of communication of the umpires. In-situ verbalised data thus provided a way to capture real time decision-making.

9.3.3 RESULTS AND LIMITATIONS

The studies provided three kinds of results: (a) a split between the three variations of the RPD model; (b) by-products used to recognise the situation as typical; and (c) decisions driven by adaptation in the course of action or anticipation.

9.3.3.1 The Split between the Three Levels of the RPD Model

Analysis of the decision-making moments showed that the experts' decisions followed a 60–81%/13–28%/3–24% split between the three levels of the RPD model (see Table 9.1). This split suggests that the majority of the experts' decisions comprised simple match decisions. Experts sized up the pattern of situations, recognised

TABLE 9.1
Studies Using the RPD Model in Sports

References	Study aims	Method	Main results
Macquet (2009)	Test the RPD model with regard to volleyball and elicit relevant information taken into account by athletes during the decision-making process	Seven professional male volleyball players Self-confrontation interviews from a competition match Constant comparative method (Corbin & Strauss, 1990)	The athletes presented an 81%/13%/6% split between the three levels of the RPD model. Pattern-matching was based on four by-products: expectancies, relevant cues, plausible goals, and typical action.
Bossard, De Keukelaere, Cormier, Pasco, and Kermarrec (2010)	Explain the underlying process of decision-making in dynamic situations with ice-hockey counter-attack players	Six professional male ice-hockey players Self-confrontation interviews from a training match Inductive and deductive analysis	The athletes presented an 80%/17%/3% split between the three levels of the RPD model. Pattern-matching based on five by-products: expectations, knowledge, relevant cues, plausible goals, and typical action.
Mulligan, McCracken, and Hodge (2012)	Explain the use of situation familiarity in expert decision-making	23 expert and non-expert ice-hockey players Self-confrontation interviews from a competition match Inductive and deductive analysis	Experts presented an 85%/15% split between familiar (simple match) and unfamiliar situations (diagnose). Recognition related to rule-based situations and event-based situations. Experts perceived situations as familiar twice often as non-experts.
Kermarrec and Bossard (2014)	Describe decision-making by defensive soccer players and how decisions were made using the RPD model	Four professional football players Self-confrontation interviews from a competition match Inductive and deductive analysis	Athletes presented a 60%/16%/24% split between the three levels of the RPD model. Pattern-matching is based on five by-products: expectancies, knowledge, relevant cues, plausible goals, and typical action
Macquet and Lacouchie (2015)	Explain how expert judokas made their decisions under stress	Two top elite female athletes Critical Decision-Making method conducted after an international competition	Decisions were based on a recognition process with regard to four by-products and experience. Under stress, the judokas favoured the use of their favourite technique in two ways: one-off decisions made by adapting the decision to the situation and two-stage decisions made by manipulating the situation to adapt it to the decision to be implemented. Two-stage decisions were more effective than one-off decisions Decisions were driven by two temporalities: immediate and anticipated.

(Continued)

TABLE 9.1 (CONTINUED)
Studies Using the RPD Model in Sports

References	Study aims	Method	Main results
Macquet and Skalej (2015)	Explain how elite athletes manage their time to perform in sports competitions and academic exams. More specifically, describe how they made sense of situations and what made them adapt time management decisions	12 elite athletes Semi-directive interviews conducted to elicit comments on athletes' planning and concomitant time management activity over a day/week/year Constant comparative method (Corbin & Strauss, 1990)	Athletes presented a 72%/28% split between level 1 and 2 of the RPD model. They recognised the situation by comparing the time-frames required for a specific activity and time allocated by the training centre. When athletes noticed an anomaly (i.e. conflict in time-frames), they enriched the frame using a diagnose process. Decisions were driven by two temporalities: immediate and anticipated.
Milazzo et al. (2015)	Determine skill-based differences in decision-making Examine the contribution of situational probability information to the decision-making process	14 elite and 14 novice karate fighters Combination of video-based analysis, eye-movement recording and verbal reports of thoughts when making decisions against a standardised expert opponent	Expert fighters made decisions faster than novices and their decision were more precise than those of novices Experts were shown to have greater perceptual and cognitive skills in comparison to novices. Experts generated more verbal reports than novices. They most frequently reported relevant cues and typical actions.
Abraham and Collins (2015)	Explain decision-making by coaches in time-limited situations Explain whether coaches used formalistic rules rather than substantive heuristics to make decisions	12 coaches in horizontal jumps Semi-structured interviews using video sequences	Decisions were proved intuitive using a pattern-recognition process (level 1 of the RPD model). Decisions were made after diagnosing the problem based on experience, and applying heuristic problem-solving procedures and intuitive mental models (level 2 of the RPD model).
Neville and Salmon (2017)	Identify whether umpires' decision-making uses the RPD model	Eight male field umpires Identify sequences of communication related to decision moments in the game Coding of each sequence as one of the three-level decisions of the RPD model	The umpires presented a 78%/18%/4% split between the three levels of the RPD model. The low proportion of mental simulation may be explained by the lesser ability of umpires to modify the existing decision.

their typicality rapidly, and implemented a course of action. This is consistent with previous research on RPD (e.g. Klein, 1997).

Results showed significant differences between ACRs across sports and players' roles within the game. They indicated that the split was less reliant on level 1 for defensive players in other sports, suggesting than defence players were more engaged in retroactive strategies in comparison to counter-attacking and attacking players (Kermarrec & Bossard, 2014). Referees were shown to exhibit a 78%/18%/4% split between the three levels of the RPD model, suggesting that experts used similar strategies in fast sports, whatever their role (Neville & Salmon, 2017).

As the RPD model predicts, when the situations were not at first perceived as typical, the experts diagnosed the situation. They had to wait for useful information relating to situation development (i.e. what, where, and when the action occurred) Bossard et al., 2010; Kermarrec & Bossard, 2014; Macquet, 2009; Mulligan et al., 2012), and assessed the situation when they noticed an anomaly (Macquet, 2009; Macquet & Skalej, 2015). Results showed that the split was less reliant on diagnosing the situation for athletes managing their time (Macquet & Skalej, 2015), suggesting that when time was available, athletes could identify an anomaly and reframe the situation using additional data or connecting data differently. Reframing requires time; ACRs lack sufficient time to reframe in fast sports.

As predicted by the RPD model, the decision-maker considers one course of action and mentally simulates the likely consequences of the course of action to determine whether or not it will work. Mental simulation takes the form: 'If I do this, then x will happen'. Kermarrec and Bossard (2014) showed that soccer defenders commented more frequently on mental simulation than volleyball and ice-hockey players (see Table 9.1). They showed that when defenders were distant from the ball, they had more time for situation assessment and mental simulation of possible situation development and risks.

9.3.3.2   The By-Products Used to Recognise the Situation as Typical

In sports studies, recognition was based on specific information relating to four by-products: expectancies, relevant cues, plausible goals, and typical actions (see Table 9.2). Results showed that expectancies directed decision-makers' attention to possible situation development and relevant cues in relation to experience and knowledge, consistent with the RPD model and schema theory (Neisser, 1976). Schemata built from learning and experience direct perception to identify elements selectively as either to be noticed or to be ignored. For example, coaches used intuitive rules relating to their experience and knowledge of coaching to assess athletes' strengths and weaknesses (Abraham & Collins, 2015). Although expectations and knowledge refer to expectancies within the RPD model, Bossard et al. (2010) and Kermarrec and Bossard (2014) distinguished knowledge from expectations using a fifth by-product. This is shown in Table 9.1.

Relevant cues were context-specific (see Table 9.2) and were mostly concerned with visual perception (see Table 9.2). Judokas also reported on auditory perception: they noticed their opponent's breathing pattern, which enabled them to assess her energy levels. In sports, perception relates to all the senses. A limited number of research studies have provided evidence of experts' use of all their senses (Macquet, 2016).

TABLE 9.2
Information Commented by Decision-Making in Relation to the By-Products of the RPD Model

References	Expectancies	Relevant cues	Plausible goal	Typical action
Macquet (2009)	Expectations about the players' actions Abilities and tendencies of their opponents, teammates and themselves 10% of total by-products	Teammates' and opponents' positions and actions Own position Trajectory of the ball 52% of total by-products	Number of decisions made 20% of total by-products	Rules in the form 'if … then' Preceding event 18% of total by-products
Bossard, De Keukelaere, Cormier, Pasco, and Kermarrec (2010)	Expectations about the situation evolution Abilities and tendencies of their opponents, teammates and themselves 20% of total by-products	Teammates' and opponents' positions and actions Own position Trajectory of the puck 39% of total by-products	10% of total by-products	Ten typical actions related to opponents, teammates, the player and shooting 31% of total by-products
Mulligan, McCracken, and Hodge (2012)	NI	NI	NI	NI
Kermarrec and Bossard (2014)	Expectations about opponents' and teammates' actions Abilities and tendencies of their opponents, teammates and themselves 13% of total by-products	Teammates' and opponents' positions and actions Trajectory of the ball 44% of total by-products	Few goals 3% of total by-products	Defence actions 40% of total by-products

(Continued)

TABLE 9.2 (CONTINUED)
Information Commented by Decision-Making in Relation to the By-Products of the RPD Model

References	Expectancies	Relevant cues	Plausible goal	Typical action
Macquet and Lacouchie (2015)	Opponents' abilities and tendencies Level of expertise in a specific situation and experience with this opponent	Opponent's energy level and involvement Participant's energy level and involvement Score	Number of decisions to be implemented in the course of action	Favourite technique
Macquet and Skalej (2015)	Own sports and academic projects Ability to cope with fatigue, stress, boredom and injury	Time-frame allocation	NI	Time management strategy
Milazzo et al. (2015)	Experts: Mean = 3.2 (SD = 2.0) Novices: Mean = 1 (SD = 1.5) Expectancies non detailed	Experts: Mean = 6.4 (SD = 2.6) Novices: Mean = 3.1 (SD = 2.8) Relevant cues non detailed	Experts: Mean = 0.7 (SD = 0.05) Novices: Mean = 0.2 (SD = 0.4) Plausible goal non detailed	Experts: Mean = 5.1 (SD = 2.2) Novices: Mean = 1.2 (SD = 1.5) Typical action non-detailed
Abraham and Collins (2015)	Rationale according to coaches' experience and knowledge	Perceived athlete profile	NI	Perspectives for coaching
Neville and Salmon (2017)	NI	NI	NI	NI

Note: NI= Non Identified

Exploration of which information is prioritised in the situation assessment could be a worthwhile avenue for research, and could provide insights into situation awareness requirements in relation to the level of expertise of decision-makers and contexts, and the senses used to perceive information (Macquet, 2016).

These studies present some limitations. First, they did not show how participants take their emotions into account when they make their decisions. Little is known about the effects of emotions in decision-making in sports including when the high stakes are high. Macquet (2016) suggested that emotions can lead to a distortion of situation understanding. Exploring the role and effects of emotions on decision-making would be of interest to elicit a bigger picture of the decision-making in situations involving high stakes.

Second, they did not compare the findings with those of other participants under the same conditions. Exploring decision-making processes with a broader sample of participants could open up possibilities for future research in sports and other contexts.

Third, they did not explain collective decision-making. Decision-making was considered from an individual perspective. Studying team decision-making would be a challenging perspective for future research.

9.3.3.3   Decisions Driven by Adaptation in the Course of Action or Anticipation

Results showed that most decisions were made in the course of action (see Table 9.1). Decision-makers waited for relevant cues before making decisions. Results also showed that when relevant cues were unavailable rapidly, experts anticipated them and planned a decision in order to meet the time pressure. Before implementing the decision, they checked whether the current situation had developed as expected from the anticipated situation. If it matched, they undertook the planned decision, if not, they changed the decision. For example, volleyball players prepared to block the hit at the centre of the net moved to the side of the court when the setter set the ball there (Macquet, 2009). The RPD model does not account for planned decisions and adaptation of planned decisions in the course of action.

McLennan and Omodei (1996) showed that while monitoring events, umpires engaged in mental simulation of what was likely to occur on the field based on their knowledge of the way the game was frequently played. As an event occurred, they checked whether the event matched what they had anticipated. Such anticipation enabled them to make very rapid decisions. Macquet and Lacouchie (2015) showed that under stress, judokas planned a decision (i.e. favourite technique) and manipulated the situation so that they could implement their decision. They adjusted the situation conditions to make them resemble those required to undertake their planned decision (i.e. first decision) and then implemented the planned action (i.e. second decision; Macquet & Lacouchie, 2015, 2017). The RPD model does not explain planned decisions and manipulation of the situation to enable a decision to be imposed.

These results all suggest that decisions were driven by two temporalities: immediate and anticipated. The RPD model accounts for immediate decisions made in the course of action. However, it gives insufficient consideration to the continual involvement in sense-making from anticipated information when information is not available (McLennan & Omodei, 1996) and planned action (Macquet & Pellegrin, 2017).

To explain how experts make sense of situations, Klein et al. (2007) developed the data-frame theory of sense-making to describe the process of 'fitting data into a frame and fitting the frame around the data' (p. 120). To account for the way experts re-plan during execution, Klein (2007) conceived the flexecution model. The flexecution process consists of preparing the plan, questioning the goals in order to detect anomalies, and reframing the goals to adapt to changes. Flexecution requires time that experts do not have in fast sports. These considerations led to the need to revisit the RPD model in order to take into account the mental process that precedes the potential situation development prior to the activation of pattern-matching on encountering the actual situation development (Macquet & Pellegrin, 2017).

Macquet and Pellegrin (2017) suggested that typical situations and associated typical actions are anticipated before the situation develops, and that the decision-maker compares the pattern of the anticipated situation with that of the current situation. If they match, decision-makers carries out the planned action; if not, they change it. Decision-makers can also mentally simulate whether the action will work. Macquet and Pellegrin (2017) revisited the RPD model to explain rapid decisions and planned actions.

9.4   THE REVISITED RPD MODEL AND DISCUSSION

9.4.1   Two More Assertions

Macquet and Pellegrin (2017) extended the RPD model with two further assertions. The fourth assertion is that time pressure does not impact performance because experts anticipate possible situation evolution and plan a decision before the situation develops. Experts compare the pattern of the anticipated situation with that of the current situation. If they match, the experts carry out the planned action; if not, they change it. Mental simulation enables experts to make rapid decisions and implement them very swiftly. For example, a badminton player anticipated that his/her opponent was going to be late hitting the shuttle coming close to the net, so the opponent would have to lift it, enabling him/her to plan an attack. He/she noticed that the opponent lifted the shuttle, as anticipated. Then he/she implemented his/her planned decision and hit the shuttle to half-court with a steep and powerful trajectory.

The fifth assertion is that time pressure does not affect pattern-matching because experts are guided towards relevant cues. Expectancies direct perception, while perception itself is active (Gibson, 1969). Experts use schemata to assess the situation (Neisser, 1976). Schemata direct perception and specify the information to be noticed and that to be ignored in order to build an overview of the situation. In sports, results showed that perception was guided by knowledge about opponents (see Table 9.2). To perceive relevant information, experts used filtering strategies in the way Gibson (1969) highlighted in explaining direct perception. For example, the referee aimed to keep the game and players safe. He/she was attuned to possible risky player behaviours, such as a tackle in football, and ignored peaceful behaviours, such as a run in a free zone.

9.4.2   The Revisited RPD Model

Macquet and Pellegrin (2017) revisited the RPD model to account for planned decisions in sports (e.g. Macquet & Fleurance, 2007), medicine (Pellegrin, Gaudin, Bonnardel, & Chaudet, 2010), and ship handling (Chauvin & Lardjane, 2008). When time pressure and uncertainty are high, decision-makers anticipate what might happen and plan, rather than waiting for the situation to evolve. Anticipation enables adaptation. Anticipation is based on current information about events and possible situation development in relation to experience and schemata. Macquet and Pellegrin (2017) enriched the RPD model by adding a function referred to as forcing the situation to change. Forcing of the situation was initiated when decision-makers noticed a difference between the current situation and the situation required to implement the planned action. The required situation is based on experience and schemata. Forcing of the situation aims to make the features of the actual situation resemble those of a situation appropriate for the planned action. When the features match, the decision-maker implements his/her planned action (see Figure 9.2).

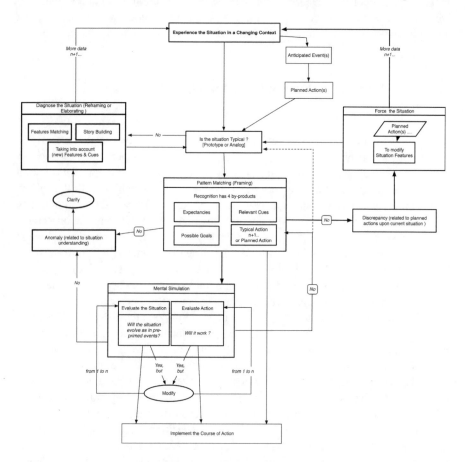

FIGURE 9.2   The revisited RPD model (Macquet & Pellegrin, 2017).

If not, he/she changes the action. The decision-maker can also mentally simulate whether the course of action will work before deciding to implement it. Forcing of the situation is a two-step decision. It enables the context to be adapted to the decision. Decisions that are adapted to the context and made in the course of action are one-step decisions.

9.4.3 Empirical Support for the Revisited RPD Model and Discussion

The studies have demonstrated empirical support for the RPD model. McLennan and Omodei (1996) highlighted that referees engaged in mental simulation to anticipate how the play was likely to unfold and what action to implement in 90% of instances they recalled. Referees made very rapid decisions on the basis of simulation of what was likely to occur. Macquet and Kragba (2015) showed that basketball players systematically anticipated a pattern of coordination, which was shared with teammates in order to achieve positive outcomes. They then compared whether the situation had evolved as expected from anticipated events. If it did not match, they reconsidered the anticipated frame and enriched it with information that had become available in the situation and adapted their course of action accordingly. Belling, Suss, and Ward (2015) showed that baseball players anticipated relevant situational cues in order to recognise the type of pitch and locate the pitch in advance of crossing the plate. Such results suggest that relevant cues and actions are over-learnt, enabling specific responses to be implemented (Baber, Chen, & Howes, 2015). Anticipation appears to be more a matter of automatic filtering than conscious framing.

Macquet and Lacouchie (2015, 2017) and Macquet and Fleurance (2007) showed that elite athletes reported manipulating the situation in order to use their favourite techniques. A favourite technique could be used in two ways: a one-off decision involving adapting the decision to the situation, and a two-stage decision involving manipulating the situation to create the conditions that resemble the typical situation associated with the favourite technique (i.e. first decision), and then implementing it (i.e. second decision). In judo, favourite techniques have been seen to win contests (Macquet & Lacouchie, 2015, 2017). In badminton, a favourite technique was reported in 12% of situations by players (Macquet & Fleurance, 2007). The revisited RPD model connects immediacy, temporality and anticipation. It enables both fast and slow decisions to be accounted for.

Future research in sports and other domains may provide more empirical support for this model. More specifically, it may provide support for the way ACRs manipulate situation conditions in order to adapt them so that they may implement their planned action. It could be a worthwhile avenue to elicit the kind of decisions made (i.e. anticipated versus planned decisions) in relation to the match score and time. It could also be fruitful to study the effects of emotions in decision-making to get a bigger picture of decision-making. It might be expected that emotions influence the use of relevant cues and expectancies. Positive and negative emotions could have different impacts. Moreover, it could be a worthwhile avenue to study the effects of physiological and mental fatigue in decision-making. If someone is tired, for example, his/her recognition process may be altered.

9.5   PRACTICAL APPLICATION IN SPORT

As a practical application, Macquet (2016) suggested developing programmes to help athletes guide perception and favour rapid pattern-matching. She suggested the use of video sequences of competitions from the athlete's perspective, paused at specific points, and encouragement of athletes to notice relevant cues and assess the situation in order to make a decision. In the same way, Kermarrec (2015) proposed small-sided controlled games in soccer in which game space, time, the number of players, and equipment are controlled to provide players with specific opportunities to act. Coaches proposed that players assess the situation and share their assessment with them and other athletes when deciding what decision to implement. Another per-spective for coaching consists of presenting athletes with a short checklist of what should be noticed. Such programmes may enable athletes to: (a) notice relevant cues more precisely; (b) develop expectancies by building knowledge about opponents and games, and anticipating what the opponent could do depending on his/her exper-tise and tendencies; (c) associate a typical action with a typical situation recognised from relevant cues and expectancies. Coaches may also encourage athletes to men-tally simulate the consequences of a possible course of action and possible situation development, and decide whether to implement the action or change it.

When relevant cues were not available rapidly, decision-makers anticipated them or waited for them to appear. Situation assessment is progressive. Anticipation of events could be a worthwhile avenue for application in sports, when time pressure is high. Training in recognition processes as a perceptual-cognitive skill could be used to anticipate key information (i.e. relevant cues). For example, Fadde (2015) devel-oped occlusion sports training programmes based on video simulation to improve early perception and situation anticipation in baseball players. Anticipation enables the complexity of situations to be reduced.

Development of programmes to train athletes in two-step decision-making could be another worthwhile avenue to explore. In boxing, for example, the coach could encourage the boxer to drive his/her opponent to the corner of the ring in order to restrict the opponent's space and possible options, and then implement his/her favou-rite technique.

9.6   CONCLUSION

The RPD model has been used to explain decision-making in sports for the past ten years. Decision-making refers to a rapid recognition process using experience in similar situations. Pattern-matching enables adaptation to the context in accordance with the time pressure and uncertainty of events. It is based on active perception guided by expectancies and goals. The data support the view that pattern-matching, and more specifically 'simple match', plays a key role in decision-making in sports, leading ACRS to adapt efficiently. Results also showed that experts use mental simu-lation to anticipate possible situation development where relevant cues are not avail-able early, consider a course of action and assess its workability.

The RPD model accounts for decisions made in the course of action to adapt them to the situation's evolution. However, it does not explain planned actions. Macquet and Pellegrin (2017) revisited the RPD model to address the issues of situation assessment, mental simulation, and planned actions. The revisited RPD model connects immediacy and anticipation temporalities. It suggests the role of filtering relevant information when decision-makers are required to anticipate action to adapt to high time pressure. It accounts for two kinds of decisions: one-off decisions and two-stage decisions. One-off decisions aim to adapt decisions to the situation, whereas two-stage decisions aim to adapt the situation to the planned decision.

This model supplements other macro-cognitive models and theories such as the Data-Frame theory of sense-making (Klein et al., 2007) and the flexecution model (Klein, 2007) which are concerned with adaptive skills in natural and complex environments. The continued study of decision-making in sports, including identifying the effects of emotions, required situation awareness and coordination among decision-makers will improve our understanding of cognitive processes when experts make individual and collective decision-making in sports.

9.7   ACKNOWLEDGMENTS

The author thanks Liliane Pellegrin for her constructive comments on this chapter.

REFERENCES

Abraham, A., & Collins, D. (2015). Professional judgement and decision-making in sport coaching: To jump or not to jump. *Proceedings of the International Conference on Naturalistic Decision-Making*, McLean, VA.

Amalberti, R. (2001). La maîtrise des situations dynamiques [control of dynamic situations]. *Psychologie Française*, 46, 107–118.

Baber, C., Chen, X., & Howes, A. (2015). (Very) rapid decision-making: Framing or filtering? *Proceedings of the International Conference on Naturalistic Decision-Making*, McLean, VA.

Belling, P., Suss, J., & Ward, P. (2015). The effects of time constraint on anticipation, decision-making, and option generation in complex and dynamic environments. *Cognition, Technology and Work*, 17, 355–366.

Bossard, C., De Keukelaere, C., Cormier, J., Pasco, D., & Kermarrec, G. (2010). L'activité décisionnelle en phase de contre-attaque en hockey sur glace (Decision-making activity during turnover phases in ice hockey). *Activités*, 7:1, 42–61.

Chauvin, C., & Lardjane, S. (2008). Decision making and strategies in an interaction situation: Collision avoidance at sea. *Transportation Research Part F: Traffic Psychology and Behavior*, 11:4, 259–269.

Corbin, J. & Strauss, A. (1990). *Basics of Qualitative Research: Grounded Theory Procedures and Techniques*. Newbury Park, CA: Newbury Park.

Fadde, P. (2015). From lab to cage: Turning the occlusion research method into a sports training program. *Proceedings of the International Conference on Naturalistic Decision-Making*, McLean, VA.

Gibson, J. J. (1969). *Principles of Perceptual Learning and Development*. New York: Appleton-Century-Crofts.

Hoc, J.-M., & Amalberti, R. (2007). Cognitive control dynamics for reaching a satisfying performance in complex dynamic situations. *Journal of Cognitive Engineering and Decision Making*, 1, 22–55.

Kanheman, D. (2011). *Thinking Fast and Slow*. New York, NY: Farrar, Strauss and Giroux.

Kermarrec, G. (2015). Enhancing tactical skills in soccer: Advances from the naturalistic decision-making approach. *Procedia Manufacturing*, 3, 1148–1156.

Kermarrec, G., & Bossard, C. (2014). Defensive soccer players' decision-making: A naturalistic study. *Cognitive Engineering and decision-making*, 8:2, 187–199. doi: 10.1177/1555343414527968.

Klein, G. A. (1997). The recognition-primed decision (RPD) model: looking back, looking forward. In: C. E. Zsambok, G. A., *Naturalistic Decision Making* (pp. 285–292). Mahwah, NJ: Lawrence Erlbaum Associates, Publishers.

Klein, G. A. (1998). *Sources of Power.* Cambridge, MA: MIT Press.

Klein, G. A. (2003). *The Power of Intuition.* New York, NY: Random House Inc.

Klein, G. A. (2007). Flexecution, part 2: Understanding and supporting flexible execution. *IEEE Intelligent Systems*, 22:6, 108–112.

Klein, G. A. (2008). Naturalistic decision-making. *Human Factors*, 50(3), 456–460. doi: 10.1518/001872008X288385.

Klein, G. A. (2009) *Streetlights and Shadows: Searching for the Keys to Adaptive Decision-Making.* Cambridge, MA: Massachusetts Institute of Technology.

Klein, G. A. (2015). Whose fallacies? *Journal of Cognitive Engineering and Decision Making*, 9:1, 55–58. doi: 10.1177/155543414551827.

Klein, G. A., Calderwood, R., & Clinton-Cirocco, A. (1986). Rapid decision making on the fireground. *Proceedings of the Human Factors and Ergonomic Society 30th Annual Meeting* (Vol. 1, pp. 576–580).

Klein, G. A., & Crandall, B. W. (1995). The role of mental simulation in problem solving and decision making. In: P. Hancock, J. Flach, J. Caird, & K. Vicente (eds.) *Local Applications of the Ecological Approach to Human-Machine Systems* (pp. 324–358). Hillsdale, NJ: Lawrence Erlbaum Associates, Publishers.

Klein, G. A., Moon, B., & Hoffman, R. (2006a). Making sense of sense-making 1: Alternative perspectives. *IEEE Intelligent Systems*, 21, 70–73.

Klein, G. A., Moon, B., & Hoffman, R. (2006b). Making sense of sense-making 2: A mACRsocognitive model. *IEEE Intelligent Systems*, 21, 88–92.

Klein G. A., Philipps, J. K., Rall R. L., & Peluso, D. A. (2007). A data frame theory of sense-making. In: R. R. Hoffman (ed.), *Expertise Out of Context* (pp. 113–155). New York, NY: Lawrence Erlbaum Associates.

Macquet, A.-C. (2009) Recognition within the decision-making process: A case study of expert volleyball players. *Journal of Applied Sport Psychology*, 21, 64–79. doi: 10.1080/10413200802575759.

Macquet, A.-C. (2016). De la compréhension de la situation à la distribution des informations: la prise de decision en sport de haut niveau [From situation understanding to information distribution: Decision-making in high level sports]. Unpublished thesis for the ability to conduct researches, Université de Bretagne Sud – INSEP.

Macquet A.-C., & Fleurance, P. (2007) Naturalistic decision-making in expert badminton players. *Ergonomics*, 50, 1433–1450. doi: 10.1080./00140130701393452.

Macquet, A.-C., & Kragba, K. (2015). What makes basketball players continue with the planned play or change it? A case study of the relationships between sense-making and decision-making. *Cognition, Technology and Work*, 17:3, 345–353. doi: 10.1007/s10111-015-0332-4.

Macquet, A.-C., & Lacouchie, H. (2015). What is the story behind the story? Two case studies of decision-making under stress. *Proceedings of the International Conference on Naturalistic Decision-Making*, McLean, VA.

Macquet, A.-C., & Lacouchie, H. (2017). Two case studies of decision-making under stress. *Journal of Sports Sciences*, 35:1, 72. doi: 10.1080/02640414.2017.1378421.

Macquet, A.-C., & Pellegrin, L. (2017). Putting the recognition primed-decision (RPD) model into perspective: The revisited RPD model. In: J. Gore & P. Ward (eds.), *Naturalistic Decision-Making Under Uncertainty* (pp. 118–124). Bath, UK: University of Bath.

Macquet, A.-C., & Skalej, V. (2015). Time management in elite sports: How do elite athletes manage time under fatigue and stress conditions? *Journal of Occupational and Organizational Psychology*, 88:2, 341–363. doi: 10.1111/joop.12105.

McLennan, J., & Omodei, M. M. (1996). The role of prepriming in recognition-primed decision-making. *Perceptual and Motor Skills*, 82, 1059–1069.

Milazzo, N., Farrow, D., Ruffault, A., & Fournier, J. F. (2015). Do karate fighters use situational probability information to improve decision-making performance during on-mat tasks? *Journal of Sports Sciences*, 34:16, 1547–1556. doi: 10.1080/02640414.2015.1122824.

Mulligan, D., McCracken, J., & Hodges, N. J. (2012). Situational familiarity and its relation to decision quality in ice-hockey, 10:3, 198–4210. doi: 10.1080/1612197X.2012.672009.

Neisser, U. (1976). *Cognition and Reality: Principles and Implications of Cognitive Psychology*. San Francisco, CA: Freeman.

Neville, T., & Salmon, P. (2017). Look who's talking – in game communications analysis as an indicator of recognition primed decision-making in elite Australian rules football empires. *Proceedings of the International Conference on Naturalistic Decision-Making*, McLean, VA.

Pellegrin, L., Gaudin C., Bonnardel N., & Chaudet H. (2010). Collaborative activities during an outbreak early warning assisted by a decision-supported system (ASTER), *International Journal of Human and Computer Interaction*, 26:2–3, 262–267.

Peräkylä, A., & Ruusuviori, J. (2011). Analyzing talk and text. In: N. K., & Y. S. Lincoln (eds.), *The Sage Handbook of Qualitative Research* (pp. 529–543). Thousand Oaks, CA: Sage Publications.

Piaget, J. (1954). *The Child's Conception of the World*. New York, NY: Harcourt Press.

Rasmussen, J. (1983). Skills, rules, and knowledge; signals, signs and symbols and other distinctions in human performance models. *IEEE Transaction Systems, Man, and Cybernetics*, 13, 257–266.

Simon, H. A. (1996). *The Sciences of Artificial*. Cambridge, MA: MIT.

Sparkes, A. C. (2009). Ethnography and the senses: Challenges and possibilities. *Qualitative Research in Sport and Exercise*, 1, 21–35.

Tversky, A., & Kahneman, D. (1974). Judgment under uncertainty: Heuristics and biases. *Science*, 185:4157, 1124–1131.

Weick, K. E. (1995). *Sensemaking in Organizations*. Thousand Oaks, CA: Sage Publications.

Zsambok, C. E. (1997). Naturalistic Decision-Making: Where are we now? In C. E. Zsambok, & G. A., Klein (eds.), *Naturalistic Decision Making* (pp. 3–16). Mahwah, NJ: Lawrence Erlbaum Associates.

10 Distributed Situation Awareness in Australian Rules Football Officiating

Timothy Neville

CONTENTS

10.1 INTRODUCTION

Borne out of the study of fighter pilots (Taylor, 1990), situation awareness (SA) has been an ongoing research effort in Human Factors and Ergonomics (HFE) for well over 30 years. SA is the concept applied to understand how individuals, teams, and systems know 'what is going on' (Endsley, 1995); how agents can understand the cues, information, and stimuli in the environment and how these factors affect task performance (Endsley, 1995; Salmon et al., 2008). SA is closely related to decision making, with prominent models of SA either establishing acquisition of SA as a precursor to decision making, or framing SA as a theory to be applied in naturalistic decision-making contexts (Stanton, 2009).

When considering the study of SA in Sport, Neville, and Salmon (2016) suggest the type and complexity of the sport should determine the specific focus of the research. Applying an individual perspective on SA is likely to aid understanding of how SA informs the decision making of a soccer midfielder or goalkeeper. Similarly applying a whole-of-system approach, such as Distributed Situation Awareness (DSA; Stanton et al., 2006), would be suitable for understanding the SA of an NFL defence. Application of SA theory in sport has provided insights into the decision making dynamics in squash (Murray et al., 2018) as well as the level of consistency in the coach/athlete relationship in training and competition (Macquet & Stanton, 2014). In both studies, SA theory was able to identify that the role and tasks being conducted impacted the SA of each actor in the system. Whether that be a coach or athlete or the strategies of two opposing squash players.

To demonstrate the utility of studying SA in sport, this chapter describes a case study in which the DSA model was applied to show how the introduction of technology impacts the knowledge, tasks, and interactions of a sports officiating team.

10.1.1 TECHNOLOGY AND OFFICIALS IN SPORT

Sports officiating can be viewed as a system comprising people and technologies working together toward the shared goal of regulating the sporting contest. Increasingly, multiple officials conduct both technical and non-technical processes in a dynamic environment in order to maintain the integrity of billion-dollar sports industries. When the performance of an official is questioned it is almost exclusively based on their ability to make an accurate decision under pressure in a high-stakes situation. In recent times technological support has been added to support officials in various sports. As a result, criticism has expanded from the decision itself to include critique of the performance, interpretation, or use of technology (see Table 10.1).

Whenever technologies are used to augment performance, it is important that the technology insertion process is appropriate (Clegg, 2000). Part of this involves understanding how the technology will impact performance and how training and education can be used to enhance positive impacts and remove negative impacts. SA theory, specifically DSA, has the ability to understand the impact of technology on a new system, with previous applications of this kind occurring for example in the military (Salmon et al., 2009). Through describing how new or different information is added and used or how additional tasks are completed, DSA can be used to explore

TABLE 10.1

A Brief Survey of the Media's Criticism of Officials Using Technology

Professional sports league (and match)	Technology system introduced (year)	Media criticism or issues
Football (Soccer): World Cup 2014 – France v Honduras (Lawton, 2014)	*Goal Line Technology* (2014)	Confusion over no-goal then goal
Cricket: Australia v West Indies – Umpire walks out of match (Newman, 2009); England v Australia 2013 – Stuart Broad (Lenten, 2013)	*Decision Review System* (2009)	Umpire resigned due to technology undermining confidence. Player refused to 'walk' after technology could not be used
Baseball: Boston Red Sox v New York Yankees 12 April 2014 (Caostello, 2014)	*Instant Replay* (2014)	Not all footage was available to the decision makers
AFL: Collingwood v West Coast Semi-Final 1 2012 (Channel-7, 2012)	*Score Review* (2012)	Procedures not followed; human decision overturned

how awareness could be enhanced, how cognitive load might increase, or how work behaviour changes with the addition of technology.

DSA's capacity to investigate the potential impacts of technology on an officiating system will be demonstrated through investigating the impact of adding radio communications technology to an Australian Rules Football (AFL) umpiring team.

10.1.2 THE GAME OF AUSTRALIAN RULES FOOTBALL

AFL is a fast-paced ball sport where two teams of 18 players compete to score points (through goals and behinds) on an oval-shaped field that is between 135–185 metres in length and 110 m and 155 m in width (AFL, 2014). Players are required to kick an oval-shaped ball through goal posts to score 6 points, if they miss to either side of the goals (known as behinds) or if the ball is not kicked directly through the goals or behinds (i.e. comes off a player's body) a point is scored. Play occurs over four quarters of 20 minutes playing time. There is no off-side rule and, while players hold notional positions as forwards (six players), midfielders (six players), and defenders (six players), the players and ball move dynamically across all areas of the field. The field is divided, through ground markings, into 'zones' known as 'attacking 50m arc', 'centre square', and 'defending 50m arc'. For the umpires, these zones are expanded slightly and termed the midzone (MZ) and right endzone (REZ) and left endzone (LEZ), as depicted in Figure 10.1.

AFL is officiated by a team of nine umpires – three central field umpires – the primary decision makers, four boundary umpires, and two goal umpires. At the elite professional level, there is also an emergency field and goal umpire, with all field umpires able to communicate using radio communication. Throughout the game,

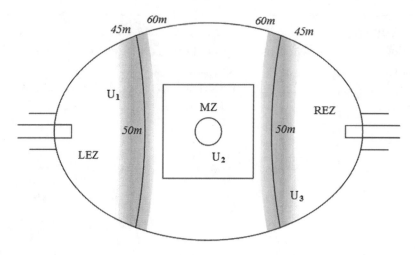

FIGURE 10.1   AFL field with umpiring zone demarcation. LEZ – Left Endzone, REZ – Right Endzone, MZ – Midzone, U1 – Umpire 1, U2 – Umpire 2, U3 – Umpire 3.

the three field umpires occupy one of the three zones on the field. As the game progresses the field umpiring team will dynamically change their positions and active decision-making role through communicating specific phrases and executing specific actions (Neville et al., 2019). For example, after a goal is kicked the two umpires furthest away from where the goal was scored will swap positions, with the umpire moving into the midzone assuming primary decision-making authority.

Previous research has investigated the role of communication in AFL Field umpires (Neville et al., 2017, 2019). When communicating to players, umpires will predominately communicate interventions (termed *free kicks*) or non-decision *play on* calls (Neville et al., 2017). Intra-team communication is used to maintain team effectiveness over the course of a match (Neville et al., 2019). Due to the geographic distribution of the umpiring team, communicating within the team can be difficult without radio communication technology, a condition which exists predominately in the lower semi-professional and amateur league. Thus, as umpires progress from amateur leagues to semi-professional and finally professional leagues, their use of communication technology increases as the level of competition (and scrutiny) increases. With the subtle but important addition of technology as umpiring proficiency increase, there is a need to understand how technology influences interactions, tasks, and, ultimately, the SA held by the officiating system.

10.2   METHOD

10.2.1   STUDY DESIGN

This naturalistic observational cross-sectional study was designed to understand how radio communication impacts DSA and teamwork in an AFL field umpiring team, using the Event Analysis of Systematic Teamwork (EAST) (Stanton et al., 2013, 2018). The study involved an analysis of umpiring teams during nine games of

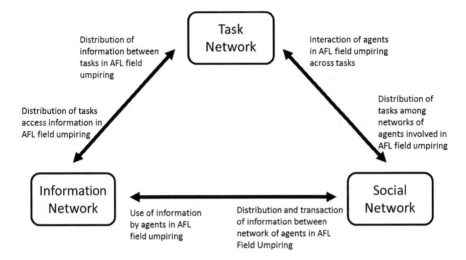

FIGURE 10.2 Relationship between tasks, information and social networks for AFL field umpiring, adapted from Stanton (2014b).

AFL – six games (four umpiring teams) at the sub-elite State League level and three games (three teams) at the elite level. Data was captured prior to a team of umpires using radio communications (Pre-Radio Group – PRG), at an umpiring team's first exposure to radio communication (Novice Radio Group – NRG), and when an umpiring team had sustained exposure to radio communication (Experience Radio Group – ERG). EAST, which has been used extensively to model and analyse DSA in different domains (see Stanton et al., 2018), was used to model the DSA of AFL field umpiring teams. The EAST process generated a conceptual base-line model of three networks representing the social interactions, information, and tasks being performed in the umpiring team prior to the introduction of radio communication (see Figure 10.2).

Following the development of the baseline social, task, and information networks of the umpiring team, an updated EAST model was developed based on the additional tasks and social interactions which occurred following the introduction of radio communication.

10.2.2 PROCEDURE

10.2.2.1 Data Collection

Live audio/visual (a/v) recordings of AFL umpiring occurred at all nine games. In each game the umpires wore radio communications equipment which allowed all verbal communication to be recorded. Within the AFL games the umpire wore broadcast-quality match communication equipment supplied by Murray Tregonning & Associates. The sub-elite umpires wore RefAudio™ provided by L&W Sports Communications. The a/v data transcribed from the nine games was approximately 16 hours in duration. Data were also collected from the grey literature document

Laws of Australian Football (Australian Football League, 2014). The *Laws* provide the declarative knowledge required to adjudicate the game. Observation notes and informal discussions were also conducted to provide greater context to the a/v recordings and *Laws*. The notes related to the specifics of how an umpire (or umpiring team) will go about umpiring a game of AFL.

All verbal communication made by the umpires (umpire-to-umpire and umpire-to-player) were transcribed and notes specific to the type of umpiring event occurring at the time were recorded. The transcripts were also informed by the corresponding video, allowing details of the tasks and information communicated to be examined. Meta-data for each communication instance included:

- the type of task being conducted,
- the time and location of the communication,
- any decision associated with the communication (i.e. free kick, play on); and,
- an indication if the communication between an umpire and the players or umpire-to-umpire.

10.2.2.2 Hierarchal Task Analysis

The transcription data, grey literature, and observation notes were used to develop a Hierarchal Task Analysis (HTA; Annett et al., 1971; Stanton, 2006) of Australian Rules Football Umpiring, with the specific goal – goal 0 of 'Officiating AFL within the spirit of the laws of the game'. HTA is used to describe tasks in terms of their goals, operations, and plans and is arguably the most commonly applied task analysis method within HFE. The HTA was subsequently used as the input for the EAST analysis, specifically the task and information networks. Developing the HTA involved decomposing the goals and tasks that field umpires conduct as they umpire the game into meaningful sub-goals, sub-tasks, and operations. Data for the HTA was sourced through the transcripts, the *Laws of Australian Rules Football*, subject matter experts and observational analysis. Branches of the hierarchal sub-goals were augmented with additional notes describing how the goals and operations occur based on the familiarisation process and analysed data.

The HTA identified 12 Level-1 sub-goals. The largest branch of the HTA was the decomposition for the sub-goal 'Adjudicate decision moment'. An example section of the HTA with associated description notes is provided in Table 10.2.

10.2.2.3 Development of the Task Network

The task network was generated from a combination of the HTA and observations from the a/v data. Initially, a task network was generated, which depicted the 12 sub-goals and their interrelations as described in the HTA (Figure 10.3). From here a set of 14 sub-tasks which AFL field umpires are required to conduct throughout the course of a game were identified and decomposed based on both the observations made from the a/v data and the collection of lower-level sub-goals in the HTA. The sub-tasks were identified from 9 of the 12 Level-1 sub-goals. Three of the Level-1 sub-goals: pre- and post-game routines and end game/quarter operations were removed as they were deemed outside the scope of the study aims of describing

TABLE 10.2

Example from the Intermediate HTA Model with Analyst Notes for the Tasks Required to Determine a 'Mark'

Decomposition Level					Sub-Goals	Analyst Notes
1	2	3	4	5		
4					Adjudicate Decision Moment	
	4.1				Adjudicate Contest	
		4.1.1			*Determine if contest is a mark*	A <u>kick</u> is determined to be a kick if the <u>ball</u> has come off a <u>player's</u>
			4.1.1.1		*Determine if ball has been kicked*	<u>leg below</u> the <u>knee</u>. From a kick, the ball must travel at least <u>15 m</u>
				4.1.1.1.1	*Determine if ball has left the hands*	before it can be considered as a <u>legitimate mark</u>, a mark occurs,
				4.1.1.1.2	*Determine if ball has made contact with the kickers' leg below the knee*	therefore, when a ball travels 15 m without <u>touching</u> the <u>ground</u> and is then <u>caught</u> by a player
			4.1.1.2		*Determine if ball has travelled 15 m*	without another player <u>touching</u> the ball.

Italicized tasks appear in task network, underlined words appear in information network.

the DSA of AFL field umpiring. Figure 10.4 represents the relationship between the Level-1 sub-goal and sub-tasks.

As an example of the relationships between the Level-1 goals and sub-tasks, the sub-task 'Conduct reporting procedure' is connected to the Level-1 sub-goals of 'Adjudicate decision moments', 'Manage players', and 'Manage match'. The purpose of identifying the sub-tasks was to increase the fidelity and understanding of the tasks field umpires are required to conduct. Furthermore, by identifying connections from sub-tasks to multiple high-level sub-goals, the relationships between different tasks become non-linear, better reflecting the complex nature of umpiring.

Subsequently, 14 sub-task networks were developed, 1 for each of the 14 sub-tasks. Within each sub-task network, tasks were linked based on an identified relationship. These can be causal (i.e. task B occurs as a consequence of task A), temporal (an umpire will do task B at the same time, or after, task A) or contingent (if the umpire does task A then task B will also occur. For instance, through observations of the video it was identified that when an umpire determined a free kick had occurred the whistle was blown (*blow whistle* task), a hand signal was made (*signal free kick*), and the umpire was required to communicate the free kicks to the players (*communicate to players*). Each sub-task network contained common nodes. For example, tasks such as *blow whistle* or *communicate to players* would occur in both the sub-task networks for an *adjudicate illegal tackle* and *conduct ball up procedure*. The common nodes allowed an overall task network for AFL field umpiring to be developed by the individual sub-task networks.

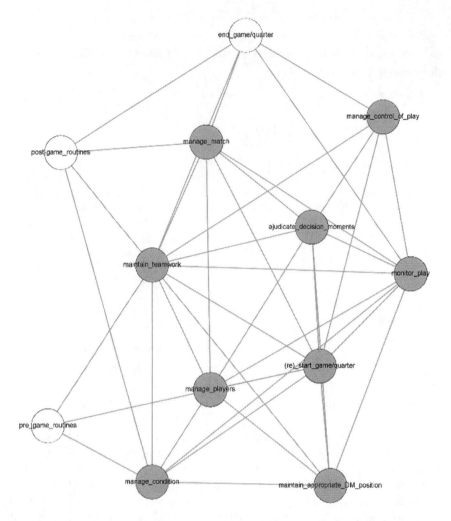

FIGURE 10.3 High level task network with blue nodes representing nodes which informed the sub-tasks networks.

10.2.2.4 Development of the Information Networks

The information networks were generated following the development of the task and sub-task networks. A similar process was followed to the development of the task network, with 21 'sub-information' networks generated. Each sub-information network was related to a sub-tasks network; however, there was not necessarily a one-to-one relationship between a sub-task and sub-information networks. For instance, the 'adjudicate illegal tackle' sub-task network had three associate sub-information networks describing content and connectedness of the information required for *illegal tackle*, *hands/push in the back (H/PITB)*, and *high tackle*. This was as each form of an illegal tackle has its own unique set of information.

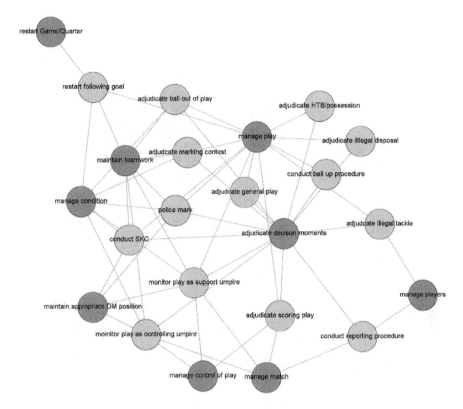

FIGURE 10.4 Relationship between Level-1 sub-goals (blue) and the sub-tasks conducted by AFL field umpires.

To identify the information nodes present in each sub-information network, a combination of the information acquired through the domain familiarisation, the observations, transcripts, *Laws of Australian Rules Football*, and the notes accompanying the intermediate HTA was used. To identify the links within each sub-information network, questions such as 'what knowledge of *y* is required for task *x*' were posed. For example, 'goal, ball, foot, player, attacking' to answer the question 'what information of the ball is required for the "determine goal" task?'. In sentence form this would be represented as:

The ball must come off the attacking player's foot for a goal to be scored.

Table 10.2 provides further examples of the information concepts identified by the intermediate HTA (information concepts are underlined in the analyst notes column).

As the example in Figure 10.5 demonstrates, sub-information networks were integrated based on their common nodes. The common nodes are coloured dark grey. In this way, the 21 sub-networks were combined to generate the overall information network.

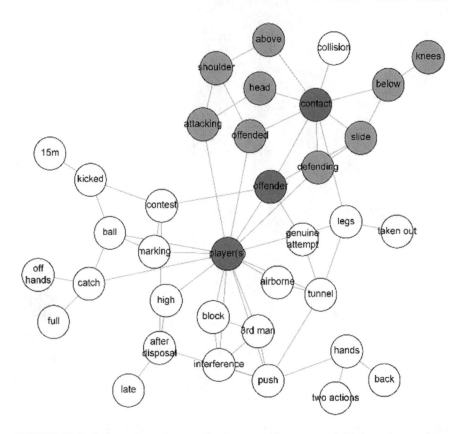

FIGURE 10.5 Information sub-networks for a marking contest (white) – when a player catches the football after it has been kicked and illegal tackle (light grey) and common nodes (dark grey).

10.2.2.5 Development of Social Network

The social network was generated through observing interactions between agents captured in the a/v recordings. The agents included human agents such as umpires and players, as well as non-human agents such as goal post, line markings, and flags. As the type of interaction between the umpires and players was consistent regardless of the specific player and as player-to-player interactions were outside of the scope of this research players were abstracted to a single node for each team (i.e. T1 and T2).

10.2.2.6 Network Validation

All networks were validated by a subject matter expert (SME), a field umpire coach within the elite Australian Football League umpiring department. At the time of the study the SME was an umpiring coach at the national level and had over 20 years of experience in umpiring ranging from the sub-elite to elite level, including four as either a state-league or national coach of field umpires.

Network validation involved the SME reviewing the social network and each of the sub-information and sub-task networks in a single review session. To start, the

SME was briefed on the network generation process and on how to read each network. Each sub-network was then presented to the SME on an A3 sheet of paper. The SME, without prompting, either agreed, removed, or added new links or nodes for each network. If sought, clarification was given to the SME on any ambiguity in the networks.

Table 10.3 presents a summary of the changes made to the initial sub-task, sub-information, and social networks. As indicated in Table 10.3, following the validation process each network had a number of additional nodes and/or links were added. The changes generally reflected instances where one or two information or task nodes were initially omitted from a sub-network. With the addition of one node, however, several additional links were inevitably required to link the additional node to existing nodes. In total, the initial networks represented over 85% of the final set of validated networks.

10.2.2.7   Network Analysis

Social network analysis (SNA, see Chapter 13) techniques were used to understand the properties of each network. SNA is used to understand the structure of networks via describing and analysing the connections between entities in the network. SNA metrics are typically used to interrogate and interpret all three EAST networks (Stanton et al., 2018). Three 'whole-of-network' measures were used to understand how the entire networks changed with the addition of radio communication. These included – Ave. Degree – the average number of nodes any one node is connected to, Density – the ratio of links and nodes in the network, and Average Path Length – the number of steps along the shortest path for all two-node pairs in the network.

For individual nodes in the network, three 'node-level' measures were used to determine how central nodes are in the network – Degree – the number of other nodes connected to a specific node in the network, Eigen Centrality – a value which considers both degree and the relative position of the node in the network, and Betweenness – a measure of how often specific nodes acts as a connecter between two other nodes. Both sets of network metrics enable analysis of the connectedness of a network; providing insight into how readily information, tasks or agents can be activated in the system (for more information please see Bounova & de Weck, 2012; MIT, 2011).

TABLE 10.3

Differences between Initial Task and Information Networks and Validated Model after SME Validation

	Information		Task		Social		Total
	Nodes	**Links**	**Nodes**	**Links**	**Nodes**	**Links**	
Initial Network	224	539	158	307	15	77	1320
Accepted Network	258	638	174	364	15	84	1533
Difference	+34	+99	+16	+57	+0	+7	213
% Change	**15.18%**	**18.37%**	**10.13%**	**18.57%**	**0**	**8.33%**	**13.89%**

10.3   RESULTS

Figures 10.6 and 10.7 present the Social and Task networks for AFL Field Umpiring following the addition of radio communication. Additional social agents and tasks identified as a consequence of the introduction of radio communication are shaded grey and were added to the network following analysis of the NRG and ERG data. Figure 10.8 presents the information network.

10.3.1   ADDITIONAL AGENTS AND TASKS

The analysis demonstrates that the introduction of radio communication introduced the following two additional agents into the system:

- Radio Comms – the radio equipment used to transmit the verbal communication between the team.
- EMG Ump – The data from the three ERG games included an emergency umpire. The addition of an emergency umpire is unique to games in the professional Australian Football League.

The updated information network included an additional nine tasks, four of which were as a direct consequence of the addition of radio communication:

- *Communicate about position* – discussion about each umpire's position relative to the other umpires.
- *Communicate about control* (of play) – discussion about how to change the decision making control from one umpire to the next.
- *Communicate about adjudications* – discussion about previous or current adjudications made by the field umpiring team.
- *Maintain morale* – communication specific to the morale of the team, such as 'good work' or 'keep it going'.

An additional five tasks were included due to the existence of a video score review process which existed only in the ERG games.

10.3.2   WHOLE-OF-NETWORK METRICS

Table 10.4 provides the whole-of-network metrics for the Task, Social, and Information Networks.

10.3.3   NODE-LEVEL METRICS

Node-level metrics offer two specific purposes. The first, presented in Table 10.5, shows the seven most highly connected and central nodes in each network, effectively the most influential social agents, information, and tasks conducted while umpiring Australian Rules Football. The second, presented in Table 10.6, presents the nodes with the greatest change in metrics following the addition of radio communication into the network.

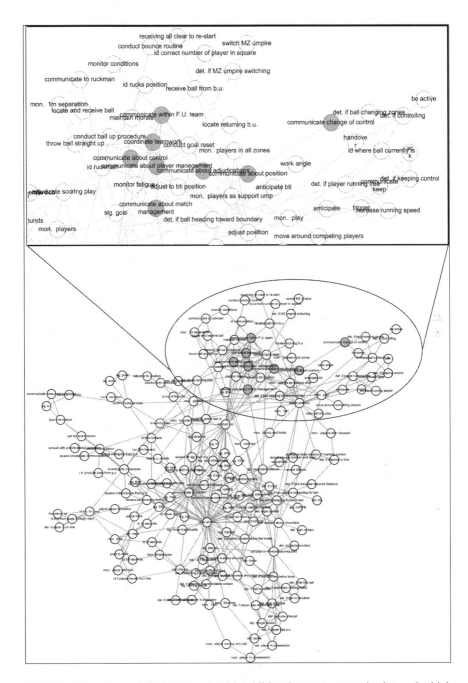

FIGURE 10.6 Updated Task Network with additional team-communication tasks highlighted. Inset-selection of the network showing the frequency of team-based communication tasks.

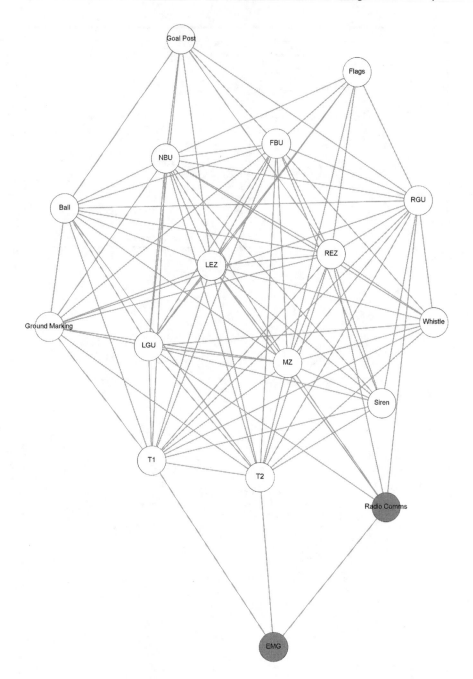

FIGURE 10.7 Updated Social Network – T1/2 – players from team 1 or 2, R/LGU, Right or left of screen goal umpire, N/FBU – near side or far side boundary umpires, R/L right or left of screen end zone umpire, MZ – midzone umpire EMG – Emergency Umpires. Dark nodes represent the additional social agents

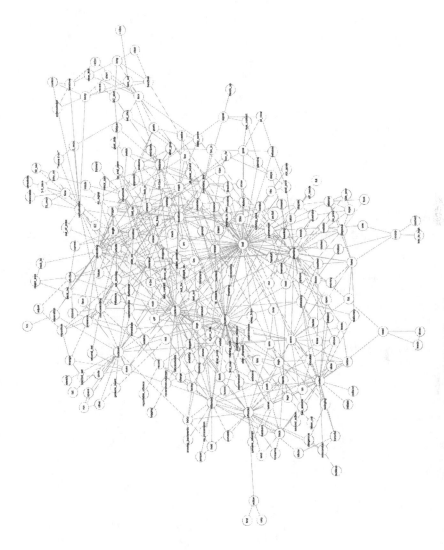

FIGURE 10.8 Validated Information Network for AFL Field Umpiring.

TABLE 10.4

Whole-of-Network Metrics for the Updated Networks (Under Three Conditions – PRG, NRG and ERG) with Comparison to the Baseline Model Ave

Network	Nodes	Links	Density	Ave. Deg.	APL
		Task			
Baseline	169	356	0.025	4.213	3.601
Updated	178	403	0.026	4.528	3.472
		Social			
Baseline	15	84	0.800	11.20	1.2
Updated	17	92	0.676	10.823	1.338
Information	258	638	0.019	4.946	3.281

Deg – Average Degree, Diam. – Network Diametre. APL – Average Path Length.

TABLE 10.5

Measures of Most Centre Nodes in Social, Task, and Information Networks

Social Agent	Deg.	Btwn.	Centr.
Far Boundary Umpire (FBU)	14	0.0261	0.304
Near Boundary Umpire (NBU)	14	0.0261	0.304
Left Endzone (LEZ)	13	0.0245	0.284
Right Endzone (REZ)	13	0.0245	0.284
Midzone (MZ)	13	0.0159	0.291
Left Goal Umpire (LGU)	12	0.0237	0.261
Right Goal Umpire (RGU)	12	0.0237	0.261
Tasks			
blow whistle	43	0.301	0.483
communicate to players	40	0.322	0.436
adjust position	20	0.266	0.062
call play on if necessary/ required	13	0.117	0.112
conduct set kick control	12	0.014	0.198
call play on	11	0.055	0.134
mon. players attempting to catch ball	10	0.074	0.141
Information			
ball	62	0.344	0.409
player(s)	40	0.197	0.286
contest	39	0.167	0.336
offender	26	0.116	0.153
attacking	26	0.100	0.205
contact	23	0.077	0.161
controlling	20	0.056	0.136

Deg. – Degree, Btwn- Betweenness, Centr. – Centrality.

TABLE 10.6
Measure of Social Agents Tasks with Largest Increase in Degree When Radio Communication Added

Social Agent	Deg.	Difference/Change in Btwn.	Centr.
Far Boundary Umpire (FBU)	0	–0.0045	–0.00785
Near Boundary Umpire (NBU)	0	–0.0045	–0.00785
Left Endzone (LEZ)	1	0.0113	0.00196
Right Endzone (REZ)	1	0.0113	0.00196
Midzone (MZ)	1	0.0101	0.00234
Left Goal Umpire (LGU)	1	0.0116	0.00217
Right Goal Umpire (RGU)	1	0.0116	0.00217
Radio Comms	*6*	*0.0176*	*0.12312*
Task	**Deg.**	**Btwn.**	**Centr.**
Conduct goal reset	8	–0.0225	0.0560
Conduct SKC	8	0.0844	0.0356
Adjust to BTI position	7	0.0081	0.0406
Maintain morale	7	0.0172	0.0519
Communicate about adjudications	6	0.0136	0.0500
Communicate about control	6	0.0052	0.0513
Conduct ball up procedure	6	0.0062	0.0338

Between baseline and updated task networks, with the difference in measure value between the baseline and updated task networks shown for the following measures Deg. – Degree, Btwn –Betweenness, Centr. – Centrality.

10.4 DISCUSSION

The aim of this chapter was to demonstrate how DSA and associated HFE methods can be used to explore the impacts of technology insertion in sport. Specifically, EAST was used to examine the impact on tasks, interactions, and information usage during AFL umpiring. The findings are discussed below in relation to each of the three EAST networks.

10.4.1 KEY TASKS, AGENTS, AND INFORMATION

The social network had a noticeably different structure to the task and information networks, with a lower number of nodes and links and a higher level of connectivity than the task and information networks. The different umpiring agents (boundary, field, and goal umpires) were the most connected to all human agents and a number of non-human agents, including the *ball* and *ground markings* (see Table 10.6). The most connected nodes were the two boundary umpires (FBU and Near Boundary Umpire (NBU)) which were connected to all other agents in the network. The node-specific metrics identified an interesting finding with respect to the midzone umpire (MZ). While the MZ's connectedness (betweenness) was less than the other field umpires, the centrality and average neighbour degree of the MZ was higher. This result reflects the geographic disposition of the MZ as an umpire in the middle of the ground, central to the operation of the game.

Perhaps unsurprisingly, the most connected non-human agent was the *ball*. From a social agent perspective, players and umpires were required to constantly interact with the ball, for example, if the ball was kicked the human agents needed to interact with the ball first through the physical act of kicking and second through adjusting their positions in reference to the geographical position of the ball.

10.4.2   Task Network

Within the task network, the most connected nodes were the *blow whistle* task followed by the *communicate to players* and *adjust position* tasks (see Table 10.6). Of these three tasks, *blow whistle* and *communicate to players* were also the most central tasks, with the *conduct set kick control* task the third most central task. In considering the connectedness (degree) and centrality, the task network showed that the key tasks used for transacting awareness around the system were *blow whistle*, *communicate to players*, *adjust position*, and *conduct set kick control*. Importantly, these tasks involve transactions between human and non-human agents, such as the umpires, whistle, and ground markings.

10.4.3   Information Network

The most connected nodes in the information network were *ball*, *player(s)*, and *contest*. These nodes were also the most central nodes with the highest betweenness value (i.e. their frequency of being between two adjacent nodes was the highest of all nodes). Note that the ball appeared in both the information and social networks as it acted as a social agent, for umpires and players to interact with, and also as an information node. Many other information nodes were connected to the *ball*, *player(s)*, and *contest*. For example, connected nodes to *contest* include *ruck*, *blocking*, *ball*, *location*, *stopped*, *player(s)*, *after disposal*, *marking*, *taggers*, *charge*, *20–25 m*, *angle*, and *future*. Exploring the list of nodes with the largest degree further (Table 10.6), the nodes most connected in the network include information such as *contact* (where one player makes contact with another player), *arms* (the limb which makes most of the *contact*), and the role of specific *players(s)* (*attacking*, *offended*, *offender*, and *ruckman*). In relation to DSA, the analysis suggests that the *ball*, *player(s)* (including the roles they identify as, e.g. *offender*, *defender*, *attacking*, and *ruckman*), and *contest* were key concepts that underpin SA and its distribution across the AFL field umpiring system. For instance, an umpire, when *adjudicating a marking contest* was required to activate information about the *ball* and the *offender* and *offended player(s)* in order for the umpiring system to be aware of 'what is going on'.

10.4.4   Changes in Network Characteristics in Updated Models

The differences in the network metrics between the baseline and updated task and social networks identified that radio communication technology changes the structure of the networks, and, therefore, the way SA is distributed across the umpiring system. The additional social agent – *radio communication* – changes the structure of the social interactions of the system. This change allows for additional opportunities for interaction between the agents who use the radio communication, and therefore an additional mechanism to transact awareness.

In the task network, the presence of radio communication technology increased the number of communication tasks within the system. The additional tasks, introduced as a consequence of the radio communications, are connected to a range of existing tasks in the network (i.e. *conduct set kick control* and *conduct ball up procedure*). In essence, these additional tasks have a second-order effect on how the game is umpired, supplementing the adjudication role of the umpires with their ability to communicate and operate as a team.

10.5 CASE STUDY – HIGH TACKLE CASE STUDY

The following case study will demonstrate the second-order effect of radio communication on the umpiring of AFL Football.

10.5.1 SCENARIO

As a full-contact sport, AFL umpires often intervene when the defending player makes contact above the shoulders while in the act of tackling the player with the ball. The resulting adjudication of *high tackle* results in the attacking player being provided with a free kick. The free kick requires the controlling umpire to control the offending and offended players (termed set kick control) while the other umpires position themselves to monitor the players in their zone. The case study incorporates two similar situations of a high tackle free kick, one from a game where the umpires did not have radio communication, the other when they did. In both examples, the free kick occurred with the middle of the playing area, with the resultant kick travelling into the attacking teams' forwards zone.

When radio communication was not present, the only communication came from the controlling umpire communicating the following:

> 'whistle', 'high', 'come around mate when you get it', 'mate you've got to come around', 'just here mate', 'thanks', 'thanks mate', 'there's the mark', 'double whistle', 'play on' (Umpire 1 – PRG1, Q1 1:46–2:00).

In the example when radio communication was present all three umpires were able to communicate, with the following communication transcribed:

> Controlling Umpire – [Whistle] high, yep, just there, [player] you're alright where you are, right there, mark's there.

> Left (Defensive) End Zone Umpire – head's up [Attacking EZ Umpire], make sure you're side on, work, work, work.

> Right (Attacking) End Zone Umpire – ah sorry [Defensive EZ umpire], yeah I know.

10.5.2 IMPACT OF RADIO COMMUNICATION

Figures 10.9, 10.10, and 10.11 show the different nodes that were activated between the no radio communication situation and the radio communication situation.

The radio communication enabled the LEZ to communicate information about a previous occurrence in the game in order to improve another umpire's future performance. The communication occurred over a distance of greater than 100 m, and also occurred while the third umpire was controlling the players during a SKC situation. Comparing the two HC/SKC situations, the free kick occurred at a similar position on the ground; therefore the relative position of each umpire was similar in both situations; the severity of the HC was similar, as was the time it took to complete the tasks. The main difference between the two situations was the intra-team communication via radio that was able to occur between the umpires during the SKC. Consequently, the radio communication enabled the verbal transactions to occur where previously this was not possible. This is reflected through the additionally activated nodes in each of the three networks.

Within the updated task network, the additional communication tasks and additional connectedness of tasks increased the number of paths and loops along which different tasks can transact information through the system. The task networks became denser and the average path lengths became shorter. With the addition of radio communication, field umpires communicated in more diverse situations. Indeed, Table 10.7 and Figure 10.7 show that that radio communication increases the connectedness of tasks, which subsequently allows information to be transacted by the field umpiring in more situations than had previously been possible.

10.6 CONCLUSIONS

10.6.1 IMPACT FOR AFL OFFICIATING

This study has a number of key findings concerning the DSA of AFL umpiring and the impact of radio communication. First, the radio communication technology allows for greater opportunities to communicate within the game. The radio communication provides increased access to information; social agents are able to transact information which enhances teamwork and coordination within the umpiring team. Overall, the findings suggest that radio communication has a positive effect on the field umpires SA. Results of the DSA analysis suggest that radio communication has very little to no impact directly on an umpire's primary role – to make decisions. Nevertheless, when combined with the study of umpire decision-making (Neville et al., 2016) and their intra-team communication (Neville et al., 2019, the findings do suggest that radio communication allows additional information to be activated within in the team which is likely to impact, albeit tangentially, the decision making of an umpire.

10.6.2 IMPACT FOR SPORT

By exploring the second and third-order impact of technology insertion, the DSA of AFL umpiring provides a model that can be replicated in other sporting domains.

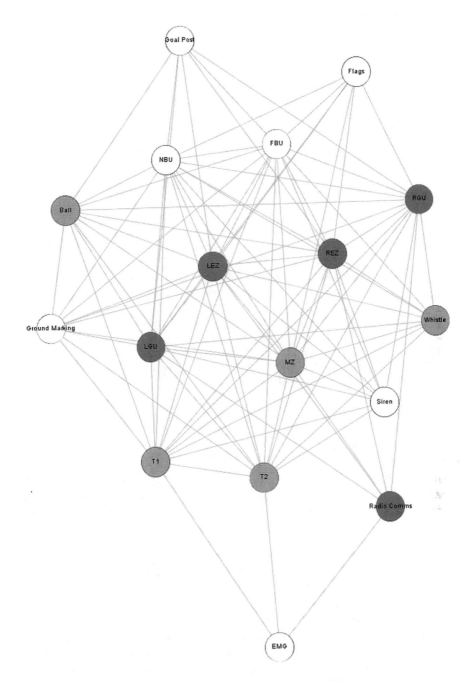

FIGURE 10.9 Activated tasks for the HC-SKC example of field umpire DSA when using radio communication –light grey nodes represent agents which would be activated without the use of radio communication; dark grey nodes represent additionally activated agents when radio communication is introduced to the system.

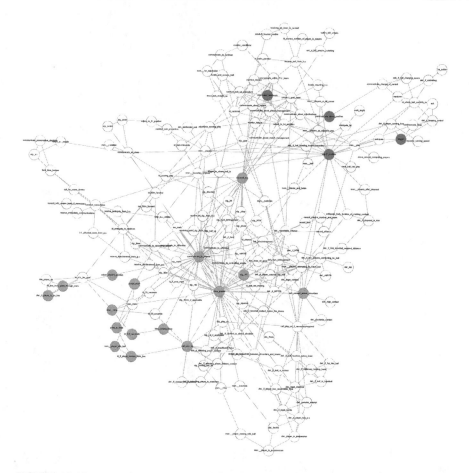

FIGURE 10.10 Activated tasks for the HC-SKC example of field umpire DSA when using radio communication – light grey nodes represent task which would be activated without the use of radio communication, dark grey nodes represent additionally activated tasks when radio communication is introduced to the system.

It is quite common for sports to trial modifications to the officiating system before the modification is permanently implemented. However, despite the trials, there have been numerous occasions when the system failed to quickly adjust to the change (i.e. goal-line technology in soccer, Lawton, 2014;, or video review in the MLB, Caostello, 2014). The current research has demonstrated the benefit of using multiple analytical methods, such as content analysis and on the system under trial conditions. The analysis provided a level of understanding of the potential effects on the system, which may limit the initial disruption of the introduced technology.

Specific to AFL, the DSA analysis has the ability to inform future changes to the AFL umpiring system. For example, the permanent introduction of a fourth field umpire to the system. The change has been recently trialled but not implemented (Edwards, 2016). The DSA model has the capacity to be modified to include an extra field umpire and associated tasks. The model can then be analysed, similar to the

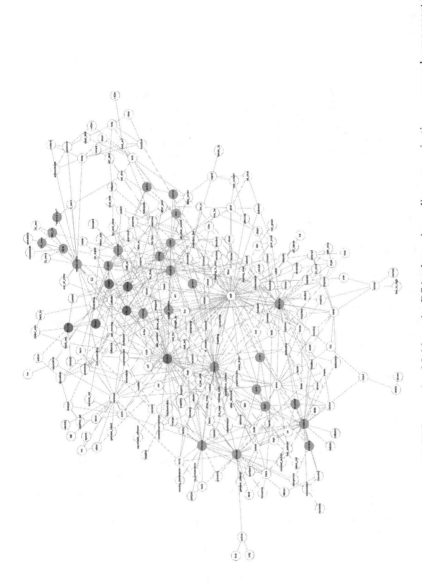

FIGURE 10.11 Activate information for the HC-SKC example of field umpire DSA when using radio communication – grey nodes represent information which would be activated without the use of radio communication orange nodes represent additionally activated information when radio communication is introduced to the system.

way the model was analysed with the introduction of radio communication, in order to understand the impact of a fourth umpire. Furthermore, such approaches to the analysis of the work of officials in sports, not just their adjudication, could benefit sporting organisations and leagues aiming to introduce decision aid technologies to aid officiating.

10.6.3 Conclusion

Finally, when considering the theory of SA, and the application of HFE methods to understand 'what is going on', due attention needs to be applied to sport. Competitive advantages can undoubtedly be made through tracking and exploring exactly what is known at any given situation in a sporting contest; a line of enquiry SA theory can be used to explore. Indeed, SA theory and methods have the capability to enable enhanced understanding of teamwork, coordination and synchronisation and decision making in systems, teams, and individuals in all sports.

REFERENCES

Annett, J., Duncan, K. D., Stammers, R. B., & Gray, M. J. (1971). *Task analysis*. Department of Employment Training Information Paper 6. London, UK: Her Majesty's Stationary Office (HMSO).

Australian Football League. (2014). Laws of Australian Football 2014. In: A. F. League (ed.). Melbourne, Victoria: Australian Football League.

Bounova, G., & de Weck, O. (2012). Overview of metrics and their correlation patterns for multiple-metric topology analysis on heterogeneous graph ensembles. *Physical Review E*, 85:1, 016117.5.

Caostello, B. (2014). MLB replay officials botch call on Anna double in eight. *New York Post*. Retrieved from: http://nypost.com/2014/04/12/mlb-replay-officials-botch-call-on-anna-double-in-eighth/.

Channel-7 (Writer). (2012). 2012 AFL Premiership Finals Series – Semi Final 1 Collingwood v West Coast. Melbourne, Victoria.

Clegg, C. W. (2000). Sociotechnical principles for system design. *Applied Ergonomics*, 31:5, 463–477.

Edwards, N. (2016). Four umpires on hold as AFL continues trial. Retrieved from: http://www.afl.com.au/news/2016-02-03/four-umpires-on-hold-as-afl-continues-trial.

Endsley, M. R. (1995). Toward a theory of situation awareness in dynamic systems. *Human Factors*, 37:1, 32–64. doi: 10.1518/001872095779049543.

Lawton, M. (2014). FIFA to review goal-line technology after mass confusion over Honduras own goal. *Daily Mail*. Retrieved from: http://www.dailymail.co.uk/sport/worldcup2014/article-2659180/FIFA-review-goal-line-technology-mass-confusion-Honduras-goal.html.

Lenten, L. (2013). Media misses point on cricket's decision review system. Retrieved from https://theconversation.com/media-misses-point-on-crickets-decision-review-system-16086.

Macquet, A. C., & Stanton, N. A. (2014). Do the coach and athlete have the same «picture» of the situation? Distributed situation awareness in an elite sport context. *Applied Ergonomics*, 45:3, 724–733. doi: 10.1016/j.apergo.2013.09.014.

MIT. (2011). *Matlab Tools for Network Analysis*. Cambridge, MA: Massachusetts Institute of Technology. Retrieved from: http://strategic.mit.edu/downloads.php?page=matlab_networks.

Murray, S., James, N., Perš, J., Mandeljc, R., & Vučković, G. (2018). Using a situation aware-ness approach to determine decision-making behaviour in squash. *Journal of Sports Sciences*, 36:12, 1415–1422.

Neville, T. J., & Salmon, P. M. (2016). Never blame the umpire – a review of Situation Awareness models and methods for examining the performance of officials in sport. *Ergonomics*, 59:7, 962–975. doi: 10.1080/00140139.2015.1100758.

Neville, T. J., Salmon, P. M., & Read, G. J. (2017). Analysis of in-game communication as an indicator of recognition primed decision making in elite Australian rules football umpires. *Journal of Cognitive Engineering and Decision Making*, 11:1, 81–96.

Neville, T. J., Salmon, P. M., & Read, G. J. (2019). Radio Gaga? Intra-team communication of Australian rules football umpires–effect of radio communication on content, structure and frequency. *Ergonomics*, 61:2, 313–328.

Newman, P. (2009). Mark Benson consigned to county cricket as ICC claim English umpire couldn't handle the job' following walkout Down Under. *Daily Mail*. Retrieved from: http://www.dailymail.co.uk/sport/cricket/article-1233930/Mark-Benson-consigned-county-cricket-ICC-claim-English-umpire-handle-job-following-walkout-Down-Under.html.

Salmon, P. M., Lenne, M. G., Walker, G. H., Stanton, N. A., & Filtness, A. (2014b). Using the Event Analysis of Systemic Teamwork (EAST) to explore conflicts between different road user groups when making right hand turns at urban intersections. *Ergonomics*, 57:11, 1628–1642. doi: 10.1080/00140139.2014.945491.

Salmon, P. M., Stanton, N. A., Walker, G. H., Baber, C., Jenkins, D. P., McMaster, R., & Young, M. S. (2008). What really is going on? Review of situation awareness models for individuals and teams. *Theoretical Issues in Ergonomics Science*, 9:4, 297–323. doi: 10.1080/14639220701561775.

Salmon, P. M., Stanton, N. A., Walker, G. H., & Jenkins, D. (2009). *Distributed Situation Awareness: Advances in Theory, Measurement and Application to Teamwork*. Aldershot, UK: Ashgate.

Stanton, N. A. (2006). Hierarchical task analysis: Developments, applications, and extensions. *Applied Ergonomics*, 37(1), 55–79.

Stanton, N. A. (2009). Situation awareness: where have we been, where are we now and where are we going? *Theoretical Issues in Ergonomics Science*, 11:1–2, 1–6. doi: 10.1080/14639220903009870.

Stanton, N. A. (2014a). East: A method for investigating social, information and task net-works. Paper presented at the Contemporary Ergonomics and Human Factors 2014.

Stanton, N. A. (2014b). Representing distributed cognition in complex systems: How a submarine returns to periscope depth. *Ergonomics*, 57:3, 403–418. doi: 10.1080/00140139.2013.772244.

Stanton, N. A., Salmon, N. A., & Walker, G. H. (2018). *Systems Thinking in Practice: The Event Analysis of Systemic Teamwork*. Boca Raton, FL: CRC Press.

Stanton, N. A., Salmon, P. M., Rafferty, L., Walker, G. H., Jenkins, D. P., & Baber, C. (2013). *Human Factors Methods: A Practical Guide for Engineering and Design*. Aldershot, UK: Ashgate.

Taylor, R. (1990). Situational Awareness Rating Technique (SART): The development of a tool for aircrew systems design. *AGARD, Situational Awareness in Aerospace Operations*, 17p. (no. 90-28972 23-53).

Walker, G. H., Gibson, H., Stanton, N. A., Baber, C., Salmon, P., & Green, D. (2006). Event Analysis of Systemic Teamwork (EAST): a novel integration of ergonomics methods to analyse C4i activity. *Ergonomics*, 49:12–13, 1345–1369. doi: 10.1080/00140130600612846.

Walker, G. H., Stanton, N. A., Baber, C., Wells, L., Gibson, H., Salmon, P., & Jenkins, D. (2010). From ethnography to the EAST method: a tractable approach for represent-ing distributed cognition in air traffic control. *Ergonomics*, 53:2, 184–197. doi: 10.1080/00140130903171672.

11 Cognitive Load in Sport

Suzanna Russell, Vincent G. Kelly, Shona L. Halson, and David G. Jenkins

CONTENTS

11.1 INTRODUCTION

11.1.1 HUMAN FACTORS AND COGNITION IN SPORT

The interaction of an athlete with their sporting environment involves a multitude of human factors (Hulme et al., 2019). Such factors contribute to sporting outcomes and achievements that are often described as being 'unexplainable'. Part of such unforeseeable outcomes or actions is a result of humans being unpredictable, and subsequently unreliable. Despite the rapid advancement and use of technology in sport to inform and increase control, including statistical quantification of performance and monitoring of load, human factors will continue to influence outcomes in sport. Thus, the human factors approach of advancing knowledge on cognitive load through understanding how an individual athlete interacts with a task in their sporting environment has the potential to lead to performance optimisation. Measuring, quantifying, and understanding load informs the design and monitoring of training programs (Halson, 2014). Historically, however, the majority of measures used in sport have only assessed physical load, and not the brain and associated networks (Halson, 2014).

Nevertheless, awareness of the importance of assessing cognitive load in sport is increasing. An example of this is the recommended inclusion of 'an assessment of cognitive function' in the '11 key features of a sustainable monitoring program' (for understanding load and fatigue in athletes) (Halson, 2014). Cognitive load is gaining the recognition and attention it deserves as a vital component relating to human factors that influence performance in sport.

11.1.2 DEFINITION OF COGNITIVE LOAD

Cognitive load has previously been defined as 'a multidimensional construct representing the load that performing a particular task imposes on the learner's cognitive system' (Paas, Tuovinen, Tabbers, & Van Gerven, 2003). In other words, cognitive load is complex, interactive, and dependant on the individual. It is important to note that in the related sport-specific literature, excessive cognitive load has the potential to lead to cognitive fatigue – also commonly referred to as mental fatigue (MF) (Van Cutsem et al., 2017c). For the purpose of this chapter, cognitive and MF will be considered as being the same. It has previously been stated that cognitive load should be defined in terms of an interaction between the task itself and the individual performing the task (Hancock & Chignell, 1988).

Consequently, in the sports setting, the cognitive load experienced by an athlete, coach, or official is dependent not only on the task at hand, but individual factors, such as expertise and environmental dynamics (Durantin, Gagnon, Tremblay, & Dehais, 2014). The assessment of cognitive load should therefore not only be estimated by the properties of the task, but also the mental effort or cognitive capacity allocated and required to accommodate the demands imposed by the task (Paas et al., 2003). For example, the cognitive load imposed by the execution of kicking a soccer ball may depend on several task properties, including the position it was received, proximity and pressure of the nearest defender, ground conditions such as firmness, and distance to the target (Oppici, Panchuk, Serpiello, & Farrow, 2017). However, the cognitive load required may also differ between individuals, with novice players requiring more cognitive capacity to accommodate for these variables and execute the skill proficiently than an expert with increased familiarity. Thus, cognitive load in sport is about the cognitive demands placed upon athletes as a result of the environmental requirements and task constraints, and the interaction with an individual's capacity to accommodate such loads. In contrast to the relative lack of research specifically investigating cognitive load as a human factor in the sporting environment, cognitive load has been more extensively examined in other settings. The examination of cognition in the fields of transport, military, nuclear power, education, and the medical field inform us of the potential negative impact of cognitive load in the sporting domain (Klaassen et al., 2016; Lal & Craig, 2001; Weeks, McAuliffe, DuRussel, & Pasquina, 2010). Such evidence demonstrates how the understanding and assessment of cognitive fatigue can benefit both the mitigation of risk and optimisation of performance. Accordingly, how we manage cognitive load, avoid cognitive fatigue, and maintain athlete performance during competition is an area of growing interest.

11.1.3 OVERVIEW OF COGNITIVE DEMANDS IN SPORT

Though the assessment and programming of physical capacity arguably remain the present focus of high-performance and development sporting programs, the role of cognitive capacity, both in isolation from, and in interaction with, physical load, is becoming recognised (Pageaux & Lepers, 2018; Russell, Jenkins, Rynne, Halson, & Kelly, 2019b; Smith et al., 2018). The brain shapes behaviour, response, and subsequent adaptation to stimuli and there is a large body of research showing the benefit of exercise on brain function (Voss, Nagamatsu, Liu-Ambrose, & Kramer, 2011). Assessment of brain activity which has typically targeted the pre-frontal cortex, indicates exercise tasks to be associated with changes in cortical patterns (Brümmer, Schneider, Abel, Vogt, & Strueder, 2011). Sporting performance requires optimisation of a combination of physical, technical, tactical, and psychological components (Joyce & Lewindon, 2014), all of which, require elements of cognition. Perception is known to influence physical performance outcomes, with perception of effort being related to a number of physiological markers of performance and fatigue (Coutts, Rampinini, Marcora, Castagna, & Impellizzeri, 2009; Eston, 2012).

Additionally, neurotransmitters and the reward system of the brain influence motivation to perform physical work (Meeusen & Roelands, 2017). Therefore, the brain assumes a key role in determining the volume and/or intensity an athlete undertakes and performs. The learning and execution of technical skill, such as a swimming start, hockey reverse hit, tennis slice, or a slower ball in cricket, also requires significant cognitive activity. A large amount of complexity and cognitive demand is required to coordinate segments of the body, activate muscles at precisely the right time, in the correct sequence, and to an appropriate degree. Further, optimal technical execution often involves adjustment for environmental constraints, e.g. position of an opposing player. Although athletes often make this process appear simple, the cognition required to perform a skill is complex (Nielsen & Cohen, 2008). Consequently, the brain holds a key role in technical performance, ranging from, for example, the accuracy of passing between team-mates or closeness of foot position to the front of the long jump take-off board. Further, the execution of tactics, or tactical discussion are also critical components of sporting performance (Rein & Memmert, 2016) that rely heavily on the brain. Brain activation for effective tactical execution in team sports influences the creation of tactical plans and the effectiveness of their execution, or lack thereof. In individual sports, such as swimming or running, decision-making and tactical execution are also required, in the form of pacing. This may involve the execution of a pre-planned race strategy, in addition to the need to adapt and respond to emerging new information, as the race unfolds (Konings & Hettinga, 2018). Additionally, the use of psychological techniques in enhancing sporting performance (Meyers, Whelan, & Murphy, 1996) draws clear links to the demands of the brain. Strategies such as visualisation, self-talk, and relaxation techniques are commonly encouraged by sports psychologists (Hammermeister & VonGuenthner, 2005), demonstrating additional cognitive demand of high-performance sport.

Despite the examples discussed, in practice, the value of cognition in sport has largely been overlooked. This is likely attributed, at least in part, to the complexity of assessing and managing cognitive load. Qualitative research informs us that coaches and athletes perceive understanding MF and the variability in athlete susceptibility to MF to be challenging (Russell et al., 2019b) and at present the use of valid objective assessment methods is limited by the constraints of the elite sporting context (Russell, Jenkins, Smith, Halson, & Kelly, 2019c). However, with the margins of difference between winning and losing in elite sport being minuscule, optimising cognitive load in sporting practice is an area that has the potential to equip athletes and teams with a winning edge.

11.1.4 WHY COGNITIVE LOAD IN SPORT?

Monitoring an athlete's physical load is beneficial for informing future training or competition stress and in identifying how athletes are responding to training (Kelly & Coutts, 2007). Load may be increased or decreased to induce adaptation or reduce fatigue, respectively. Individual athletes respond differently to exposure to the same stressor or training session, and coaches adjust the training dose to optimise individual responses (Joyce & Lewindon, 2014). A similar approach can be used to understand and manage cognitive load. Whilst it is not possible to completely remove the unpredictability and unreliability of humans in performance sport, understanding cognitive load may minimise the risk of adverse events, such as those resulting from poor execution or response, and better inform training and competition practices. In light of this, the following section will discuss specifics of cognitive load through a human factors perspective. Practical recommendations for the assessment and management of cognitive load will be provided.

11.2 DISCUSSION

11.2.1 COGNITIVE FATIGUE AS A RESULT OF COGNITIVE LOAD

Cognitive fatigue is defined as; a psychobiological state caused by prolonged periods of demanding cognitive activity (Desmond & Hancock, 2001; Job & Dalziel, 2001). The resulting cognitive fatigue experienced by an athlete in response to the stimulus of cognitive load can aid in contextualising the importance of cognitive load in sport (Smith et al., 2018). Further, cognitive fatigue, or the ability to cope with fatigue, is a human factor particularly relevant to sport. Whilst there has been only limited research investigating cognitive fatigue in the competition or elite sport setting, cognitive fatigue has been shown to negatively influence a number of physical performance tasks, including time to exhaustion, self-selected power output, and completion time (Van Cutsem et al., 2017a), as well as aspects of sporting performance such as skilled motor performance and decision-making (Russell et al., 2019c; Smith et al., 2018). As previously discussed, factors contributing to performance success include physical, technical, tactical, and psychological components – each of which are influenced by cognitive fatigue (Russell et al., 2019c). Evidence suggests a preceding mentally fatiguing task results in a detrimental impact on endurance-based exercise performance (Van Cutsem et al., 2017c). However, evidence relating to the impact on all-out, or explosive type activities is less convincing (Van Cutsem et al., 2017c). An example of the detrimental impact of MF on a physical task regularly used in the elite sporting environment is a 16.3 ± 5.1% decrease in distance covered on the Yo-Yo Intermittent Recovery Test Level 1 (Yo-Yo IR1) found with recreational soccer players following a 30-min modified Stroop task (Smith et al., 2016). A significantly lower Yo-Yo IR1 completion distance was also found with mentally fatigued elite cricketers, together with a significant increase in completion times for the run-two test, which tests acceleration, running at maximal speed, decelerating, and completing a 180° turn (Veness, Patterson, Jeffries, & Waldron, 2017). Again, a 30-min Stroop was used as the mentally fatiguing stimuli. Technical performance has also been shown to be negatively influenced by preceding cognitive load; increased passing and ball control errors and decreased accuracy have been reported and attributed to the interaction of cognitive fatigue and task time-constraints (Smith, Fransen, Deprez, Lenoir, & Coutts, 2017). Further, an increased number of ball control errors, and a lower percentage of technical involvements which resulted in either a positive outcome, possessions, accurate passes, and/or successful tackles, was demonstrated in small-sided soccer games, where prior inducement of cognitive fatigue was applied (Badin, Smith, Conte, & Coutts, 2016). Tactically, there is evidence to support the impact of cognitive fatigue on how individuals interact with environmental information. Similarly, tactical behaviour has been shown to be negatively influenced by a state of cognitive fatigue. For example, using positional data Coutinho et al. (2017) reported that with MF the synchronisation of movement in the lateral direction between players was impaired and speed of contraction (distance between players when ball won or lost) was slower, comparative to a control condition. Psychological constructs such as ego-depletion, response-inhibition, and motivation have been discussed in relation to cognitive fatigue (Aitken & MacMahon, 2019; Van Cutsem et al., 2017c). Given that cognitive fatigue is a psychobiological state (Desmond & Hancock, 2001; Job & Dalziel, 2001), psychological factors are clearly involved in the processes influencing changes in behaviour or performance as a result of cognitive fatigue. The increased rating of perceived exertion (RPE) seen when completing an endurance exercise under MF (Van Cutsem et al., 2017c) evidences this influence of perceptual or psychological factors. The processes and theories by which cognitive load may be induced by sporting activities and experienced in the applied sporting environment will be discussed below. Figure 11.1 provides a visual representation of perceived factors and inducers of cognitive fatigue as reported by elite athletes, coaching, and support staff. This figure was devised utilising a qualitative research approach of focus group discussions lasting 27.38 ± 7.82 minutes. Participants included 15 coaches, sports scientists, or medical personnel (>five years of elite experience), and 17 elite athletes (>two years of elite experience). Athletes and coaches were involved in sports, including athletics, Australia Rules Football, cricket, netball, rugby, basketball, hockey, and soccer. Participants were asked guiding questions on their perceptions to inform the descriptors and associated symptoms and the potential impact and causes of MF in the applied sporting environment.

FIGURE 11.1   Word cloud presenting the nature of perceived factors and inducers of cognitive fatigue as reported by elite athletes, and coaching and support staff.

11.2.2   Typical Cognitive Loading Tasks

As discussed, cognitive load has the potential to influence an individual's ability to successfully perform both cognitive and physical tasks. Traditionally, cognitive load is experienced when an individual undertakes a task requiring mental activity. Numerous examples of such tasks exist in the cognitive and physical performance literature (Van Cutsem et al., 2017c). These tasks require the allocation of cognitive resources, and consequently, when undertaken for a prolonged duration, cognitive fatigue can result. Tasks that have been used preceding a physical performance task have included; the Stroop task (Rozand, Pageaux, Marcora, Papaxanthis, & Lepers, 2014; Schücker & MacMahon, 2016; Smith et al., 2016), AX-Continuous Performance task (AX-CPT) (Marcora, Staiano, & Manning, 2009; Martin, Thompson, Keegan, Ball, & Rattray, 2015; Smith, Marcora, & Coutts, 2015), 0-back and 2-back test (Tanaka et al., 2012), Continuous cognitive activity task (Brownsberger, Edwards, Crowther, & Cottrell, 2013), Switch task paradigm (Budini, Lowery, Durbaba, & De Vito, 2014; Lorist et al., 2000), Concentration grids (Duncan, Fowler, George, Joyce, & Hankey, 2015; Greenlees, Thelwell, & Holder, 2006), Psychomotor vigilance task (Xiao et al., 2015), Go/No-Go task (Guo, Ren, Wang, & Zhu, 2015; Kato, Endo, & Kizuka, 2009), Oddball task (Zhao, Zhao, Liu, & Zheng, 2012), four-choice reaction time task (Xiao et al., 2015), The 'National Aeronautics and Space Administration Task Load Index' (Head et al., 2016), and Flanker task (Van Cutsem et al., 2017a). The categorisation of these cognitive loading protocols as either a task used to induce cognitive fatigue, or a test to assess cognitive performance varies across the literature. Adding to the complexity, response time, and accuracy (i.e. percentage of correct or incorrect answers) during cognitively demanding processes are often reported as objective markers indicating the level of load or fatigue experienced by the individual. Though the effort to complete each task varies, all require rapid information processing from visual, spatial, or audio information – eliciting stress on memory capacity. A number of key behaviours required for these tasks translate across to demands in the competitive sporting environment. The response-inhibition or counter-intuitive components may be required be inhibit a response to a dummy, fake, or feint in invasion type sports, and select the appropriate response to target or non-target stimuli. Similarly, information recall may be required to bring to mind the favoured action of an opponent or previously discussed game or pacing tactics. The component of sustained attention for vigilance and concentration is needed to avoid the detrimental effects of 'lapses in concentration' as commonly discussed by commentators. Athletes are also required to cope with random inter-stimuli intervals, information switching demands, resistance to distraction, and demand anticipation capacity. These examples demonstrate how a number of the components of typical cognitive loading tasks resemble the demands that contribute to cognitive load in sport.

11.2.3 INFORMATION PROCESSING AND WORKING MEMORY

The acquisition, recording, organisation, retrieval, display, and dissemination of information is generally referred to as information processing. In other words, the process of taking in to appropriately responding to information. The previously discussed cognitive tasks used to induce cognitive fatigue require significant information processing (Brownsberger et al., 2013). The closely linked central cognitive mechanistic concepts of long-term and working memory are also related to information processing (Cowan, 2008) which sports performance requires as one of many higher-order cognitive capacities. Whilst it is acknowledged that many basic and commonly executed tasks in sport may be completed autonomically, the unpredictable and complex scenarios presented require instantaneous processing of information that places a high demand on working memory capacity. For example, whilst the simple execution of a pass between two players may be autonomous for an experienced basketballer, the attentional and cognitive demands of accounting for a multitude of situational factors, including opponent positioning, time restrictions, and score differential, demand working memory. Similarly, performing a technically correct freestyle stroke cycle may be an autonomous task for an elite swimmer. However, processing physiological feedback to internally gauge pace, the positioning or apparent pacing strategy of a competitor as the race unfolds (McGibbon, Pyne, Shephard, & Thompson, 2018), and coordinating the body to execute the rotational stroke with optimal power and timing may require higher levels of working memory and subsequent cognitive load. In addition to the cognitive load required for internal information acquisition and organisation, athletes listen to, comprehend, respond to, and immediately action verbal feedback. Both coaches and team-mates offer sources of feedback which may be in the form of a simple instruction such as calling a team-mate left or right to intercept a ball or block opposition player movement. More sophisticated responses to verbal demands may also be needed; a captain may call a number which corresponds to a strategic play to be implemented. At an even higher level of complexity a coach may direct a series of instructions at an athlete, either during competition or perhaps a short time-out period. Whilst coaches are trained in the art of communicating to athletes, cognitive demand may not always be a consideration when delivering tactics or instruction, and emotion plays a significant role in the communication process. Further, not all coaches have awareness of an appropriate level of information and the optimal delivery method for each of their individual athletes (Millar, Oldham, & Donovan, 2011).

11.2.4 Response-Inhibition

Athletes are often required to regulate their emotions during competition (Lane, Beedie, Jones, Uphill, & Devonport, 2012). For example, managing decisions made by officials that they disagree with. Such situations contribute to cognitive load. Similarly, when a team-mate makes a mistake that negatively impacts the team, or an opponent employs a tactic deemed as unsporting or inappropriate, this requires response-inhibition and adds to the cognitive load of the athlete. In the present age of extensive telecommunications and social media, athletes may need to moderate their instinctive responses and behaviour in a specific manner to attract sponsorship prospects and ultimately benefit the athlete financially. Social media is known as a powerful tool for fan–athlete interaction and subsequent financial and fringe benefits (Filo, Lock, & Karg, 2015). Yet, athletes must represent themselves in a positive light to meet contractual requirements and maintain a healthy public perception (Geurin-Eagleman & Burch, 2016). The consequences of negative social media interactions have been made apparent in traditional media following scenarios in sports including, tennis, rugby union, and national rugby league. Many team codes of conduct and sponsorship deals have clauses relating the appropriate use of social media. Athletes have lost both playing and sponsorship contracts, e.g. Israel Folau and the Australian Rugby Union in 2019, following subsequent contractual disputes. The sharing of inappropriate media content may more likely be made following periods of high cognitive demand. Furthermore, even when posting positive content, the emotional processing and response-inhibition demands may be high, with athletes opening themselves up to negative responses or potential cyberbullying when sharing content. Examples reported in the media include Arsenal footballer Granit Xhaka, British speed skater Elise Christie, and tennis player Rebecca Marino. Research has also evidenced that 30-mins of smartphone application use on social networking apps (WhatsApp, Facebook and Instagram) can impair passing decision-making performance in male soccer athletes (Fortes et al., 2019). Accordingly, response-inhibition demand appears as a concurrent consequence of social media use, a medium that is commonplace in elite sport. Dealing with crowd behaviour and feedback from coaches also adds to the cognitive load experienced by athletes. Real-life sporting examples exist where athletes have responded inappropriately with outburst to spectator behaviour including Eric Cantona in 1995, Sven Nys in 2012, and Nick Kyrgios in 2019. Elite staff and athletes have linked MF to a diminished ability to control emotional responses and decreased impulse control (Russell et al. 2019b). Self-regulation refers to the act of exerting control over one's behaviour and has been described as the capacity to override impulses and habitual responses, including controlling thoughts, emotions, desires, and behaviours and is noted to be a seemingly limited capacity (Martin, Thompson, Keegan, & Rattray, 2019). The cognitive demands required during other aspects of sport may increase one's MF and reduce the subsequent capacity to self-regulate in response to crowds during competition.

Selecting and training athletes to tolerate and cope with MF may improve an athlete's ability to remain on task and respond appropriately (i.e. not be influenced by any distractions) to the crowd, subsequently improving athlete image, and potentially performance outcome. Whilst constructive feedback is critical for improvement, coaches need to balance negative and positive feedback to athletes appropriately. The most successful coaches read situations carefully to manage cognitive demand. Indeed, athletes suggest that a lack of staff cohesion contributes to cognitive fatigue (Russell et al., 2019b). Conflicting messages or tension between staff may again require the cognitively demanding process of response-inhibition, or at least, an adjustment of response, likely combined with internal frustration. Media responsibilities are another example which involves response-inhibition and cognitive load increases. Athletes are often questioned on topical and provoking questions when conducting media interviews. Further, information in the professional sporting environment is sensitive to maintain a competitive advantage; unwittingly disclosing this during media interviews would be inappropriate. Consequently, it is not uncommon for an athlete to be required to inhibit their natural response or personal opinion to provide a media-appropriate statement in response to provoking questions. Such interactions may be likely to occur immediately prior competition and thus cognitively load an athlete just prior to their performance. Similarly, post-match media will test an athlete's resistance to cognitive fatigue to employ such appropriate responses following the cognitive load experienced during competition.

11.2.5 INFORMATION RECALL

Information recall is another cognitively demanding task; the recalling of instructions to appropriately apply. For example, the modified incongruent Stroop task requires an individual to choose the colour in which a word is presented in; the word may, for example, be 'blue' but be written in the colour green. The individual has to continually remember to ignore the words and to choose the colours shown (Schücker & MacMahon, 2016). Further, tasks such as the '2-back test' require participants to recall the target letter presented two displays prior to a subsequent display (Tanaka et al., 2012). Again, this concept of information recall is not unfamiliar in sporting competition. Athletes are required to recall a variety of information relevant to their performance. For some sports, such as soccer, tennis, or swimming, this may be an opponent's typical pattern of play or race plan, passing preference, or technical weakness. For individual endurance performance, this could include recalling aspects of a previously analysed competition track to help guide appropriate pacing or intake of fluid and nutritional aids. Recalling the impact of weather conditions on a ball flight or path may also be required by a kicker or golfer to anticipate the appropriate adjustments, just prior to executing a set shot or penalty kick. Information recall is also required in the training setting. An example is an athlete who is completing a strength session not under direct supervision of their physical performance coach. In the absence of their coach to optimise the technical execution and maximise the resulting physical benefits the recall of previously provided information may be required, commonly assisted by utilisation of cues. Whilst the physical demands of such sessions are easily quantified and monitored, the cognitive load induced by the information recall to successfully execute compound lifts or perform exercises with optimal technique is seldom considered. Nonetheless, it is evident that information recall is required for the effective execution of technical and tactical components, two factors previously discussed to hold critical impact on the outcome of competition.

11.2.6 SUSTAINED ATTENTION

Sustained attention is the ability to focus on a task or stimulus for an extended period of time. This is associated with a number of the aforementioned cognitive loading tasks, some lasting 30 minutes or longer. To optimally perform in training and competition, athletes are required to sustain their attention, often on a variety of stimuli relevant to the one task. Vigilance is required to detect stimuli appearance and concentration to continually focus on such stimuli. In the sporting context, a split second of attentional lapse may result in an opponent overtaking and gaining a more optimal racing position, a dropped ball, or loss of a wicket; ultimately determining competition outcome. Motorsport is another sport-specific example where sustained attention and vigilance is crucial (Brown, Stanton, & Revell, 2019). Racing drivers experience arguably the most demanding of cognitive challenges amongst athletes with the consequences of loss of sustained attention extending beyond impaired performance to threatening preservation of life. Whilst outside of the competition environment, in the training setting, the consequences of sustained attention demands may be less severe, the nature of activity makes the task equally, and on occasions potentially more, cognitively demanding. The repetitive nature and required attention to continuously execute the same training drills has been reported by athletes to result in cognitive fatigue (Russell et al., 2019b), suggesting a high cognitive load of such activities. With the numerous cognitively demanding aspects of sport, it is challenging for athletes to sustain their attention, yet the implications of a lapse of concentration are significant. Multiple distractions may further challenge the ability of an athlete to sustain attention on the specific task at hand. Competing thoughts such as contract negotiations and sponsorship requirements, have previously been identified by athletes to be a factor contributing to cognitive fatigue (Russell, Jenkins, Rynne, Halson, & Kelly, 2019a). A multitude of further distractions may also be present. A deliberate distraction, such as a 'dummy' lead in a team sport, fake pass, or opposition shouting out deliberate irrelevant information as a distraction in an individual race pursuit scenario may also be present in competition. In such situations, athletes are required to sustain their attention on the task at hand and avoid the temptation to respond to the deliberately competing distracting stimuli. As mentioned earlier, spectators may provide a deliberate or non-purposeful distraction. The entertainment value of sport with theatrical components such as lighting, music, and commentary can provide further stimuli which an athlete has to adequately manage, in order to sustain their attention on the appropriate task.

11.2.7 SUMMARY

The examples discussed above of sport-specific cognitive load differ in nature from those of typically cognitively demanding task, such as the Stroop or AX-CPT. However, regardless of the superficial differences, striking similarities exist with respect to the nature and causes of such cognitive demand. Aspects of information processing, limitation of working memory capacity, response-inhibition, information recall, and sustained attention, are similar. Reflecting on the mechanistic equivalences, it is reasonable to conclude that the resulting physiological responses and subsequent behavioural and subjective consequences following such demands are also similar.

11.3 PRACTICAL APPLICATIONS

In the sections above, situations that increase cognitive load and can lead to cognitive fatigue in athletes were provided. Evidence was also considered that demonstrates the potential for cognitive fatigue, or a tired brain to impair physical, technical, tactical, and psychological components of sporting performance (Russell et al., 2019c). In light of this, the question emerges of how best we can identify activities demanding cognitive load and manage the cognitive load of individual athletes. Improved understanding of how an individual interreacts with tasks in their sporting environment, specifically in regard to cognitive load, may lead to improved performance. Harnessing the concept of cognitive load as a human factor may in turn improve the likelihood of athletes achieving the ultimate goal in competitive sport; to win.

11.3.1 Assessment of Cognitive Load

Given the limited available evidence, it is clear that assessing and quantifying cognitive load in athletes during training and competition has the potential to improve outcomes for both athletes and teams. However, the complexity of cognitive load makes this more challenging than assessing and quantifying physical load (Halson, 2014). First and foremost, practitioners must remember that cognitive load is defined in terms of an interaction between a task and the individual performing the task. Accordingly, an individual's ability to self-regulate may indeed be a moderating or protective factor for resilience to cognitive load (Martin et al., 2019). Previous and current experiences, personality, and genetic factors will all influence how an individual experiences cognitive load as a result of a specific task (Russell et al., 2019b). As such, anticipating and indeed assessing cognitive load is not straight forward. Nevertheless, awareness and adaptation for heterogeneity in response to cognitive load is potentially beneficial (Martin et al., 2019). Research has investigated an array of physiological measures to assess their sensitivity to changes in cognitive load or effort. Electroencephalogram (EEG), in particular beta-band activity, has previously been proposed as an indicator of cognitive fatigue in response to cognitively demanding tasks (Balasubramanian, Adalarasu, & Gupta, 2011; Barwick, Arnett, & Slobounov, 2012). In the sporting environment, however, the setup time and resource requirements, cost, and potential signal noise make even mobile EEG currently impractical for use with elite athletes (Russell et al., 2019c). A range of other physiological indicators include pupil dilation and blink rate (Marquart, Cabrall, & de Winter, 2015), accelerated plethysmography (Mizuno et al., 2011), and functional near-infrared spectroscopy (McKendrick, Parasuraman, & Ayaz, 2015); most show some promise as potential assessment methods but further research is required. Identification of a practical yet robust physiological marker reflecting changes in cognitive load suitable for use with athletes in the applied setting is still being sought (Russell et al., 2019c; Smith et al., 2018). Accounting for the interactive effect between task and individual, the subjective measure of a 100-mm visual analogue scale of cognitive effort or load is currently used by some researchers in applied sports settings to assess, for example, mental (cognitive) fatigue (Russell et al., 2019c). It is essential that this scale is appropriately anchored, e.g. none at all (0) to maximal (100), and it is clearly explained to the participant surrounding what is being subjectively questioned (Smith et al., 2018). Routine use of a subjective scale used in context with physical load can indicate an athlete's experience of cognitive load which in turn can be used to differentiate the origin of fatigue as physical or mental (Russell, Jenkins, Halson, & Kelly, 2020). Such information can guide best-practice management of the athlete, i.e. preventing scenarios such as over-training – in both the physical and cognitive domains.

11.3.2   MANAGEMENT OF COGNITIVE LOAD

As mentioned in an earlier section, it should be remembered that the cognitive load experienced by an athlete may originate from both within and/or outside the sporting environment. Commitments, including study, emotionally challenging work, and management of personal relationships, can contribute to cognitive fatigue (Russell et al., 2019a). Assessing cognitive load using, for example, a visual analogue scale enables coaching and support staff to appropriately manage or adapt the load experienced by an athlete at a specific time-point. This, much like physical load and recovery, will depend on the intended outcome, i.e. adaptation to, or reduction of, fatigue.

A number of benefits may result by reducing cognitive load, or by providing additional support during periods of high cognitive load and subsequent cognitive fatigue. Whilst also directly improving performance, there will also be indirect benefits. These include, but are not limited to, reduced levels of distractibility and disengagement, being less prone to error, increased time on task, and greater attention to detail (Russell et al., 2019a, 2019c). In essence, management of cognitive load may benefit day-to-day so-called 'one-percenters' believed to influence performance. Furthermore, monitoring of cognitive load and documenting such information to coaches, performance analysts, strength and conditioning coaches, etc. can help practitioners make appropriate adjustments. Such examples may be a coach with improved awareness of the limited capacity of a specific athlete to take on new information (working memory capacity), or process negative feedback (response-inhibition) and fine-tuning communication accordingly.

Some may argue such adaptations to practice may seem counter-intuitive by reducing load or omitting communication of important information. However, principles of optimising dose or stimulus to elicit a desired response are a common practice and can be transferred to cognitive, not just physical, load. As such, a practitioner can structure and deliver training sessions or key performance messages at a level most likely to result in a benefit. Of course, this process is not simple and to optimise this, coaches and support staff must be educated on cognitive load and the abundance of related research such as learning style, information processing, etc.

A very simple example, adapted from other human factor recommendations is the implementation of a pre-game checklist (Hales & Pronovost, 2006). Such checklists, as utilised in aviation, aeronautic, product manufacturing, and health care, provide a list of action items or criteria, presented in a systemic manner (Hales & Pronovost, 2006). The result is that an athlete limits their cognitive load to only the most important commitments relating to their performance. Clear communication of roles, responsibilities, and planned action for each individual in scenarios which may unfold, may also reduce the likelihood of divided attention and improve efficiency. The brain and cognition in sport is important and should be considered as another component that can aid in the search for a competitive edge.

11.3.3 Environment and Nutrition

With regards to the daily training environment, environmental and nutritional practices will potentially influence cognitive load. A related and noteworthy example for applied practitioners is heat stress. A negative impact of heat stress on cognition, including deterioration in information processing and working memory, has been demonstrated. For example, increased brain activity whilst compeleting the AX-CPT with elevated core temperature has been reported (Hocking, Lau, Silberstein, Roberts, & Stough, 2000). Accordingly, practitioners should consider the impact of training with increased thermal strain on cognitive load; through either intentional acclimation training, i.e. heat chamber, or that experienced outdoors in a natural field setting through ambient environmental temperature. Evidence also supports the timely and appropriate intake of glucose, caffeine, and creatine to manage or acutely mitigate cognitive fatigue as a result of cognitive load (Van Cutsem et al., 2017b; Van Cutsem et al., 2019). A caffeine-maltodextrin mouth rinse (0.3 g/25 ml caffeine, 1.6 g/25 ml maltodextrin) has been found to acutely reduce self-reported mental fatigue and improve Stroop task accuracy, comparative to a placebo (Van Cutsem et al., 2017b). Seven-days of creatine supplementation, with 20 g of creatine administered per day, demonstrated improved strength endurance and prolonged Stroop task accuracy (Van Cutsem et al., 2019). Accordingly, such strategies are likely to acutely support the achievement of cognitive performance, perhaps most effectively when an individual is particularly challenged or fatigued. The role of glucose as a cerebral fuel and hence adequate intake of carbohydrate cannot be ignored as a factor influencing cognitive exertion in response to load. Combined caffeine and carbohydrate ingestion have been shown to reduce subjective MF during prolonged mental exertion (Kennedy & Scholey, 2004). For caffeine and creatine intake, optimal dosages will depend on the task at hand and the individual athlete. Previous evidence has found caffeine (5 mg/kg) intake to benefit endurance performance in those previously exposed to 90-minutes of the AX-CPT, a typically cognitively demanding task (Azevedo, Silva-Cavalcante, Gualano, Lima-Silva, & Bertuzzi, 2016).

11.3.4 Recovery From Cognitive Load

In circumstances where high cognitive load cannot be avoided, yet a reduction in cognitive fatigue is sought (e.g. pre or between-competition), strategies to accelerate 'cognitive recovery' can be attempted (Loch, Ferrauti, Meyer, Pfeiffer, & Kellmann, 2019). Limited research has been undertaken to examine cognitive recovery following sporting activity; there are generally perceptual benefits of those same methods used by athletes to improve their physical recovery and well-being (Kellmann, 2002). For example, many elite athletes currently engage in biofeedback breathing and mindfulness to reduce stress and anxiety; these practices may also benefit cognitive function, particularly if used regularly during a competitive season. Indeed, resonant frequency biofeedback, a heart-rate variability training protocol to promote autonomous nervous system balance through slow breathing rate, has been shown to improve some cognitive functions (Sutarto, Wahab, & Zin, 2010). Whilst investigated in the industrial operational environment, where human factors are better understood, the concept is transferable to the applied sporting environment. More needs to be done to inform evidence-based guidelines to derive the benefits of such recovery methods on cognition in sport. Further, as perceptions from elite athletes inform us, helping athletes to have periods where they can avoid thinking about their sport (Russell et al., 2019b) is likely to reduce chronic cognitive fatigue. Of course, this depends on the athlete not replacing sports or training commitments with other cognitively demanding activities, and taking the allocated time removed from the setting to cognitively rest their brain.

11.3.5 Cognitive Training

Aligning with the principle that physical training improves an individual's ability to tolerate physical load, specific cognition training has the potential to help an athlete better manage cognitive load and thus delay the development of cognitive fatigue during competition. As such, in periods where adaptation is sought (e.g. the pre-season phase), cognitive loading may be used to an athlete's advantage, by harnessing the principles of brain endurance training. Research has demonstrated brain endurance training (BET) combined with traditional physical cycling training to be highly effective in reducing perception of effort whilst improving endurance performance (Staiano, Merlini, & Marcora, 2015). The BET protocol used by Staiano et al. (2015) involved three sessions per week for 12 weeks; in each session participants cycled for 60 minutes at 65% of VO_2 max whilst completing the AX-CPT (Staiano et al., 2015). Future research needs to provide proof-of-concept specifically in elite athletic populations, who may display superior inhibitor control and resistance to cognitive fatigue (Martin et al., 2016), potentially due to their regular exposure to cognitive load. However, performing BET under the additional stress of a physical stimuli can more closely mimic the combination of cognitive and physical load experienced by athletes during competition. In turn this may help to develop a resilience to cognitive fatigue that will benefit competition performance. Considering that there is often a limited capacity to further increase physical load with elite athletes, enhancing cognitive capacity and tolerance to cognitive load is an underutilised area where significant gains in performance may be possible.

11.3.6 Practical Applications Summary

In summary, a number of practical changes can be made to improve the management of cognitive load in athletes. Use of valid methods to assess cognitive load

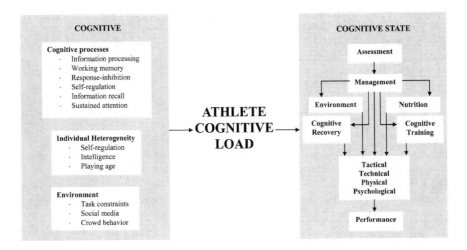

FIGURE 11.2 Summary diagram of cognitive demand and state in athletes.

to improve awareness and inform appropriate training and competition practices for athletes should be implemented. Consideration of the impact of heat stress on cognition and strategic use of nutritional and ergogenic aids should be employed to manage cognitive fatigue. Lastly, the employment of recovery methods which target cognition, and the carefully periodised use of BET to build resilience to cognitive fatigue may all be of benefit.

11.4 CONCLUSION

Cognitive load is a complex yet important factor that, from the available evidence, clearly influences an athlete's performance, as summarised in Figure 11.2. Theoretical concepts of information processing, working memory, response-inhibition, self-regulation, information recall, and sustained attention are all relevant to sport-specific tasks. Better understanding how to monitor and limit cognitive load in athletes has the potential significantly benefit athletic performance.

REFERENCES

Aitken, B., & MacMahon, C. (2019). Shared demands between cognitive and physical tasks may drive negative effects of fatigue: A focused review. *Frontiers in Sports and Act Living*, 1, 45. doi: 10.3389/fspor.

Azevedo, R., Silva-Cavalcante, M. D., Gualano, B., Lima-Silva, A. E., & Bertuzzi, R. (2016). Effects of caffeine ingestion on endurance performance in mentally fatigued individuals. *European Journal of Applied Physiology*, 116:11–12, 2293–2303.

Badin, O., Smith, M., Conte, D., & Coutts, A. (2016). Mental fatigue: impairment of technical performance in small-sided soccer games. *The International Journal of Sports Physiology and Performance*, 11:8, 1100–1105.

Balasubramanian, V., Adalarasu, K., & Gupta, A. (2011). EEG based analysis of cognitive fatigue during simulated driving. *International Journal of Industrial and Systems Engineering*, 7:2, 135–149.

Barwick, F., Arnett, P., & Slobounov, S. (2012). EEG correlates of fatigue during administration of a neuropsychological test battery. *Clinical Neurophysiology*, 123:2, 278–284. doi: 10.1016/j.clinph.2011.06.027.

Brown, J., Stanton, N., & Revell, K. (2019). *A Review of the Physical, Psychological and Psychophysiological Effects of Motorsport on Drivers and Their Potential Influences on Cockpit Interface Design*. Cham.

Brownsberger, J., Edwards, A., Crowther, R., & Cottrell, D. (2013). Impact of mental fatigue on self-paced exercise. *International Journal of Sports Medicine*, 34:12, 1029–1036. doi: 10.1055/s-0033-1343402.

Brümmer, V., Schneider, S., Abel, T., Vogt, T., & Strueder, H. K. (2011). Brain cortical activity is influenced by exercise mode and intensity. *Medicine and Science in Sports and Exercise*, 43:10, 1863–1872.

Budini, F., Lowery, M., Durbaba, R., & De Vito, G. (2014). Effect of mental fatigue on induced tremor in human knee extensors. *Journal of Electromyography and Kinesiology*, 24:3, 412–418. doi: 10.1016/j.jelekin.2014.02.003.

Coutinho, D., Gonçalves, B., Travassos, B., Wong, D. P., Coutts, A. J., & Sampaio, J. E. (2017). Mental fatigue and spatial references impair soccer players' physical and tactical performances. *Frontiers in Psychology*, 8, 1645.

Coutts, A. J., Rampinini, E., Marcora, S. M., Castagna, C., & Impellizzeri, F. M. (2009). Heart rate and blood lactate correlates of perceived exertion during small-sided soccer games. *Journal of Science and Medicine in Sport*, 12:1, 79–84.

Cowan, N. (2008). Chapter 20 What are the differences between long-term, short-term, and working memory? In: W. S. Sossin, J.-C. Lacaille, V. F. Castellucci, & S. Belleville (eds.), *Progress in Brain Research* (Vol. 169, pp. 323–338). Elsevier.

Desmond, P., & Hancock, P. (2001). Active and passive fatigue states. In: P. Hancock (ed.), *Stress, Workload, and Fatigue* (pp. 455–465). Mahwah, NJ: Lawrence Erlbaum Associates Publishers.

Duncan, M. J., Fowler, N., George, O., Joyce, S., & Hankey, J. (2015). Mental fatigue negatively influences manual dexterity and anticipation timing but not repeated high-intensity exercise performance in trained adults. *Research in Sports Medicine*, 23:1, 1–13.

Durantin, G., Gagnon, J.-F., Tremblay, S., & Dehais, F. (2014). Using near infrared spectroscopy and heart rate variability to detect mental overload. *Behavioural Brain Research*, 259, 16–23.

Eston, R. (2012). Use of ratings of perceived exertion in sports. *International Journal of Sports Physiology and Performance*, 7:2, 175. doi: 10.1123/ijspp.7.2.175.

Filo, K., Lock, D., & Karg, A. (2015). Sport and social media research: A review. *Sport Management Review*, 18:2, 166–181. doi: 10.1016/j.smr.2014.11.001.

Fortes, L. S., Lima-Junior, D., Nascimento-Júnior, J. R., Costa, E. C., Matta, M. O., & Ferreira, M. E. (2019). Effect of exposure time to smartphone apps on passing decision-making in male soccer athletes. *Psychology of Sport and Exercise*, 44, 35–41.

Geurin-Eagleman, A. N., & Burch, L. M. (2016). Communicating via photographs: A gendered analysis of Olympic athletes' visual self-presentation on Instagram. *Sport Management Review*, 19:2, 133–145. doi: 10.1016/j.smr.2015.03.002.

Greenlees, I., Thelwell, R., & Holder, T. (2006). Examining the efficacy of the concentration grid exercise as a concentration enhancement exercise. *Psychology of Sport and Exercise*, 7:1, 29–39. doi: 10.1016/j.psychsport.2005.02.001.

Guo, W., Ren, J., Wang, B., & Zhu, Q. (2015). Effects of relaxing music on mental fatigue induced by a continuous performance task: Behavioral and ERPs evidence. *PLoS One*, 10:8, e0136446.

Hales, B. M., & Pronovost, P. J. (2006). The checklist – a tool for error management and performance improvement. *Journal of Critical Care*, 21:3, 231–235. doi: 10.1016/j.jcrc.2006.06.002.

Halson, S. L. (2014). Monitoring training load to understand fatigue in athletes. *Sports Medicine*, *44*(2) Supplement 2, 139–147. doi: 10.1007/s40279-014-0253-z.

Hammermeister, J., & VonGuenthner, S. (2005). Sport psychology: Training the mind for competition. *Current Sports Medicine Reports*, 4:3, 160–164.

Hancock, P. A., & Chignell, M. H. (1988). Mental workload dynamics in adaptive interface design. *IEEE transactions on Systems, Man, and Cybernetics*, 18:4, 647–658.

Head, J., Tenan, M., Tweedell, A., Price, T., LaFiandra, M., & Helton, W. (2016). Cognitive fatigue influences time-on-task during bodyweight resistance training exercise. *Frontiers in Physiology*, 7:373. doi: 10.3389/fphys.2016.00373.

Hocking, C., Lau, W. M., Silberstein, R., Roberts, W., & Stough, C. (2000). *The Effects of Thermal Strain on Cognition*. Melbourne, Australia: Defence Science and Technology Organisation Melbourne.

Hulme, A., Thompson, J., Plant, K. L., Read, G. J. M., McLean, S., Clacy, A., & Salmon, P. M. (2019). Applying systems ergonomics methods in sport: A systematic review. *Applied Ergonomics*, 80, 214–225. doi: 10.1016/j.apergo.2018.03.019.

Job, R., & Dalziel, J. (2001). Defining fatigue as a condition of the organism and distinguishing it from habituation, adaptation, and boredom. In: P. A. Hancock (ed.), *Stress, Workload, and Fatigue*. Mahwah, NJ: Lawrence Erlbaum Associates Publishers.

Joyce, D., & Lewindon, D. (2014). *High-Performance Training for Sports*. Champaign, IL: Human Kinetics.

Kato, Y., Endo, H., & Kizuka, T. (2009). Mental fatigue and impaired response processes: event-related brain potentials in a Go/NoGo task. *International Journal of Psychophysiology*, 72:2, 204–211.

Kellmann, M. (2002). *Enhancing Recovery: Preventing Underperformance in Athletes*. Champaign, IL: Human Kinetics.

Kelly, V. G., & Coutts, A. J. (2007). Planning and monitoring training loads during the competition phase in team sports. *Strength and Conditioning Journal*, 29:4, 32.

Kennedy, D. O., & Scholey, A. B. (2004). A glucose-caffeine 'energy drink'ameliorates subjective and performance deficits during prolonged cognitive demand. *Appetite*, 42:3, 331–333.

Klaassen, E. B., Plukaard, S., Evers, E. A. T., de Groot, R. H. M., Backes, W. H., Veltman, D. J., & Jolles, J. (2016). Young and middle-aged schoolteachers differ in the neural correlates of memory encoding and cognitive fatigue: A functional MRI study. *Frontiers in Human Neuroscience*, 10:148. doi: 10.3389/fnhum.2016.00148.

Konings, M. J., & Hettinga, F. J. (2018). Pacing decision making in sport and the effects of interpersonal competition: A critical review. *Sports Medicine*, 48:8, 1829–1843. doi: 10.1007/s40279-018-0937-x.

Lal, S. K. L., & Craig, A. (2001). A critical review of the psychophysiology of driver fatigue. *Biological Psychology*, 55:3, 173–194. doi: 10.1016/S0301-0511(00)00085-5.

Lane, A. M., Beedie, C. J., Jones, M. V., Uphill, M., & Devonport, T. J. (2012). The BASES expert statement on emotion regulation in sport. *Journal of Sports Sciences*, 30:11, 1189–1195. doi: 10.1080/02640414.2012.693621.

Loch, F., Ferrauti, A., Meyer, T., Pfeiffer, M., & Kellmann, M. (2019). Resting the mind–A novel topic with scarce insights. Considering potential mental recovery strategies for short rest periods in sports. *Performance Enhancement and Health*.

Lorist, M. M., Klein, M., Nieuwenhuis, S., De Jong, R., Mulder, G., & Meijman, T. F. (2000). Mental fatigue and task control: Planning and preparation. *Psychophysiology*, 37:5, 614–625.

Marcora, S. M., Staiano, W., & Manning, V. (2009). Mental fatigue impairs physical performance in humans. *Journal of Applied Physiology*, 106:3, 857–864.

Marquart, G., Cabrall, C., & de Winter, J. (2015). Review of eye-related measures of drivers' mental workload. *Procedia Manufacturing*, 3, 2854–2861. doi: 10.1016/j.promfg.2015.07.783.

Martin, K., Staiano, W., Menaspà, P., Hennessey, T., Marcora, S., Keegan, R., Thompson, K. G., Martin, D., Halson, S., & Rattray, B. (2016). Superior inhibitory control and resistance to mental fatigue in professional road cyclists. *PLoS One*, 11:7, e0159907.

Martin, K., Thompson, K. G., Keegan, R., Ball, N., & Rattray, B. (2015). Mental fatigue does not affect maximal anaerobic exercise performance. *European Journal of Applied Physiology*, 115:4, 715–725.

Martin, K., Thompson, K. G., Keegan, R., & Rattray, B. (2019). Are individuals who engage in more frequent self-regulation less susceptible to mental fatigue? *Journal of Sport and Exercise Psychology*, 1–9. doi: 10.1123/jsep.2018-0222.

McGibbon, K. E., Pyne, D. B., Shephard, M. E., & Thompson, K. G. (2018). Pacing in swimming: A systematic review. *Sports Medicine*, 48:7, 1621–1633. doi: 10.1007/s40279-018-0901-9.

McKendrick, R., Parasuraman, R., & Ayaz, H. (2015). Wearable functional near infrared spectroscopy (fNIRS) and transcranial direct current stimulation (tDCS): Expanding vistas for neurocognitive augmentation. *Frontiers in Systems Neuroscience*, 9:27. doi: 10.3389/fnsys.2015.00027.

Meeusen, R., & Roelands, B. (2017). Fatigue: Is it all neurochemistry? *European Journal of Sport Science*, 1–10.

Meyers, A. W., Whelan, J. P., & Murphy, S. M. (1996). Cognitive behavioral strategies in athletic performance enhancement. *Progress in Behavior Modification*, 30, 137–164.

Millar, S.-K., Oldham, A. R. H., & Donovan, M. (2011). Coaches' self-awareness of timing, nature and intent of verbal instructions to athletes. *International Journal of Sports Science and Coaching*, 6:4, 503–513. doi: 10.1260/1747-9541.6.4.503.

Mizuno, K., Tanaka, M., Yamaguti, K., Kajimoto, O., Kuratsune, H., & Watanabe, Y. (2011). Mental fatigue caused by prolonged cognitive load associated with sympathetic hyperactivity. *Behavioral and Brain Functions*, 7:1, 17.

Nielsen, J. B., & Cohen, L. G. (2008). The Olympic brain. Does corticospinal plasticity play a role in acquisition of skills required for high-performance sports? *Journal of Physiology*, 586:1, 65–70. doi: 10.1113/jphysiol.2007.142661.

Oppici, L., Panchuk, D., Serpiello, F. R., & Farrow, D. (2017). Long-term practice with domain-specific task constraints influences perceptual skills. *Frontiers in Psychology*, 8, 1387–1387. doi: 10.3389/fpsyg.2017.01387.

Paas, F., Tuovinen, J. E., Tabbers, H., & Van Gerven, P. W. (2003). Cognitive load measurement as a means to advance cognitive load theory. *Educational Psychologist*, 38:1, 63–71.

Pageaux, B., & Lepers, R. (2018). The effects of mental fatigue on sport-related performance. *Progress in Brain Research*, 240, 291–315. doi: 10.1016/bs.pbr.2018.10.004.

Rein, R., & Memmert, D. (2016). Big data and tactical analysis in elite soccer: Future challenges and opportunities for sports science. *SpringerPlus*, 5:1, 1410–1410. doi: 10.1186/s40064-016-3108-2.

Rozand, V., Pageaux, B., Marcora, S. M., Papaxanthis, C., & Lepers, R. (2014). Does mental exertion alter maximal muscle activation? *Frontiers in Human Neuroscience*, 8:755. doi: 10.3389/fnhum.2014.00755.

Russell, S., Jenkins, D., Halson, S., & Kelly, V. (2020). Changes in subjective mental and physical fatigue during netball games in elite development athletes. *Journal of Science and Medicine in Sport*.

Russell, S., Jenkins, D., Rynne, S., Halson, S., & Kelly, V. (2019a). What is mental fatigue in elite sport? Perceptions from athletes and staff. *European Journal of Sport Science*.

Russell, S., Jenkins, D., Rynne, S., Halson, S. L., & Kelly, V. (2019b). What is mental fatigue in elite sport? Perceptions from athletes and staff. *European Journal of Sport Science*, 1–10.

Russell, S., Jenkins, D., Smith, M., Halson, S., & Kelly, V. (2019c). The application of mental fatigue research to elite team sport performance: New perspectives. *Journal of Science and Medicine in Sport*, 22:6: 723–728. doi: 10.1016/j.jsams.2018.12.008.

Schücker, L., & MacMahon, C. (2016). Working on a cognitive task does not influence performance in a physical fitness test. *Journal of Sport and Exercise Psychology*, 25, 1–8.

Smith, M. R., Coutts, A. J., Merlini, M., Deprez, D., Lenoir, M., & Marcora, S. M. (2016). Mental fatigue impairs soccer-specific physical and technical performance. *Medicine and Science in Sports and Exercise*, 48:2, 267–276.

Smith, M. R., Fransen, J., Deprez, D., Lenoir, M., & Coutts, A. J. (2017). Impact of mental fatigue on speed and accuracy components of soccer-specific skills. *Science and Medicine in Football*, 1:1, 48–52.

Smith, M. R., Marcora, S. M., & Coutts, A. J. (2015). Mental fatigue impairs intermittent running performance. *Medicine and Science in Sports and Exercise*, 47:8, 1682–1690. doi: 10.1249/mss.0000000000000592.

Smith, M. R., Thompson, C., Marcora, S. M., Skorski, S., Meyer, T., & Coutts, A. J. (2018). Mental fatigue and soccer: Current knowledge and future directions. Sports Medicine, 48:7, 1525–1532. doi: 10.1007/s40279-018-0908-2.

Staiano, W., Merlini, M., & Marcora, S. (2015). A randomized controlled trial of brain endurance training (BET) to reduce fatigue during endurance exercise. Paper presented at the ACSM Annual Meeting, San Diego, CA.

Sutarto, A. P., Wahab, M. N. A., & Zin, N. M. (2010). Heart rate variability (HRV) biofeedback: A new training approach for operator's performance enhancement. *Journal of Industrial Engineering and Management*, 3:1, 176–198.

Tanaka, M., Shigihara, Y., Ishii, A., Funakura, M., Kanai, E., & Watanabe, Y. (2012). Effect of mental fatigue on the central nervous system: an electroencephalography study. *Behavioral and Brain Functions*, 8, 48. doi: 10.1186/1744-9081-8-48.

Van Cutsem, J., De Pauw, K., Buyse, L., Marcora, S., Meeusen, R., & Roelands, B. (2017a). Effects of mental fatigue on endurance performance in the heat. *Medicine and Science in Sports and Exercise*. doi: 10.1249/mss.0000000000001263.

Van Cutsem, J., De Pauw, K., Marcora, S., Meeusen, R., & Roelands, B. (2017b). A caffeine-maltodextrin mouth rinse counters mental fatigue. *Psychopharmacology*, 1–12.

Van Cutsem, J., Marcora, S., De Pauw, K., Bailey, S., Meeusen, R., & Roelands, B. (2017c). The effects of mental fatigue on physical performance: A systematic review. *Sports Medicine*, 47:8, 1569–1588. doi: 10.1007/s40279-016-0672-0.

Van Cutsem, J., Roelands, B., Pluym, B., Tassignon, B., Verschueren, J., De Pauw, K., & Meeusen, R. (2019). Can creatine combat the mental fatigue-associated decrease in visuomotor skills? *Medicine and Science in Sports and Exercise*. doi: 10.1249/mss.0000000000002122.

Veness, D., Patterson, S. D., Jeffries, O., & Waldron, M. (2017). The effects of mental fatigue on cricket-relevant performance among elite players. *Journal of Sports Science*, 1–7. doi: 10.1080/02640414.2016.1273540.

Voss, M. W., Nagamatsu, L. S., Liu-Ambrose, T., & Kramer, A. F. (2011). Exercise, brain, and cognition across the life span. *Journal of Applied Physiology*, 111:5, 1505–1513. doi: 10.1152/japplphysiol.00210.2011.

Weeks, S. R., McAuliffe, C. L., DuRussel, D., & Pasquina, P. F. (2010). Physiological and psychological fatigue in extreme conditions: the military example. *Pm&r*, 2:5, 438–441.

Xiao, Y., Ma, F., Lv, Y., Cai, G., Teng, P., Xu, F., & Chen, S. (2015). Sustained attention is associated with error processing impairment: evidence from mental fatigue study in four-choice reaction time task. *PLoS One*, 10:3, e0117837.

Zhao, C., Zhao, M., Liu, J., & Zheng, C. (2012). Electroencephalogram and electrocardiograph assessment of mental fatigue in a driving simulator. *Accident Analysis and Prevention*, 45, 83–90. doi: 10.1016/j.aap.2011.11.019.

12 Using Constraints in a Complex Sports System
Modern Day Training for Modern Day Cricket

Vickery W., McEwan K., Clark M. E., and Christie C.J.

CONTENTS

12.1   INTRODUCTION

Our ability to coordinate our movements is an integral part of life and is a process which develops over a long period of time. We need only to look at how a young child learns the movements associated with a new sport and how long this can take to master. We might also consider how an elite athlete recovers from a season-ending injury to realise what developing a new set of movement patterns or relearning a previous skill involves. Training, or skill acquisition, is the internal process that brings about relatively permanent changes in an individual's movement capabilities (Schmidt & Wrisberg, 2004). This process typically requires the individual to use and adapt to the information available to them during training, and previous experience of a particular movement, and perform the motor skill to their best of their ability in a repetitive manner. The process should result in a coordinated movement pattern that has the ability to adapt to the continually changing environment in which we work, play, and compete (Davids, Button, & Bennet, 2008).

In practice, understanding the way in which an individual learns and develops a new skill or movement pattern is vital for professions such as occupational therapy, childcare, or sports coaching (Haubenstricker & Seefeldt, 1986; Hodges & Williams, 2019; Poole, 1991; Steel, Harris, Baxter, & King, 2013). It is also important to consider how these practitioners develop and deliver these training sessions or learning programmes. Recently, there has been an increase in the information relating to the most effective environment for an individual to acquire skills which focus more on ecologically valid environments (Hodges & Williams, 2019; Renshaw, Chow, Davids, & Hammond, 2010c). Further, some have suggested extending this further by using methods offered by the discipline of Human Factors and Ergonomics (HFE) to incorporate more systems thinking methods (McLean, Salmon, Gorman, Read, & Solomon, 2017). Although these methods have mostly been applied to contexts such as the military, aviation, and transportation industries, to ensure safe and more effective systems, they can be applied to various sports. The game of cricket can be viewed as a complex socio-technical system as there are many interacting human and non-human components operating within a dynamic and constantly changing environment (McLean et al., 2017).

Despite this, most cricket coaching is still done by isolating the disciplines and parts thereof and providing one-on-one feedback coaching, which is similar in many sports. Subsequently, descriptive feedback is provided to the learner to make small adjustments to their movement patterns (Cushion & Jones, 2001; Potrac, Jones, & Armour, 2002; Smith & Cushion, 2006). Using a method such as Cognitive Work Analysis (CWA; Vicente, 1999), a method used in systems theory, can assist in investigating complex socio-technical systems (McLean et al., 2017), such as the game of cricket.

Therefore, the aim of this chapter is to highlight the importance of ecologically valid training for the complex game of cricket and how, using tools from HFE, we can optimise cricketing performance.

12.2   THE GAME OF CRICKET

Cricket is a bat and ball game played between two teams of 11 players each. It is arguably more complex than other team-based sports. Typically cricket researchers and practitioners draw from the sport of baseball and from previous experience to inform coaching practice. However, it is contended that cricket is a lot more complex than other team-based sports, including baseball. This is because many more factors determine the outcome of a game. For example, the opening batters lay the foundation of an innings by seeing off the new ball and playing a sheet anchor role. The lower-middle, depending on the score, try to make as many runs as possible or try to keep their wickets.

In addition, what makes it more complex is that bowlers are required to bat once the top order batters are dismissed. Further, sometimes batters are used as part-time bowlers depending on the game situation. The team batting first tries to score as many runs as possible while the other team bowls and fields trying to get the batter's out. At the end of the innings, the teams switch between batting and fielding requiring most team members to bat, bowl, and field with one person required to keep wicket for the fielding side. Two batters bat in partnership and the opening partnership is responsible for facing the 'new ball' and scoring runs. The middle-order batters consolidate the innings. The outcome of a match depends on the batting partnerships as large partnerships (high combined scores) not only add runs but also exhaust the tactics of the fielding side. The concept of partnerships become vital if only one recognised batter remains. It is thus important to identify the key players in a team by constructing a network of batting partners. Pitch conditions, weather, time of day, and type of game played as well as the opposition they are playing against are just some of the factors which feed into the complexity of the game.

12.3   SKILL ACQUISITION WITHIN SPORT

Much debate has surrounded the basis for superior performance, from a physiological perspective, as well as a motor skill perspective, with theoretical ideologies divided between genetics and deliberate practice (Ericsson, Krampe, & Tesch-Römer, 1993; Baker & Davids, 2005; Davids & Baker, 2007; Tucker & Collins, 2012). On the one hand, it is argued that elite performance is the result of innumerable environmental influences and experiences (nurture), and that genetics (nature) has no role in the development of elite performance (Davids & Baker, 2007; Tucker & Collins, 2012). Some believe that all individuals are born equal and that subsequent performance is developed through deliberate practice and experiences (Ericsson et al., 1993; Davids & Baker, 2007). It is thought that elite performance can be attained within any given domain provided that the subsequent number of hours of deliberate practice is fulfilled. The 10,000-hour theory stemmed from this, which prescribed that to achieve an elite level of performance within any given domain, individuals were required to perform a minimum of 10,000 hours (or ten years) of deliberate practice (Ericsson et al., 1993; Davids & Baker, 2007; Gladwell, 2008; Tucker & Collins, 2012). Conversely, it is believed that elite performance can be traced back to the genetic make-up of individuals; elite athletes are born to succeed due to predisposed genetic characteristics (Davids & Baker, 2007; Tucker & Collins, 2012). As a result, an individual not in possession of the specific genetic characteristics necessary for a given domain will not be able to reach an elite level, regardless of the amount of deliberate practice they perform (Davids & Baker, 2007; Tucker & Collins, 2012). The debate has primarily accepted an extremist or dualistic view, with researchers generally adopting one or the other perspective as truth.

Opposing the nature versus nurture debate, Davids and Baker (2007) and Tucker and Collins (2012) provide a theory that suggests that elite performance results from the optimal interaction of intrinsic (nature) and extrinsic (nurture) factors (Figure 12.1). The genetic ceiling model states that individuals are given an innate, limited ('ceiling') ability that can only be fulfilled through the attainment of skills developed by means of favourable and optimal environmental constraints, such as deliberate practice, coaching, experience, and exposure (Tucker & Collins, 2012). The domain of HFE has come to a similar conclusion, as traditionally skill acquisition used an 'information-processing' approach to motor learning which conceptualised

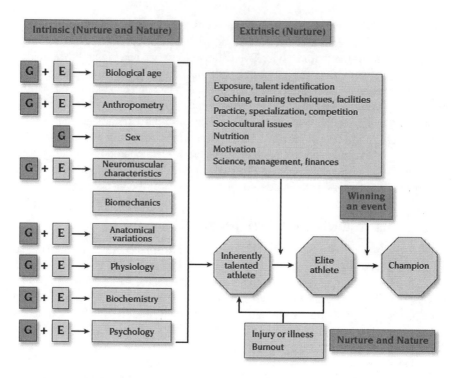

FIGURE 12.1 Breakdown of the intrinsic (genetic) and extrinsic (environment) factors that influence the acquisition of skilled performance. (Extracted from Tucker & Collins, 2012.)

the human brain as a 'capacity limited computational device' into which raw sensory information is channelled and then processed with input from prior external environmental experiences (Davids et al., 2008). HFE views individuals as an integral part of systems, which account for their abilities and limitations (both of which are defined by intrinsic and extrinsic factors) when optimising the system's overall performance (Gurses, Ozok, & Pronovost, 2012).

According to this model, the level of performance, incorporating subsequent training essential for improvement, has an upper limit or performance capacity which is determined by the genetic predisposition of each individual (Tucker & Collins, 2012). Regardless of the amount of training and practice that is performed by an individual, the upper limit of performance (i.e. the maximum potential) cannot be exceeded (Tucker & Collins, 2012). To reach this maximum potential, individuals are required to maximise the extrinsic factors (Figure 12.1) such as the amount of deliberate practice and exposure to correct coaching techniques to ensure improved performance (Tucker & Collins, 2012). Weissensteiner, Abernethy, Farrow, and Müller (2008) agree with this understanding of skill acquisition, describing that 'the perceptual skills that are known to be critical for expert performance in adults improve not with maturation or chronological age alone but with experience with, and exposure to, vast amounts of task-relevant practice' (p. 644). However, for these inherently talented athletes to achieve elite status, the extrinsic factors that influence development are required to be optimised. This includes superior training regimes, correct nutritional diets, motivation to succeed, and so forth (Tucker & Collins, 2012). In the event of injury or burnout, these athletes would lose some of the benefits that arose from optimising the extrinsic factors, returning to their previous inherently talented status (Tucker & Collins, 2012).

12.4   SKILL ACQUISITION WITHIN CRICKET

Our understanding of skill acquisition and as a result, the process of how to train for the purpose of skill development has evolved over the last few decades. Although there still remains no final consensus to how an individual acquires a motor skill (and note that this chapter does not set out to try to resolve this issue), a multidisciplinary theoretical framework for capturing and making practical use of the interaction of so many interacting factors has begun to evolve.

HFE used to have a 'reductionist' approach (i.e. separating the system into parts, attempting to optimise the parts independently, and then joining the parts back together on the assumption of improved functionality) which did not work. Nowadays, HFE employs a 'systems' approach which optimises performance based on understanding the systematic interactions between individuals and all elements of the system in which they function (i.e. the physical environment, tasks, tools/technologies, and organisational conditions). For example, in healthcare, instead of designing each care service independently (e.g. the emergency department and intensive care unit), HFE focusses on the integrated whole (i.e. the hospital) to maximise the system's overall performance (e.g. patient safely, clinicians' job satisfaction, and efficiency) (Gurses et al., 2012).

A cricketer is a human interacting in a nonlinear complex system, i.e. a network of interrelated, interacting components, whereby each of the various components influence each other (Hristovski, Balague, Daskalovski, Zivkovic, & Naumovski, 2012), which may include the interactions between teammates, the opposition, or the physical and social environment. Cricketers try and outsmart their opponents through the use of various tactics usually unknown to the opposition, which enhances the complexity of the interactions. Thus they need to adapt in a constantly changing, sometimes hostile environment, working as individual players within specific roles but also as a collective team. Developing skills in a sport like cricket is nonlinear (i.e. skill development does not move from one clear stage to another when attaining expertise) and so it is contended that we cannot control skill development by breaking the sport or position/s into its parts, but that we rather need to create more holistic and realistic training scenarios which allow players to adapt and self-organise in a multitude of scenarios while training.

This can be achieved by drawing on the constraints-led training approach, as well as some of the methods used in other disciplines, including the domain of HFE. Rasmussen (1986) considers constraints on individual behaviour using the method of CWA which is seen as formative (Naiker, 2017). So when constraints are included they limit the opportunity for action by individuals but they then create many remaining degrees of freedom for behaviour which can promote adaption by ensuring individuals get exposed to numerous unpredictable scenarios (Naiker, 2017). In cricket, constraints are numerous, dynamic, and constantly changing and these scenarios need to be replicated, as far as possible, in training sessions so that the player can react successfully to a multitude of situations. Traditional cricket training does not typically do this.

Why this would be important is explained using an example of a batter going out to bat (Adapted from Goble, 2017): From the noise of the crowd to the fall of a wicket, each batter, as they walk out to bat, will be overwhelmed by an abundance of stimuli. As batters set foot onto the field and approach the pitch, the brain continually needs to sift and process information. The crowd is often the first stimulus that batters perceive, where home and away games play an integral role in feelings of confidence or intimidation. Thereafter, and with the intention of intimidation, the 11 opposition fielders often try to talk the batter out of their comfort zone. The intimidation increases as the batter prepares to face the first delivery. At this stage the number of stimuli that need processing increase exponentially and before the batter can intercept the ball, several essential information processing sequences are completed. The ability to do this efficiently, while reducing errors, is fundamental to batting success.

First, the batter must observe the field and fielding positions for potential 'traps' that have been set by the opposition. They must then select certain areas in the field that suit their strengths and scoring areas. Second, and while cognizant of these areas, the batter must observe certain gestures of the bowler as they approach their delivery stride (point of ball release). These include, the speed of approach, the angle of approach, the position of the ball in the bowler's hand, and the height at which the ball is released. This pre-delivery information allows the batter to make anticipatory movements in the preparation of shot selection. Therefore, as early as pre-ball release, the information processing system is hard at work.

At the point of ball release, the temporal constraints placed on batters are severe. When facing a fast bowler ($\geq$120 kmph^{-1}) batters have approximately 425–530 ms from ball release to arrival at the bat. In this time batters must detect and track the ball, interpret the visual information obtained (line and length of the delivery and late deviations in ball flight), make a decision to play or leave the ball, and then programme the motor system based on this decision. When a shot is executed the batter must then reassess the situation. Here, the batter must determine the direction of the ball and whether or not the ball has entered a 'gap' in the field. The batter must then decide whether or not there is time to complete a run. In the event that a run is 'on' they must initiate a new motor programming sequence that will start the process of running. While running, the batter needs to continually observe and update the ball position, relative to the fielders, and decide whether a second or perhaps a third run is possible and if so, once again, the neuromuscular system must be reprogrammed. While this is only one delivery, batters repeat these processes hundreds of times when scoring a century. Additionally, batters also need to complete runs for their partners. This necessitates similar information processing sequences and as a result the batter's mind is never at rest.

Where this example is examined from a fast bowling standpoint, batters are also required to intercept balls that move prodigiously in the air (swing bowling) or off the pitch (spin bowling). Despite slower ball speeds, swing and spin bowling requires heightened selective attention as the deviations in flight or movement often occur late in the delivery (just before the ball arrives at the batter). In these examples, early detection of pre-delivery information is crucial, as this often indicates to the batter what type of delivery could be bowled.

It is therefore evident that successful batting requires the following traits: vigilance (filtering important from non-important stimuli), fast response times that facilitate accurate and rapid decision making (shot selection and execution), spatial awareness (shot execution and running between the wickets), and efficient executive functioning (executive control in each delivery). Batting also requires effective retrieval of information from short- term and long-term memory to allow decisions to be made and the correct shot selected. The same can be applied to all the positions on the field to varying degrees. How to create this type of 'real match' scenario within a cricketing context is thus difficult.

The high explanatory power of systems approaches and, in particular, CWA could be used in this framework to describe in-depth, socio-technical systems and the factors influencing (cricketing) performance (McLean et al., 2017). The first phase of CWA, Work Domain Analysis (WDA), can construct a description of the functional structure of the system including the purposes of the game, the objects used, the behaviours needed to perform well, and the criteria to determine between good and bad performance. This is something that requires further investigation and should be prioritised.

The following section of this chapter highlights and explores how manipulating constraints in a cricket training environment can assist in creating some of these 'real match' scenarios in order to better prepare cricketers for game situations. Further, the following section describes the use of the systems approach, adopted by HFE, through studying the systematic interaction of each component of cricket performance in order to optimise success.

12.5   WHAT ARE CONSTRAINTS?

The term constraint is typically (and unfortunately) associated with a negative orientation. Most people would see a constraint as something which hinders our ability to complete a particular objective. In fact, the *Oxford Dictionary* (2018) defines a constraint as 'a limitation or restriction', which certainly suggests a more negative than positive definition. From the perspective of skill acquisition though, Newell (1986) defined constraints as boundaries which shape the emergence of behaviour from a movement system (learner) seeking a stable state of organisation. In a more recent definition, Davids, Button, and Bennett (2008) suggested that constraints are variables that influence, limit, and enable the different behavioural trajectories (development of skill acquisition) of a complex system. Regardless of the definition used, the term constraint is very much about limiting or restricting a learner's performance by using constraints to their advantage to form a co-ordinated movement pattern.

Newell (1986) classified constraints into three distinct categories in order to provide those involved in the development of a learner's skill with a more coherent understanding of how different movement patterns may emerge (Davids, Button, & Bennett, 2008). These are:

12.5.1 Organismic Constraints

These are constraints which are unique and specific to an individual. For example, a learner's height, weight, muscle mass and distribution, genetic make-up, emotions, and behaviour, cognitions, or even as simple as whether they are right- or left-handed. These characteristics are for the most part unchangeable and, therefore, the learner (and the person/people responsible for helping to acquire a motor skill) must adapt and develop movement patterns based on these organismic constraints. For example, in cricket, physically strong batters may approach batting by muscling the ball to various areas of the playing surface, whereas physically weaker batters are typically required to manipulate the ball skilfully to accumulate runs. One also need only to consider the different movement patterns associated with those with full use of their limbs compared to those that do not to gain insight into how different organismic constraints can impact of the acquisition of skill.

12.5.2 Environmental Constraints

Environmental constraints are those which occur in both the physical and social environment which impact on the performance of a learner which are not within their control. Physical environmental constraints include factors such as surroundings (i.e. gravity, altitude, light, and noise), weather conditions, playing surfaces (i.e. location, size, and type of surface), and previous environmental settings of informal gameplay (Davids, 2010; Renshaw, Davids, & Savelsbergh, 2010b; Greenwood, 2014). Within the sport of cricket this may be the type of pitch played on, which would influence the approach used by batters to score runs; for example, a hard pitch that provides extra bounce would typically require a greater amount of back foot strokes to be played. Further, social environmental constraints such as those imposed from family and friends, societal and cultural norms, fans of the team/sport and media outlets (Haywood & Getchell, 2014), and cultural constraints (the culture of a sports club and access to high-quality coaching) also have an impact on skill acquisition.

12.5.3 Task Constraints

Task constraints are those which are task specific and concerned with the goals of a specific activity (Renshaw & Davids, 2014). These include aspects of gameplay, such as rules and regulations of the activity, the objectives and outcomes for successful performance, number of players within a team, the opposition a team or individual is competing against, and the equipment being used (Davids, 2010; Renshaw et al., 2010b; Greenwood, 2014). This is an area of gameplay which can be manipulated to shape the emergence of individual behaviours during motor learning (Renshaw et al., 2010b; Greenwood, 2014). For example, a cricket coach may alter the rules and regulations of a training session to encourage batters to change their typical batting behaviour. This may take the form of scoring constraints, whereby batters are only allowed to score runs in certain areas of the field (i.e. straight back past the bowler). Batters may be punished if they play horizontal bat strokes, or if they play the deliveries to third-man or fine-leg. A simple constraint such as this would encourage a different set of action responses from the batters to succeed with the training session.

Modifying equipment has been shown to allow the athlete to cope with or manage the large amount of information that surrounds them by restructuring the practice environment, or in other words to make things a little less challenging so that they can focus on the information which is important (Davids, Button, & Bennett, 2008). In some cases, using modified equipment is necessary for athletes starting out in a particular sport for them to develop an effective movement pattern. Consider child athletes who may have to use adult-sized equipment. Using modified equipment which is suitable for the size and strength of the younger athlete, compared to an adult, provides them with the opportunity to develop an optimal movement pattern which is suitable to the individual (Buszard, Reid, Masters, & Farrow, 2016). Stretch, Nurick, and McKellar (1998b) attempted to train cricket batters to improve their accuracy of hitting the ball in the middle of the bat; this was done through the use

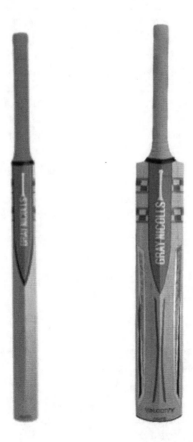

FIGURE 12.2 Modified cricket bats (left) which are narrower than a standard cricket bat (right) are now commonly used so that batters have a smaller surface area.

of a modified bat, one-third the width of regulation size (Figure 12.2). The theory behind the design was that the narrow bat would force participants to hit the ball out of the centre of the bat, therefore decreasing miss-hits and misses (Stretch et al., 1998b). Skilled batters did not benefit from the use of a narrow bat, but unskilled batters improved their performance significantly (Stretch et al., 1998b) and as such it was concluded that this form of training should be implemented to assist the development of unskilled batters.

12.6   THE CONSTRAINTS-LED MODEL

The aim of a constraints-led approach to motor learning is to identify the nature of the interacting constraints that influence the skill acquisition of an individual (Davids, 2010). Once this has been identified, learning environments can be created to further assist with the development of motor learning. A well-designed learning environment manipulating informational constraints would constrain individuals to the emergence of functional movement patterns, as well as afford an environment which could shape decision-making behaviour (Davids, 2010). Newell (1985, 1986) suggested that each of these different constraints – organismic, environmental, and task – interact not only with the learner but each other, which results in the emergence of a movement (action) to ultimately lead to the movement goal (see Figure 12.3). The manner in which a learner utilises the information (perception) gained from each of these different constraints is what may distinguish the ability of a performer.

The key principles behind this approach are that of functionality and action fidelity (Pinder, Davids, & Renshaw, 2011a). Functionality refers to the degree to which the task maintains the perception–action coupling present in real performance environments; whereas action fidelity refers to the degree to which the training environment reflects that of the real performance environment (Pinder et al., 2011a). The constraints-led approach encourages the design of a training environment that affords individuals a high level of functionality as well as a high degree of action

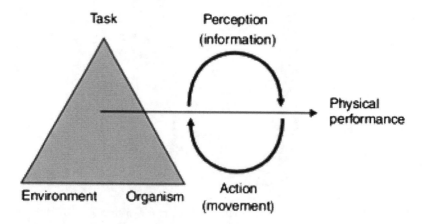

FIGURE 12.3   Newell's Constraints-Led model (1986).

fidelity. However, even this may be limited as there are many interacting factors that impact a cricketer's performance in a very complex way.

12.7   DYNAMICAL SYSTEMS THEORY

When explaining the constraints-led model (Newell, 1985, 1986) many authors have tended to base this on the theory of Dynamical Systems (for more insight refer to Anson, Elliot, & Davids, 2005). Despite originating within the areas of mathematics and physics, nowadays the Dynamical Systems theory has been adapted across a range of disciplines including HFE and that of motor skill acquisition which suggests that

> the human movement system is a highly intricate network of co-dependent sub-systems (e.g. respiratory, circulatory, nervous, skeletomuscular, perceptual) that are composed of a large number of interacting components (e.g. blood cells, oxygen molecules, muscle tissue, metabolic enzymes, connective tissue and bone). In dynamical systems theory, movement patterns emerge through generic processes of self-organization found in physical and biological systems.
>
> **(Glazier, Davids, & Barlett, 2003)**

In simple terms, the movement patterns of a learner are the result of the many interacting constraints over a period of time through practice. This practice is referred to as the process of self-organisation and is described as 'a dynamical and adaptive process where systems acquire and maintain structure themselves, without external control' (De Wolf & Holvoet, 2004). Regardless of the context in which it is used, when defining and describing the theory of Dynamical Systems a number of key factors are always acknowledged, these are:

- The interaction of numerous parts, otherwise known as degrees of freedom;
- The integration of these interacting parts into movement solutions;
- This all takes place within an ever-changing (dynamic) environment.

One point that needs to be emphasised and is influenced by these factors is that the more complex the system the greater the number of degrees of freedom and, therefore, the more difficult it is to coordinate this into an effective movement solution (Davids, Button, & Bennett, 2008). Within the context of cricket, think of the many different interacting parts that a cricket fast bowler must coordinate to get from the top of their run-up to completing their follow-through after releasing the ball towards the batters. This would include not only their limbs and body parts but the different muscles within the body which all contract at different times and serve a different purpose at different points of the bowling action, the nervous system, and the role it plays in making sure the correct muscles contract at the right time, and so forth. The fast bowler must coordinate each of these limbs and body parts into an action which will be effective and achieve a specific outcome. In addition to this, the playing conditions are constantly changing depending on the state of the game. The same fast bowler may have to change the way in which they deliver the ball from the first over of the day to the last due to factors such as fatigue, the environmental conditions, or different batters. Just from this example you can begin to understand the complexity and difficulty in ensuring that cricketers are appropriately trained. Given the number of interacting parts and the dynamic characteristics of the environments in which we all live, work, and play, this links closely as one can see with that of the constraints-led model originally proposed by Newell (1985, 1986). The theory emphasises the influence of multiple constraints on behaviour and proposes a framework through which individual performers satisfy the multiple constraints upon them in dynamic environments (Davids, Araújo, Shuttleworth, & Button, 2003). This implies that in order to promote learning, individuals should be exposed to training environments that produce dynamic informational affordances that require individuals to re-organise movement patterns according to new opportunities.

12.8 PRACTICE DESIGN

Regardless of the environment in which a coach, a physical therapist, or a parent is trying to develop a motor skill within an athlete, a patient in recovery, or an infant learning to walk, respectively, it has long been advocated that the most effective environment for this to occur is one which represents as closely as possible all the demands associated with that motor skill when performed in the intended environment (Müller & Rosalie, 2010). When we consider Dynamical Systems theory, this would suggest ensuring that the information that is available to the learner is also representative of the intended performance environment to allow the learner to adapt to the ever-changing environment.

Unfortunately, this representative training environment has not always been applied. Even within present-day skill development practice many still adopt training environments which are not always representative of the performance environment or context in which the motor skill will be regularly performed. In the case of young cricket players playing within the age range of under 10 to under 17 years, Low, Williams, McRobert, and Ford (2013) discovered that overall the majority of time during training sessions was spent performing activities which focused on isolated skills or technique (Table 12.1). Those who were only there for recreational purposes, or in other words just to have fun playing cricket, spent more time in these specific and isolated skills than those who were there for performance (Table 12.1). Gibson (1979) proposed that invariant (persistent features) and variant information can act as affordances for action, through which a performer perceives information from the environment in relation to what it offers or demands in action responses. Over time, performers become attuned to information through experience and practice in different performance environments, creating relationships between movement patterning and specific sources of perceptual information (information–movement coupling). The importance of ensuring the presence of key specifying information in practice tasks is captured by the ecological concept of representative learning design. Traditionally, representativeness has referred to the generality of task constraints in a specific research context to the perceptual variables available in actual performance settings. In sports, practice environments are the equivalent of experimental settings, suggesting that they need to be accurately designed to ensure congruence with a

TABLE 12.1

Activity of Under-10 to Under-17 Male Cricket Players (Adapted From Low, Williams, McRobert, & Ford, 2013)

Group (mean session duration)	Proportion of Training Form Activity			Proportion of Playing Form Activity		
	% Fitness	% Technique	% Skills	% Small-sided conditioned	% Full games conditioned	% Full game
Recreational-children (74, s = 32 min)	2 sessions (mean = 4 min)	96 ± 6	0	50 ± 55	2 sessions (mean = 73 min)	1 session (mean = 74 min)
Elite-children (110, s = 21 min)	7 ± 8	29 ± 34	64 ± 35	0	100 ± 0	0
Recreational-adolescents (99, s = 21 min)	7 ± 6	33 ± 12	60 ± 10	1 session (mean = 103 min)	0	0
Elite-adolescents (96, s = 33 min)	7 ± 5	45 ± 39	48 ± 38	0	0	0

Note.   Number of session (duration) is used where there are less than three practice sessions recorded. In this instance, Training Form Activity refers to specific, isolated skill development activity, whereas Playing Form Activity refers to sessions representative of the performance environment.

performance environment in which the movements will be implemented. Changing the informational constraints on action might result in less representative practice designs and changes to a performer's acquisition of functional movement control. This idea has been exemplified in cricket batting research. Pinder, Renshaw, Davids, and Kerhervé (2011b) demonstrated that batters adapted spatiotemporal characteristics of emergent action when facing a live opponent through the pickup of advanced kinematic information, in contrast to facing balls delivered via a projection machine where movements were delayed through a need to sample early ball flight to determine the bounce point of the ball.

At this point though, one may begin to wonder how this idea of using an environment which is representative of the environment or context in which a learner would need to perform a motor skill actually fits together. Remember, the constraints-led model is based on the premise that the constraints (organismic, environmental, and task) within the environment interact with the learner leading to a new movement pattern (Newell, 1985, 1986). As such coaches, for example, should design learning environments that provide controlled boundaries of exploration in dynamic settings through the provision of relevant constraints to facilitate new movement patterns. The use of this type of training suggests the learner will adapt to the various constraints and over time will be able to learn and retain the skills learned, along with transferring these learned skills into a real-world context. There are several ways in which constraints can be manipulated to provide a learner with an environment that represents that of the performance environment or context. Due to the greater ease at which task constraints can be adapted, much of the research up to this point has made this the focus when applying the constraints-led model. This is not to say though that the organismic and environmental constraints cannot be utilised for the purpose of skill development. The remaining sections of this chapter will provide insight into the different ways in which the constraints-led model has been used within the context of cricket.

12.9 CONSTRAINTS-LED TRAINING IN CRICKET: A HFE APPROACH

Cricket possesses many characteristics of a complex socio-technical system due to the multiple, components operating within a dynamic environment. As such, cricket performance is highly complex, multi-faceted, and ultimately difficult to define. Training within cricket incorporates a combination of different factors, including physiological (strength and conditioning), tactical (situational awareness and decision making), and technical (technique and execution) skill development. The fundamental basis of traditional cricket training requires batters and bowlers to refine their skills and techniques within the confinements of a netted-off pitch (Stretch, McKellar, & Nurick, 1998a; Renshaw & Holder, 2009).

Traditional cricket training (net practice) has both strengths and limitations in design: the advantage for using such a practice structure is that the net stops the ball from travelling away from the practice area making it time efficient. Net practices, with the use of live bowlers, also maintain a strong perception–action coupling between batter and bowler, as would be the case in real match situations. However, net practice has a number of limitations with regards to structure and goal-oriented outcomes (Woolmer & Noakes, 2008; Renshaw & Holder, 2009). For example, net practice typically has a strong focus on the development of individual techniques, undermining the importance of training the execution of cricket strokes, which are essential for elite performance (Renshaw & Holder, 2009). Developing individual techniques is an important facet of the sport as it provides batters with a structured 'template' in which to face deliveries of a greater difficulty (Woolmer & Noakes, 2008). However, equally important is the ability to utilise those learned skills and techniques in an optimal, functional manner to achieve maximum performance.

Two main consequences may arise from net practice: (i) batters tend to play an array of attacking strokes with the knowledge that they cannot be dismissed, and (ii) batters are lured into believing that they are batting well due to good ball-striking, as well as the execution of 'textbook' cricket strokes. Consequently, batters can leave net practices under false pretences that they are 'in form' or 'hitting the ball well', but in reality these batters may not have the knowledge of whether they would have hit any number of balls straight to opposition fielders, or whether those cricket strokes would have produced any runs at all. Therefore, when transitioning to real match situations, these same batters may struggle to score runs because many of their strokes are being hit straight to fielders due to a lack of game-specific constraints, as well as a lack of outcomes-based focus (such as runs scored) that would be seen in real match situations (Vickery, Dascombe, & Duffield, 2014).

The latest trend in cricket research has been the use of a centre-wicket protocol, termed the Battlezone, which is designed to simulate real match situations within a training environment (Renshaw, Chappell, Fitzgerald, Davidson, & McFadyen, 2010a; Vickery et al., 2014). The Battlezone is a training environment which makes use of a circular net, 1 metre in height, enclosing the 30-yard circle around the pitch (Renshaw et al., 2010a). The design of this protocol promotes an intensified training scenario within the inner ring (see Renshaw et al., 2010a; Vickery et al., 2014). Within this design, batters train against live bowlers, with live or mock fielders in place to provide holistic environmental cues (Renshaw et al., 2010a). As a result, batters are made aware of where specific cricket strokes are being placed and the resultant runs rewarded for each stroke. This training design also provides cricketers with a high-intensity exercise bout, which mimics the physical demands of real match situations (Vickery et al., 2014).

However, research on the Battlezone has focused primarily on the ecological validity of the training exercise, with regards to the physiological demands of real match situations (Renshaw et al., 2010a; Vickery et al., 2014). These studies have demonstrated that the Battlezone protocol is reflective of the physical and physiological demands over and above that of real match situations (Renshaw et al., 2010a; Vickery et al., 2014). In other words, the demands of training using the Battlezone protocol are at a greater intensity than the demands in real match situations, for specific positions within the inner ring. The technical aspect of this form of training has not been adequately studied, though with only Vickery et al. (2014) reporting on the technical demands of the Battlezone in relation to real match situations to support ecological validity (Vickery et al., 2014). Here the technical demands of batting were defined according to the number of balls faced, the number of balls hit, and the percentage of good contact shots (Vickery et al., 2014) with the authors suggesting that that the technical demands of the batting during Battlezone were representative of the demands of real match situations (Vickery et al., 2014).

Table 12.2 is an example of ideas in which a constraints-led approach can be used in training to elicit improvement in cricket fast bowlers.

Whether this constraints-led approach is sufficiently holistic can be questioned so the authors' contend that building on this idea of constraints-led training in cricket, would be to try and develop in-depth models of cricket through the application of HFE methods. The CWA method could be used to describe the constraints that influence performance within the cricketing context (Rasmussin, 1986; Naikar, 2017). CWA has also been used to support the design of systems, particularly training systems (e.g. healthcare, aviation, automotive, and rail transportation) (Stanton, Salmon, Walker, & Jenkins, 2017). To elucidate, the first phase of CWA, WDA, is used to construct an in-depth description of the functional structure of the system under analysis – in this case, for example, the cricket match. This functional structure covers the purposes of the system, the objects used, the behaviours required for successful performance, and criteria that is used to determine whether the system is achieving its functional purposes. For example, instead of using isolated performance variables for individual players (e.g. averages, economy, strike rate), WDA would consider teamwork functions as well as physical, psychological, and environmental constraints (McLean et al., 2017). Once the CWA analyses are developed, the

TABLE 12.2

Examples of Constraints to Elicit Improvement in Cricket Fast Bowlers

Constraint Classification	Type of Constraint	Intervention	Desired Objective
Task	Rules	Ball must hit desired target on pitch but bowler's feet cannot land on the same part of the popping crease during an over	Change the angle of the delivery with improved accuracy
	Playing Area	Modified centre wicket practice which replicates match scenario	Challenge accuracy when under pressure, e.g. final over of limited overs match with few runs to win for the opposing team
	Player Involvement	Include silly mid-on/silly point fielder into net-based practice	Learn to bowl line and length which is advantageous for wicket-taking opportunities for this fielder
	Equipment	Bowl using an old (40+ overs) ball	Improve ability to create wicket-taking and fewer run-scoring deliveries
		Modified bat, one-third the width of regulation size (Stretch et al., 1998b).	Improve accuracy of hitting the ball in the middle of the bat
Environment	Pitch surface	Play on dry pitch against well-established batters with the intention of dismissing them after a period of minimal runs scored	Bowl a fuller length with greater accuracy and develop greater variety of change up deliveries in batter friendly conditions
	Overcast/ humid weather conditions	Short spell in practice nets against top-order batters with new ball	Learn to create wicket-taking opportunities in short space of time
Organismic	Coordination of body segments	Shorten pre-delivery run-up distance	
		Bowler to complete shuttle-run circuit before net practice session	

CWA-Design Toolkit (CWA-DT) can be used to support the design of constraints-based training systems. The CWA-DT is a participatory design approach that has been developed specifically for generating new designs based on CWA analyses.

Cricket is a very traditional game with very traditional methods of coaching and while constraints-led training, as described in this chapter, is a new way of approaching coaching in cricket, it is our view that we need to expand this even further by including methods used in the HFE domain to better understand the complex game of cricket and the best way to train cricketers for optimal performance.

12.10   CONCLUSION

The constraints-led approach emphasises the interaction between individual, task, and environmental constraints at numerous levels within a complex human system. Co-ordinated movement patterns, or behaviour, are regulated by the relationship between perception and action systems, with the influence of specific individual, task, and environmental constraints. It is believed that specific constraints influence the information sources which regulate human behaviour in varying performance environments. Therefore, the constraints-led approach considers behaviour which emerges from the interaction between the individual and the environment. The constraints-led approach should be extended further to include methods from HFE to better understand the complexity of cricket and resulting training requirements. This will aid in developing more ecologically valid coaching sessions to better pre-pare cricketers for the complex skills required to succeed in the game. For instance, WDA (see Chapter 13) can be used to develop a model for the cricket match system, which in turn could be used to support the design of a new constraints-based cricket training system using the CWA-Design Toolkit (Read, Salmon, Goode, & Lenne, 2018).

REFERENCES

Anson, G., Elliott, D., & Davids, K. (2005). Information processing and constraints-based views of skill acquisition: divergent or complementary? *Motor Control*, 9:3, 217–241.

Baker, J., & Davids, K. (2005). Genetic and environmental constraints on variability in sport performance. *Movement System Variability*, 109–130.

Buszard, T., Reid, M., Masters, R., & Farrow, D. (2016). Scaling the equipment and play area in children's sport to improve motor skill acquisition: A systematic review. *Sports Medicine*, 46:6, 829–843.

Cushion, C. J., & Jones, R. L. (2001). A systematic observation of professional top-level youth soccer coaches. *Journal of Sport Behaviour*, 24, 1–23.

Davids, K. (2010). The constraints-based approach to motor learning: Implications for a non-linear pedagogy in sport and physical education. In: *Motor Learning in Practice: A Constraints-Led Approach* (pp. 3–16) Routledge.

Davids, K., Araujo, D., Shuttleworth, R., & Button, C. (2003). Acquiring skill in sport: A constraints-led perspective. *International Journal of Computer Science in Sport*, 2:2, 31–39.

Davids, K., & Baker, J. (2007). Genes, environment and sport performance: why the nature-nurture dualism is no longer relevant. *Sports Medicine (Auckland, N.Z.)*, 37:11, 961–980.

Davids, K., Button, C., & Bennett, S. (2008). *Dynamics of Skill Acquisition: a Constraints-Led Approach*. Champaign, IL: Human Kinetics.

De Wolf, T., & Holvoet, T. (2004). Emergence versus self-organisation: Different concepts but promising when combined. In: International Workshop on Engineering Self-Organising Applications (pp. 1–15). Berlin/Heidelberg: Springer.

Ericsson, K. A., Krampe, R. T., & Tesch-Romer, C. (1993). The role of deliberate practice in the acquisition of expert performance. *Psychological Review*, 100:3, 363–406.

Gladwell, M. (2008). *Outliers: The Story of Success*. New York, NY: Little, Brown and Company.

Gibson, J. J. (1979). *The Ecological Approach to Visual Perception*. Boston, MA: Wiley.

Glazier, P. S., Davids, K., & Bartlett, R. M. (2003). Dynamical systems theory: a relevant framework for performance-oriented sports biomechanics research. *Sportscience*, 7.

Goble, D. (2017). Physical, perceptual and cognitive demands placed on amateur batters while scoring a simulated one-day century. Doctor of Philosophy. Department of Human Kinetics and Ergonomics, Rhodes University, Grahamstown, South Africa.

Greenwood, D. A. (2014). Informational constraints on performance of dynamic interceptive actions. Unpublished Doctor of Philosophy thesis. Queensland University of Technology, Brisbane, Australia.

Gurses, A. P., Ozok, A. A., & Pronovost, P. J. (2012) Time to accelerate integration of human factors and ergonomics in patient safety. *BMJ Quality and Safety*, 21, 347–351.

Haubenstricker, J., & Seefeldt, V. (1986). Acquisition of motor skills during childhood. *Physical Activity and Well-Being* (pp. 41–92).

Haywood, K.M., & Getchell, N. (2014). *Life Span Motor Development*. 6th edition. Champaign, IL: Human Kinetics.

Hodges, N. J., & Williams, A. M. (2019). *Skill Acquisition in Sport: Research, Theory and Practice*. Routledge.

Hristovski, R., Balague, N., Daskalovski, B., Zivkovic, V., & Naumovski, M. (2012). Linear and nonlinear complex systems approach to sports. Explanatory differences and applications. *Research in Physical Education, Sport and Health*, 1, 25–31.

Low, J., Williams, A. M., McRobert, A. P., & Ford, P. R. (2013). The microstructure of practice activities engaged in by elite and recreational youth cricket players. *Journal of Sports Sciences*, *31*(11), 1242–1250.

McLean, S., Salmon, P. M., Gorman, A. D., Read, G. J., & Solomon, C. (2017). What's in a game? A systems approach to enhancing performance analysis in football. *PLoS One*, 12, e0172565. doi: 10.1371/journal.pone.0172565.

Muller, S., & Rosalie, S. (2010). Transfer of motor skill learning: Is it possible? In: M. Portus (ed.), Conference of Science, Medicine and Coaching in Cricket (pp. 109–111). Gold Coast, Australia: Sheraton Mirage, 1–3 June.

Naikar, N (2017). Cognitive work analysis: An influential legacy extending beyond human factors and engineering. *Applied Ergonomics*, 528–540.

Newell, K. M. (1985). Coordination, control and skill. In: D. Goodman, R.B. Wilberg, & I.M. Franks (eds.), *Differing Perspectives in Motor Learning, Memory, and Control* (pp. 295–317). Amsterdam, the Netherlands: Elsevier Science.

Newell, K. M. (1986). Constraints on the development of coordination. In: M.G. Wade, & H.T.A. Whiting (eds.), *Motor Development in Children: Aspects of Coordination and Control* (pp. 341–360). Dordrecht, the Netherlands: Martinus Nijhoff.

Oxford Dictionary. (2019). Oxford University Press, viewed 27 November 2018. https://en .oxforddictionaries.com/definition/constraint

Pinder, R. A., Davids, K., & Renshaw, I. (2011a). Representative learning design and functionality of research and practice in sport. *Journal of Sport and Exercise Psychology*, 33(1), 146–155.

Pinder, R. A., Renshaw, I., Davids, K., & Kerhervé, H. (2011b). Principles for the use of ball projection machines in elite and developmental sport programmes. Sports Medicine, 41, 793–800.

Poole, J. L. (1991). Application of motor learning principles in occupational therapy. *American Journal of Occupational Therapy*, 45:6, 531–537.

Potrac, P., Jones, R. L., & Armour, K., (2002). It's all about getting respect: the coaching behaviours of an expert English soccer coach. *Sport Education and Society*, 7, 183–202.

Rasmussen, J. (1986). *Information Processing and Human-Machine Interaction: an Approach to Cognitive Engineering*. New York, NY: North-Holland.

Read, G. J. M., Salmon, P. M., Goode, N., & Lenne, M. (2018). A sociotechnical design toolkit for bridging the gap between systems-based analyses and system design. *Human Factors and Ergonomics in Management and Service*, 1–15.

Renshaw, I., Chappell, G., Fitzgerald, D., Davidson, J., & McFadyen, B. (2010a). The battle zone: constraint-led coaching in action. In: M. Portus (ed.), Conference of Science, Medicine and Coaching in Cricket (pp. 181–184). Gold Coast, Australia: Sheraton Mirage, 1–3 June.

Renshaw, I., Chow, J. Y., Davids, K., & Hammond, J. (2010c). A constraints-led perspective to understanding skill acquisition and game play: a basis for integration of motor learning theory and physical education praxis? *Physical Education and Sport Pedagogy*, 1, 1–21.

Renshaw, I., Davids, K., & Savelsbergh, G. (2010b). *Motor Learning in Practice: A Constraints-Led Approach*. London, UK: Routledge.

Renshaw, I., & Davids, K. W. (2014). Task constraints. In: R.C. Eklund, & g. Tenenbaum (eds.), *Encyclopedia of Sport and Exercise Psychology* (pp. 734–737). Los Angeles, CA: SAGE Publications.

Renshaw, I., & Holder, D. (2009). The 'nurdle to leg' and other ways of winning cricket matches. In: I. Renshaw k.w. Davids, & g.j.p. Savelsbergh (eds.), *Motor Learning in Practice: A Constraints-Led Approach* (pp. 109–119). London, UK: Routledge (Taylor & Francis Group).

Schmidt, R. A., & Wrisberg, C. A. (2004). *Motor Learning and Performance*. 3rd edition. Champaign, IL: Human Kinetics.

Smith, M., & Cushion, C. J. (2006). An investigation of the in-game behaviours of professional, top-level youth soccer coaches. *Journal of Sport Sciences*, 24, 355–366.

Stanton, N. A., Salmon, P. M., Walker, G. H., & Jenkins, D. P. (2017). *Cognitive Work Analysis: Applications, Extensions and Future Directions*. Boca Raton, FL: CRC Press

Steel, K. A., Harris, B., Baxter, D., & King, M. (2013). Skill acquisition specialists, coaches and athletes: the current state of play? *Journal of Sport Behavior*, 36:3, 291–305.

Stretch, R., McKellar, D., & Nurick, G. (1998a). The measurement of the position of a ball striking a cricket bat. In: S.J. Haake (ed.), *The Engineering of Sport: Design and Development*. Oxford, UK: Blackwell Science.

Stretch, R., Nurick, G., & McKellar, D. (1998b). Improving the accuracy and consistency of shot reproduction in cricket batting. *South African Journal for Research in Sport, Physical Education and Recreation*, 21:2, 77–88.

Tucker, R., & Collins, M. (2012). Athletic performance and risk of injury : can genes explain all? *Dialogues in Cardiovascular Medicine*, 17:1, 31–39.

Vicente, K. (1999). *Cognitive Work Analysis: Toward Safe, Productive, and Healthy Computer-Based Work*. Boca Raton, FL: CRC Press.

Vickery, W., Dascombe, B., & Duffield, R. (2014). Physiological, movement and technical demands of centre-wicket Battlezone, traditional net-based training and one-day cricket matches: a comparative study of sub-elite cricket players. *Journal of Sports Sciences*, 32:8, 722–737.

Weissensteiner, J., Abernethy, B., Farrow, D., & Müller, S. (2008). The development of anticipation : A cross-sectional examination of the practice experiences contributing to skill in cricket batting. *Journal of Sport and Exercise Psychology*, 30:6, 663–684.

Woolmer, B., & Noakes, T. D. (2008). *The Art and Science of Cricket*. Cape Town, South Africa: Struik Publishers. ISBN: 978-1-77007-658-7.

13 Performance Analysis in Sport

A Human Factors and Ergonomics Approach

Scott McLean, Paul M. Salmon, Adam D. Gorman, and Colin Solomon

CONTENTS

13.1 INTRODUCTION

Performance analysis (PA) in elite level football is well established, and globally accepted as a fundamental requirement for understanding and enhancing individual and team performance (Mackenzie & Cushion, 2013; Sarmento et al., 2014). In football, there has been a substantial amount of research, using a variety of research methods aimed at identifying the performance indicators (physical, technical, tactical, and psychological) related to successful and unsuccessful performance (Mackenzie & Cushion, 2013; Sarmento et al., 2014). Research investigating PA in football has traditionally involved descriptive, comparative, or predictive analysis, or combinations of these (Bishop, 2008; McGarry, 2009; Sarmento et al., 2014). Descriptive PA involves describing the actions of players during matches, e.g. distances run, and frequencies and percentages of technical behaviours (passing, tackles, and shots, etc.) (Dellal et al., 2010; Di Salvo et al., 2009). Comparative PA has involved investigating the differences between different physical and technical behaviours of players within different leagues and competitions, successful versus unsuccessful teams, quality of opposition, and matches played home or away (Chamari et al., 2010; Lago, 2009; Taylor et al., 2008). Predictive analysis is designed to determine the most effective methods of achieving future successful performance, and includes methods of scoring goals, and analysis of tactical patterns (e.g. counter-attack versus progressive build-up) (Camerino et al., 2012; Sarmento et al., 2018). Although these approaches have enhanced our understanding of football performance, they have been criticised for being reductionist, which limits a full understanding of the nonlinearity and complexity in football (McLean et al., 2017b; Salmon & McLean, 2019).

Within the past decade, advances in computer and video aided match analysis systems have produced an increase in new methods and research used to investigate PA in football (Low et al., 2019; Sarmento et al., 2018). The emergence of new PA methods derived from large data sets using player tracking technology have provided improvements in our understanding of group tactical behaviours (Low et al., 2019). For example, variables measuring group behaviours, such as team centroids, team dispersion, pattern analysis, and team connectedness have advanced upon the traditional descriptive, comparative, and predictive methods of PA (see review by Low et al., 2019). The advantage of these modern methods over traditional PA methods is the consideration of football as a complex and dynamic system, with continual interactions in space and time between players. For example, measuring the distance of the spread of a team (length and width) is important for understanding a team's attacking and defensive tactics and how this relates to performance (Sarmento et al., 2017). Although the use of methods to understand group behaviours has improved football PA on a theoretical basis, work is needed to ensure that the information generated by such methods is able to be understood and used by coaches and practitioners.

Together, the substantial amount of football PA research has contributed to an increased understanding of football performance. In particular, descriptive and comparative analysis methods have provided information regarding the who (player), what (action), where (location), and when (time) of performance in football (McGarry, 2009). However, descriptive, comparative and predictive PA research does not evaluate the 'how' and 'why' of performance, both of which are necessary to provide a complete understanding (McGarry, 2009). Understanding how and why players perform as they do will better inform coaches to improve performance, thereby supporting the fundamental aim of PA in sport which is to improve the performance of individual players and teams (Bishop, 2008). To achieve this, PA methods need to focus on the who-what-where-when-why-and-how for all components of performance (McGarry, 2009). The extensive use of descriptive and comparative research methods in PA research is a possible explanation for why the majority of research on PA in football has only had a minimal impact on coaches translation of research into practice (Bishop, 2008; Drust & Green, 2013). Describing the who-what-where-when sequence that occurs in a football match limits what coaches are able to gain from such research (Bishop, 2008; Drust & Green, 2013; Mackenzie & Cushion, 2013; McGarry, 2009). By not evaluating the how and why of performance, current approaches define football as a series of uncomplicated steps to follow to ensure successful performance (Mackenzie & Cushion, 2013). Further, by not integrating these different elements these current approaches do not consider the complexity and numerous interactions that occur during a game of football (McLean et al., 2017b; Salmon and McLean, 2019).

Although PA is shifting more towards understanding the inherent complexity of football (Low et al., 2019), there remain key issues with current analysis methods, gaps in the knowledge base, and an apparent lack of specific future directions of PA. In addition, several longstanding issues remain (Carling et al., 2014; Drust & Green, 2013; Mackenzie & Cushion, 2013; Sarmento et al., 2014). For example, the absence of a clear conceptual base of current PA (Carling et al., 2014), including what football performance actually comprises (McLean et al., 2017a; Salmon & McLean, 2019), the lack of a set of standardised operational definitions (Sarmento et al., 2014), the suitability of PA methods and findings for use in practice, and the resistance of coaches and practitioners to accept and use new PA methods (Drust & Green, 2013). These are in addition to the commonly criticised reductionist approaches of typical notational analysis, and the lack of consideration of match context (Mackenzie & Cushion, 2013; Sarmento et al., 2014).

It is our contention that many of these critical issues can be resolved by developing and applying new forms of PA methods, both independently and in combination with existing measures. Appropriate analysis methods exist and are used in disciplines such as HFE to examine human, team, organisational, and system behaviour (see Chapter 1). Notably, in complex systems analysis disciplines, such as that of HFE, there has been a shift in the way that performance is viewed, analysed, and addressed. Central to this shift is the argument that deterministic approaches do not fully consider and are not designed to measure the inherent complexity in socio-technical systems or the full range of factors influencing behaviour (Read et al., 2013; Salmon, Walker, Stanton, Goode, & Read, 2017). Using an example from road safety, in which practitioners have typically taken a deterministic, reductionist approach to road trauma prevention, involving breaking down systems into their component parts (e.g. drivers, vehicles), examining the parts in isolation (e.g. driver errors, vehicle safety components), attempting to improve the performance of single parts, and then reintroducing those parts back into the system on the assumption that overall system behaviour will improve.

This deterministic and reductionist approach appears to be prevalent in PA in football where it is common practice to break down a match into passes, tackles, distance run, and time in specific speed zones, which are then assessed and used together to assess performance without actually integrating the measures. Whilst a detailed understanding of the selected components of performance is achieved, team and match level performance remains poorly understood. It is therefore argued that a systems HFE approach is required in PA to provide a deeper and more meaningful and informative overview of performance. In road safety research, the use of systems HFE methods have initiated a paradigm shift in road safety research (Salmon et al., 2019). It is argued here that HFE methods can be used in a similar manner to initiate a paradigm shift in PA in football. The aim of this article is to introduce and demonstrate the use of systems HFE methods that have the potential to provide a more complete and comprehensive understanding to PA in football.

13.2   SYSTEMS HUMAN FACTORS AND ERGONOMICS METHODS

As outlined in Chapter 1, HFE methods are used to study the performance of humans, teams, organisations, and overall socio-technical systems. A fundamental component and contribution of systems HFE methods is the capacity to describe complex systems and the factors that interact to determine how well the system performs. Another advantage of HFE methods over traditional PA methods is the use of Subject Matter Experts (SME's) in study design, data collection, and review and refinement of analysis outputs. This ensures that outputs are valid and relevant and specific to the topic of research. For example, if this approach were to be applied to research in football, SME's such as players, coaches, and sports scientists would be active in the design of the project, data collection, and review of the outputs to ensure validity, relevance, and importance to practitioners. Unfortunately, this is currently not the case in the majority of football PA studies which are predominantly undertaken by researchers.

As noted in Chapter 1, there are well over 100 HFE methods available (Stanton et al., 2013), many of which can be applied to sport and PA to provide a detailed and unique perspective on sports performance (Salmon, Stanton, Gibbon, Jenkins, & Walker, 2009). To demonstrate the potential utility of HFE methods in sports PA, we discuss broad categories of HF methods to advance PA in football. Specifically, we focus on the following categories of method: (1) Systems analysis and design: Cognitive Work Analysis (CWA; Vicente, 1999), (2) Teamwork assessment: Social Network Analysis (SNA) and the Event Analysis of Systemic Teamwork (EAST, Stanton et al., 2018), and (3) Cognitive task analysis: The Critical Decision Method (CDM; Klein et al., 1989). The use of these four methods, each representing different components in the context of football performance, will advance our understanding of performance in football and subsequently the way in which we assess and analyse football.

13.3   SYSTEMS ANALYSIS AND DESIGN: COGNITIVE WORK ANALYSIS

Cognitive Work Analysis is a framework which is used to support the design, analysis, and evaluation of behaviour in complex sociotechnical systems (Naikar, 2013; Vicente, 1999). CWA provides a detailed analysis of the system under investigation via five phases (Vicente, 1999). These include (1) an initial model of the system, (2) an analysis of the control tasks that are undertaken in the system, and the constraints imposed on these activities during different situations, (3) a description of how work in the system is and can potentially be achieved, (4) a description of how activities are and can be distributed among human and non-human agents in the system, and (5) an analysis of the cognitive competencies that an ideal worker should exhibit (Jenkins et al., 2008; Naikar, 2013; Vicente, 1999).

By focusing on the constraints within the system, the CWA framework is used to establish how tasks can be done, rather than focusing only upon how tasks should be done as determined using predictive PA in football, or how tasks are currently done in the system as determined using descriptive and comparative PA approaches (Naikar, 2013). Given that football possesses many of the characteristics of a sociotechnical system (i.e. human and/or technical elements working together towards a shared goal, see Chapter 2), CWA can be used to enhance how various components of football can be analysed, and evaluated (McLean et al., 2017a). Applying CWA to PA in football in conjunction with the current methods has the potential for the development of new ideas and approaches to evaluate performance. A notable feature of the CWA framework is that it can also be used to design or modify systems (Read et al., 2016). Hence, CWA offers a framework for PA that can also be used to develop coaching and tactical interventions designed to improve performance.

The potential utility of CWA for PA in football has been demonstrated (McLean et al., 2017a). McLean et al. (2017a) used the first phase of CWA, Work Domain Analysis (WDA), to develop a constraints-based model of the football 'match' system, with the specific aim of identifying the components of performance that should be analysed through PA. The study added to the current knowledge base on football PA by identifying, first, that the game of football is indeed a complex socio-technical system, with many competing functions. Second, a research-practitioner gap exists, whereby researchers do not often investigate and assess features of performance that are actually useful to coaches and practitioners. Third, novel PA measures not currently investigated by football researchers, which were deemed to be important by SMEs (McLean et al., 2017a). A summary of the WDA model of a football match, developed using the abstraction hierarchy method, is presented in Figure 13.1. The model depicts the physical objects in the system through to the processes and functions required to meet the overall purposes of the match system, as well as the measures used to assess whether the purposes are being achieved. The model shows the relationships and interactions of the different components of performance and highlights the many interacting functions and measures of performance that occur during a match.

Although McLean et al. (2017a) only used the WDA phase, one of the primary strengths of CWA is that the five-phase analysis approach allows the method to be applied for many purposes (Figure 13.2). For example, applications in the safety-critical domains have used CWA for system design (Bisantz et al., 2003), team design (Naikar et al., 2003), training program design (Naikar et al., 1999), situational awareness requirements analysis (Jenkins et al., 2010), decision requirements analysis (Mulvihill et al., 2016), and the development of performance measures (Crone et al., 2003, 2007). This flexibility is particularly appealing for potential sports applications and allows it to be applied for various purposes in football. For example, the decision ladder method used during phase 2 (Control task analysis) could be used to identify decision making and situation awareness requirements during different match scenarios such as defending set-pieces. This information in turn could be used to support the development of coaching practices. The Strategies Analysis phase could be used to identify the range of potential ways in which different match strategies could be enacted, and then to assess the different ways in which activities are undertaken during games. Social Organisation and Co-operation Analysis (SOCA), the fourth phase, could be used to identify which players should perform different activities during the game, which again could assist with training design. Finally, Worker Competencies Analysis (WCA) could be used to understand and assess the cognitive aspects of player performance. Using the full CWA framework for PA in football has the capacity to provide an in-depth analysis of the who-what-where-when-why-and-how components of performance.

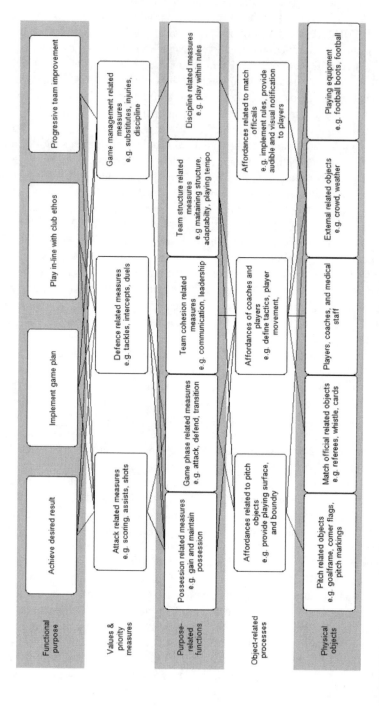

FIGURE 13.1 Summary of the WDA for a football match system, developed by McLean et al., (2017).

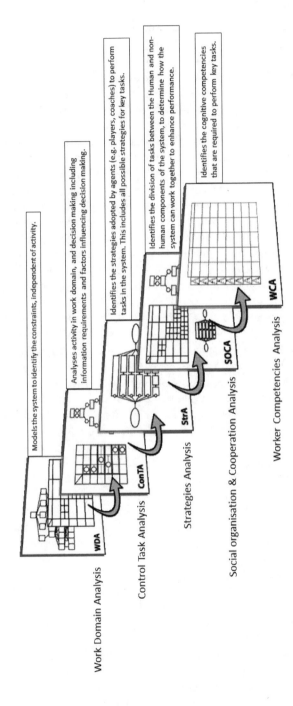

Models the system to identify the constraints, independent of activity.

Analyses activity in work domain, and decision making including information requirements and factors influencing decision making.

Identifies the strategies adopted by agents (e.g. players, coaches) to perform tasks in the system. This includes all possible strategies for key tasks.

Identifies the division of tasks between the Human and non-human components of the system, to determine how the system can work together to enhance performance.

Identifies the cognitive competencies that are required to perform key tasks.

WDA

ConTA

StrA

SOCA

WCA

Work Domain Analysis

Control Task Analysis

Strategies Analysis

Social organisation & Cooperation Analysis

Worker Competencies Analysis

FIGURE 13.2 The CWA framework describing each level of analysis. (Adapted from Jenkins et al., 2008.)

13.4 TEAMWORK ASSESSMENT: SOCIAL NETWORK ANALYSIS (SNA)

Social Network Analysis (SNA) is a method used in HFE to understand network structures via description, visualisation, and statistical modelling (van Duijn & Vermunt, 2006;Wasserman & Faust, 1994). In the past decade, SNA has been applied either on its own (Houghton et al., 2006) or as part of frameworks such as EAST (Stanton et al., 2013) across multiple domains and topics. These topics have included communications and teamwork in different systems (Baber et al., 2013; Barth et al., 2015), interactions between drivers and vulnerable road users (Salmon, Lenne, Walker, Stanton, & Filtness, 2014), and emergency service response teams (Houghton et al., 2006). SNA provides an appropriate method for assessing team performance in terms of connectivity and interactions of team members, as well as determining the most influential entities (Clemente et al., 2015a; van Duijn & Vermunt, 2006). In a football context, SNA has been used to analyse multiple performance components including match passing networks (Grund, 2012), communication networks (Mclean et al., 2018a), unsuccessful passing networks (Mclean & Salmon, 2019), the use of different playing formations (Clemente et al., 2015b), and pitch locations involved in passing networks (Mclean et al., 2018b). SNA can potentially be used in all team sports that require the co-operation and interactions of team members to collectively overcome the opposition (Clemente et al., 2015a).

When performing network analysis in football, the players represent the nodes within the network and the connections between players represent the links between the nodes (Grund, 2012). Passing has typically been considered as the primary connection between players; however, connectedness via verbal and non-verbal communication has also been assessed using SNA (Mclean et al., 2018a). Social network data is collected using a variety of methods such as questionnaires, interviews, and observations (van Duijn & Vermunt, 2006). For passing networks, the collection of data is best achieved via observation of the matches, whereas for the communication networks, post-match questionnaires have been used (Mclean et al., 2018a). SNA uses graph theory to produce quantifiable metrics to understand team connectivity, as well as multiple metrics to measure individual prominence (Table 13.1).

In addition to providing quantifiable metrics, SNA diagrams provide a visual representation of networks (Figure 13.3). This is important as coaches often prefer a quick snapshot of performance (Carling et al., 2014), and SNA diagrams are capable of providing visualisations for coaches to quickly assess if key aspects of the match plan are being accomplished or not.

Previous research on match passing networks has shown that a fundamental characteristic of successful teams, compared to less successful teams, is the high level of connectivity (density) between the players (Clemente et al., 2015a; Grund, 2012). Furthermore, high levels of passing connectivity was a characteristic of successful, compared to unsuccessful teams, at the 2014 World Cup (Clemente et al., 2015a). For the goal scoring passing networks at the 2016 European Championships (EUROs), on average the networks contained low-density passing networks comprising three players with low reciprocity over a short duration (8 s) and using two pitch zones (Mclean et al., 2018b). For SNA of intra-team communication, a professional team won more matches when communication was increased, but in matches with low

TABLE 13.1

Description of SNA Metrics and Current Uses for Performance Analysis in Football

SNA variable	Description	Uses to date
Network edge	An edge is a connection between two nodes.	Passing; communication
Network density	How connected a network is which is represented by a score between zero and one, i.e. a score of one indicates every node is connected to every other node, whereas a score of zero indicates no node is connected to any other node.	Passing; communication
Network cohesion	How reciprocal a network is, i.e. a cohesive network will have connections exchanged back and forth between nodes.	Passing; communication
Sociometric status	Identifies the most connected entities in terms of incoming and outgoing ties.	Passing; communication
In-Degree centrality	How many inbound ties a node receives from all other nodes in the network.	Passing; pitch zones; communication
Out-Degree centrality	How many outbound ties a node has to all other nodes in the network.	Passing; pitch zones; communication
Closeness centrality (CC)	How close each node is to all other nodes in the network, i.e. a node with high CC can reach many other nodes in a few steps.	Passing
Betweenness centrality (BC)	The extent to which a node lies on a path between two nodes, i.e. nodes that act like a bridge.	Passing
Eigenvector centrality	An extension of degree centrality, it gives each actor a score proportional to the scores of its neighbours, i.e. a node may be important because it has numerous ties, or it has or fewer ties to other important nodes.	Passing
Clustering coefficient	The degree to which nodes tend to cluster together, i.e. how close a node and its neighbours are to making a sub-network (clique).	Passing

possession the team communicated more compared to matches with high possession. This indicates that increased defending requires more communication for organising the defence to respond to the opposition's attacks (Mclean et al., 2018a).

The benefit of SNA over traditional notational analysis where the players are examined as independent and autonomous entities (Lusher et al., 2010) is that SNA views the entire team, including the connectivity of all team members (Clemente et al., 2015a), as well as the prominent players relative to all team members (Clemente et al., 2015b; McLean et al., 2017a). The SNA metrics produced during the analysis, provide a more detailed description, compared to the typical reductionist approach of frequencies and percentage measures, by way of identifying the interactions of all components in the system (McLean et al., 2017). A limitation of SNA for passing in football is that it is a static representation of the passing networks. Future uses of SNA for passing analysis in football would benefit from developing dynamic passing

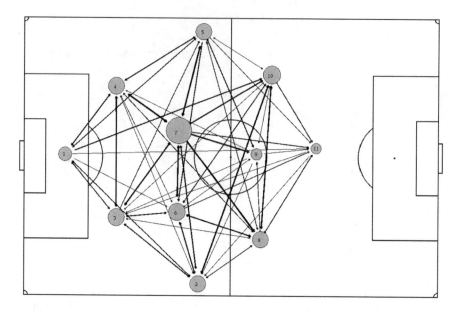

FIGURE 13.3 Social network diagram of the passing network of a professional team in a full competitive match. The direction of attack is from left to right (node #1 is the goal-keeper), the position of the players is the average position for the match recorded by GPS. The node (player) size is based on degree centrality (number of outward passes in the match). The larger nodes represent more passes made, and the edges (passes) are weighted by frequency of passes completed between players e.g. thicker lines indicate a greater number of passes.

networks that also include the player movements derived from player tracking technologies. Lastly, the use of SNA has mostly been limited to passing in football, with only one use each for communication (McLean et al., 2018a) and for pitch zone analysis (McLean et al., 2018b) (Table 13.1). This should encourage researchers to expand the use of SNA in football to new areas of PA.

13.5 TEAMWORK ASSESSMENT: EVENT ANALYSIS OF SYSTEMIC TEAMWORK (EAST)

The Event Analysis of Systematic Teamwork (EAST) (see Chapters 1 and 10) is a framework developed to assess complex sociotechnical system performance, using the integration of three network analysis-based methods including: (1) task networks, (2) social networks, and (3) information networks (Stanton et al., 2013). EAST uses these three interlinked network-based representations to describe and analyse the behaviour of team, organisational, or system behaviour (Stanton et al., 2018). (Figure 13.4)

EAST has previously been used to analyse individual, team, organisational and system performance in various domains outside of sport, including road and rail safety, aviation, land and naval warfare, air traffic control, the darknet, and emergency service response (Stanton et al., 2018).

Within the EAST framework, task networks are used to describe tasks being performed within a system (i.e. which agents, both human and non-human, do what).

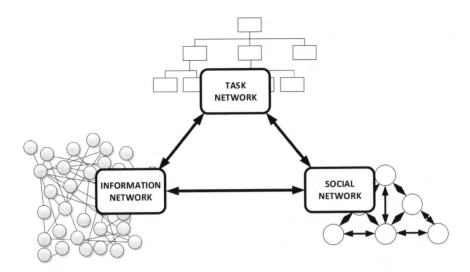

FIGURE 13.4 EAST network of networks analysis approach. The Figure shows example representations of each network including task network, social network, and information network.

In football, task networks could be built from post-match video analysis of the tasks that players perform during a match, or specific scenarios (i.e. a counter-attack). Alternatively, performance could be mapped onto pre-defined 'optimal performance' task networks to identify where requisite tasks were achieved or omitted. As described above, social networks are used to analyse the interactive structure of the system and the communications taking place between agents (i.e. who/what interacts and communicates with who/what). Moment to moment communication exchanges are yet to be captured in football; however, communications could be captured post-match using video footage and asking players to describe how they were communicating in given scenarios. In a football context, the social networks could also be used to capture the passing network (as per the above section). Information networks show what information is held and used by different agents within the system (i.e. who/what knows what at different periods in time) (Stanton et al., 2013). For football PA, information networks could be used to identify what information is required for different match scenarios and subsequently to identify what information was used by whom during the match. For example, a counterattack information network would show what information was used by the different players involved.

Task, social, and information networks are subsequently integrated to provide a composite network which shows the relationships between agents, tasks, communications (or passing), and information. By showing the relationships between task, social, and information networks, and then by analysing these networks using network analysis metrics, an in-depth understanding of performance is obtained.

To date, EAST has only been used once for sports performance analysis, despite case study analyses demonstrating its potential utility (Salmon et al., 2009). The only application of EAST to analyse elite sports performance has been to understand the teamwork and situational awareness of cyclists in a cycling peloton in elite women's road races (Salmon et al., 2017). The analysis identified the key tasks, interactions, and information required during an elite women's cycling road race. The application of EAST to football has enormous potential to advance our understanding of the situational awareness of players in certain scenarios. For example, EAST could be used to determine the required tasks, information used, and communication between players for multiple scenarios including counter-attacks, and the different phases of play (possession, transition, opposition possession). By showing the relationships between task, social, and information networks, and then by analysing these using network metrics an in-depth understanding of football performance is achieved. This approach will enable the design of training practice to replicate specific match related scenarios and enhance SA in these types of match scenarios.

13.6   COGNITIVE TASK ANALYSIS: CRITICAL DECISION METHOD

Cognitive task analysis methods are used to describe and analyse the cognitive processes underpinning performance. The Critical Decision Method (CDM) (Klein et al., 1989) is the most commonly used cognitive task analysis method within HFE. The CDM is used to examine how and why critical decisions are made in complex environments, by asking SMEs how and why they did what they did during key decision points (Crandall et al., 2006). The CDM involves conducting semi-structured interviews with the decision-maker to understand the cognitive processes involved, the courses of action available, and the information used to support decision-making at each decision point during the scenario under analysis (Crandall et al., 2006; Klein et al., 1989). CDM has previously been used in domains such as emergency services (Blandford & Wong, 2004), the military (Salmon et al., 2009), aviation (Plant & Stanton, 2013), energy distribution (Salmon et al., 2008), road transport (Walker et al., 2009), and white-water rafting (O'Hare et al., 2000) to obtain information on the decisions made regarding critical incidents. The outputs describe various facets of decision-making and performance, including the aims that influence behaviour, the information used, potential courses of action considered, and the factors influencing decision making. Findings are typically used to understand performance and aid the design of interventions to improve future performance such as training drills and programs, technological support, and new operating procedures (Hoffman et al., 1998).

There are numerous aspects of football performance where CDM could assist in understanding the cognitive processes underpinning coach and player behaviour and decision making. For example, specific match scenarios could be analysed in-depth to reveal the factors influencing decision-making and performance. These include goals scored and missed, penalties, disciplinary issues, coach's decisions, defensive errors, and saves made by goalkeepers. The outputs derived from CDM can then be used for the design of training practices to replicate these specific scenarios. Whilst decision-making research is common in sport science, including visual search strategies, perceptual cue and gaze fixation, and response time and accuracy, particularly comparing expert and novice performers (Kim & Lee, 2006; Mann et al., 2007; Vilar et al., 2013; Williams et al., 1994), less work has been conducted to understand the cognitive processes that contribute to decision-making in dynamic 'real world' sporting situations (Kermarrec & Bossard, 2013; Williams & Ericsson, 2005). This is perhaps due to the difficulty of measuring decision-making in dynamic sporting environments with sufficient levels of precision and without disrupting the performers themselves (Kermarrec & Bossard, 2013; Smith et al., 2016; Williams & Ericsson, 2005).

A further advantage of the CDM is its capacity to elicit the players' implicit, rather than explicit, knowledge (Klein et al., 1989). An example of explicit knowledge is when a player performs a certain action (routines and behaviours) based on the coaches' instructions, whereas the CDM is able to elicit the implicit knowledge (pattern recognition, mental models, judging typicality, and perceptual discriminations) that is responsible for critical decisions (Klein et al., 1989).

The detailed analysis of the scenario under investigation is performed by using predefined cognitive probes to elicit the cognitive processes underlying the decisions made by the performer. The cognitive probes for CDM are presented in Table 13.2.

Additional uses of CDM in football could be to evaluate coach decision-making (e.g. selection of tactics, substitutions) and the decisions around team selection and

---

TABLE 13.2
Critical Decision Method Cognitive Probes

Goal specification	What were your specific goals at the various decision points
Cue identification	What features were you looking for when you formulated your decision? How did you know you needed to make that decision? How did you know when to make that decision?
Expectancy	Were you expecting to make this sort of decision during the match? Describe how this affected your decision-making process?
Conceptual	Are there any situations in which your decision would have turned out differently?
Influence of uncertainty	At any stage, were you uncertain about either the reliability or the relevance of the information you had available?
Information integration	What was the most important piece of information that you used to formulate the decision?
Situation awareness	What information did you have available to you at the time of the decision?
Situation assessment	Did you use all of the information available to you when formulating the decision? Was there any additional information that you might have used to assist in the formulation of the decision?
Options	Were there any other alternatives to you other than the decision you made?
Decision blocking	Was there any stage during the decision-making process in which you found it difficult to process and integrate the information available?
Basis of choice	Do you think you could develop a rule, based on your experience, which could assist another person to make the same decision successfully?
Analogy/generalisation	Were you at any time reminded of previous experiences in which a similar/ different decision was made?

Source of CDM cognitive probes comes from O'Hare et al. (2000), and has been previously used in published articles (Salmon et al., 2009; Stewart et al., 2008; Walker et al., 2009).

player transfers. More importantly, the CDM would provide information on how and why players chose to make a specific decision during a match. Practical uses of CDM could be to investigate specific critical moments identified in match review process undertaken by coaches and analysts to obtain the players' perspectives on decisions made.

13.7 CONCLUSION

The aim of this chapter was to introduce and discuss the potential application of a series of analytical HFE methods to enhance PA in football. This argument is based on football systems having many of the characteristics of complex safety-critical systems, along with an expanding set of HF applications in sport (see Chapters 1 and 2). The inescapable conclusion when one considers existing PA measures is that, despite decades of research, we still do not fully understand football performance and the ways in which its components interact with one another. This key knowledge gap is omnipresent across most sporting domains. Contemporary HFE methods have been developed specifically to describe and analyse complexity and describe exactly these features of performance. It is our contention that further research should investigate the use of existing HFE methods in the football PA context; it is encouraging also to see many chapters in this book applying such methods in different sports for purposes outside of PA. A final research direction to highlight is the integration of HFE methods with existing PA methods to form a toolkit approach to football PA. This integration would ostensibly provide coaches and practitioners with a way of coping with the complexity of football systems.

REFERENCES

Baber, C., Stanton, N., Atkinson, J., McMaster, R., & Houghton, R. J. (2013). Using social network analysis and agent-based modelling to explore information flow using common operational pictures for maritime search and rescue operations. *Ergonomics*, 56:6, 889–905.

Barth, S., Schraagen, J. M., & Schmettow, M. (2015). Network measures for characterising team adaptation processes. *Ergonomics*, 58:8, 1287–1302.

Bisantz, A. M., Roth, E., Brickman, B., Gosbee, L. L., Hettinger, L., & McKinney, J. (2003). Integrating cognitive analyses in a large-scale system design process. *International Journal of Human-Computer Studies*, 58:2, 177–206.

Bishop, D. (2008). An applied research model for the sport sciences. Sports Medicine, 38:3, 253–263.

Blandford, A., & Wong, B. W. (2004). Situation awareness in emergency medical dispatch. *International Journal of Human-Computer Studies*, 61:4, 421–452.

Camerino, O. F., Chaverri, J., Anguera, M. T., & Jonsson, G. K. (2012). Dynamics of the game in soccer: Detection of T-patterns. *European Journal of Sport Science*, 12:3, 216–224.

Carling, C., Wright, C., Nelson, L. J., & Bradley, P. S. (2014). Comment on 'Performance analysis in football: A critical review and implications for future research'. *Journal of Sports Sciences*, 32:1, 2–7.

Chamari, K., Dellal, A., Wong, P. L., & Moalla, W. (2010). Physical and technical activity of soccer players in the French First League: With special reference to their playing position. *International Sports Medicine Journal*, 11:2, 278–290.

Clemente, F. M., Martins, F. M. L., Kalamaras, D., Wong, D. P., & Mendes, R. S. (2015a). General network analysis of national soccer teams in FIFA World Cup 2014. *International Journal of Performance Analysis in Sport*, 15:1, 80–96.

Clemente, F. M., Martins, F. M. L., Wong, D. P., Kalamaras, D., & Mendes, R. S. (2015b). Midfielder as the prominent participant in the building attack: A network analysis of national teams in FIFA World Cup 2014. *International Journal of Performance Analysis in Sport*, 15:2, 704–722.

Crandall, B., Klein, G. A., & Hoffman, R. R. (2006). *Working Minds: A Practitioner's Guide to Cognitive Task Analysis.* Cambridge, MA: MIT Press.

Crone, D., Sanderson, P., Naikar, N., & Parker, S. (2007). Selecting sensitive measures of performance in complex multivariable environments. Paper presented at the Proceedings of the 2007 Simulation Technology Conference (SimTecT'07), Brisbane, Australia.

Crone, D. J., Sanderson, P. M., & Naikar, N. (2003). Using cognitive work analysis to develop a capability for the evaluation of future systems. Paper presented at the Proceedings of the Human Factors and Ergonomics Society Annual Meeting.

Dellal, A., Wong, D., P.W, Moalla, & Chamari, K. (2010). Physical and technical activity of soccer players in the French First League – with special reference to their playing position. *International Sports Medicine Journal*, 11:2, 278–290.

Di Salvo, V., Gregson, W., Atkinson, G., Tordoff, P., & Drust, B. (2009). Analysis of high intensity activity in Premier League soccer. *International Journal of Sports Medicine*, 30:3, 205–212. doi:10.1055/s-0028-1105950.

Drust, B., & Green, M. (2013). Science and football: evaluating the influence of science on performance. *Journal of Sports Sciences*, 31:13, 1377–1382.

Grund, T. U. (2012). Network structure and team performance: The case of English Premier League soccer teams. *Social Networks*, 34:4, 682–690.

Hoffman, R. R., Crandall, B., & Shadbolt, N. (1998). Use of the critical decision method to elicit expert knowledge: A case study in the methodology of cognitive task analysis. *Human Factors*, 40:2, 254–276.

Houghton, R. J., Baber, C., McMaster, R., Stanton, N. A., Salmon, P., Stewart, R., & Walker, G. (2006). Command and control in emergency services operations: a social network analysis. *Ergonomics*, 49:12–13, 1204–1225.

Jenkins, D. P., Stanton, N. A., Salmon, P. M., Walker, G. H., & Rafferty, L. (2010). Using the decision-ladder to add a formative element to naturalistic decision-making research. *International Journal of Human–Computer Interaction*, 26:2–3, 132–146.

Jenkins, D. P., Stanton, N. A., Salmon, P. M., Walker, G. H., & Young, M. (2008). Using cognitive work analysis to explore activity allocation within military domains. *Ergonomics*, 51:6, 798–815.

Kermarrec, G., & Bossard, C. (2013). A naturalistic decision-making investigation of football defensive players: an exploratory study. Paper presented at the International conference on Naturalistic Decision Making (Vol. 11, pp.163–171).

Kim, S., & Lee, S. (2006). Gaze behavior of elite soccer goalkeeper in successful penalty kick defense. *International Journal of Applied Sports Sciences*, 18:1, 96–110.

Klein, G. A., Calderwood, R., & Macgregor, D. (1989). Critical decision method for eliciting knowledge. *IEEE Transactions on Systems, Man, and Cybernetics*, 19:3, 462–472.

Lago, C. (2009). The influence of match location, quality of opposition, and match status on possession strategies in professional association football. *Journal of Sports Sciences*, 27:13, 1463–1469.

Low, B., Coutinho, D., Gonçalves, B., Rein, R., Memmert, D., & Sampaio, J. (2019). A systematic review of collective tactical behaviours in football using positional data. *Sports Medicine*, 50, 1–43.

Lusher, D., Robins, G., & Kremer, P. (2010). The application of social network analysis to team sports. *Measurement in Physical Education and Exercise Science*, 14:4, 211–224.

Mackenzie, R., & Cushion, C. (2013). Performance analysis in football: A critical review and implications for future research. *Journal of Sports Sciences*, 31:6, 639–676.

Mann, D. T., Williams, A. M., Ward, P., & Janelle, C. M. (2007). Perceptual-cognitive expertise in sport: A meta-analysis. *Journal of Sport and Exercise Psychology*, 29(4), 457.

McGarry, T. (2009). Applied and theoretical perspectives of performance analysis in sport: Scientific issues and challenges. *International Journal of Performance Analysis in Sport*, 9:1, 128–140.

McLean, S., Salmon, P., Gorman, A., Naughton, M., & Solomon, C. (2017a). Do inter-continental playing styles exist? Using social network analysis to compare goals from the 2016 EURO and COPA football tournaments knock-out stages. *Theoretical Issues in Ergonomics Science*. doi:10.1080/1463922X.2017.1290158.

Mclean, S., & Salmon, P. M. (2019). The weakest link: a novel use of network analysis for the broken passing links in football. *Science and Medicine in Football*, 1–4.

Mclean, S., Salmon, P. M., Gorman, A. D., Dodd, K., & Solomon, C. (2018a). Integrating communication and passing networks in football using social network analysis. *Science and Medicine in Football*, 1–7.

McLean, S., Salmon, P. M., Gorman, A. D., Read, G. J., & Solomon, C. (2017b). What's in a game? A systems approach to enhancing performance analysis in football. *PLoS One*, 12:2, e0172565.

Mclean, S., Salmon, P. M., Gorman, A. D., Stevens, N. J., & Solomon, C. (2018b). A social network analysis of the goal scoring passing networks of the 2016 European Football Championships. *Human Movement Science*, 57, 400–408.

Mulvihill, C. M., Salmon, P. M., Beanland, V., Lenné, M. G., Read, G. J., Walker, G. H., & Stanton, N. A. (2016). Using the decision ladder to understand road user decision making at actively controlled rail level crossings. *Applied Ergonomics*, 56, 1–10.

Naikar, N. (2013). *Work Domain Analysis: Concepts, Guidelines, and Cases*. Boca Raton, FL: CRC Press.

Naikar, N., Pearce, B., Drumm, D., & Sanderson, P. M. (2003). Designing teams for first-of-a-kind, complex systems using the initial phases of cognitive work analysis: Case study. *Human Factors*, 45:2, 202–217.

Naikar, N., Sanderson, P. M., & Lintern, G. (1999). Work domain analysis for identification of training needs and training-system design. Paper presented at the Proceedings of the Human Factors and Ergonomics Society Annual Meeting.

O'Hare, D., Wiggins, M., Williams, A., & Wong, W. (2000). Cognitive task analysis for decision centred design and training. *Task Analysis*, 41, 170–190.

Plant, K. L., & Stanton, N. A. (2013). What is on your mind? Using the perceptual cycle model and critical decision method to understand the decision-making process in the cockpit. *Ergonomics*, 56:8, 1232–1250.

Read, G. J., Salmon, P. M., & Lenné, M. G. (2013). Sounding the warning bells: The need for a systems approach to understanding behaviour at rail level crossings. *Applied Ergonomics*, 44:5, 764–774.

Read, G. J., Salmon, P. M., Lenné, M. G., & Stanton, N. A. (2016). Walking the line: understanding pedestrian behaviour and risk at rail level crossings with cognitive work analysis. *Applied Ergonomics*, 53, 209–227.

Salmon, P. M., Clacy, A., & Dallat, C. (2017). It's not all about the bike: distributed situation awareness and teamwork in elite women's cycling teams. *Contemporary Ergonomics*, 2017, 240–248.

Salmon, P. M., Lenne, M. G., Walker, G. H., Stanton, N. A., & Filtness, A. (2014). Using the Event Analysis of Systemic Teamwork (EAST) to explore conflicts between different road user groups when making right hand turns at urban intersections. *Ergonomics*, 57:11, 1628–1642.

Salmon, P. M., & McLean, S. (2019). Complexity in the beautiful game: implications for football research and practice. *Science and Medicine in Football*, 1–6.

Salmon, P. M., Read, G. J., Beanland, V., Thompson, J., Filtness, A. J., Hulme, A., McClure, R., & Johnston, I. (2019). Bad behaviour or societal failure? Perceptions of the factors contributing to drivers' engagement in the fatal five driving behaviours. *Applied Ergonomics*, 74, 162–171.

Salmon, P. M., Stanton, N. A., Gibbon, A., Jenkins, D., & Walker, G. H. (2009). *Human Factors Methods and Sports Science: A Practical Guide*. Boca Raton: CRC Press.

Salmon, P. M., Stanton, N. A., Walker, G. H., Jenkins, D., Baber, C., & McMaster, R. (2008). Representing situation awareness in collaborative systems: A case study in the energy distribution domain. *Ergonomics*, 51:3, 367–384.

Salmon, P. M., Stanton, N. A., Walker, G. H., Jenkins, D., Ladva, D., Rafferty, L., & Young, M. (2009). Measuring situation awareness in complex systems: Comparison of measures study. *International Journal of Industrial Ergonomics*, 39:3, 490–500.

Salmon, P. M., Walker, G., Stanton, N., Goode, N., & Read, G. J. (2017). Fitting methods to paradigms: are ergonomics methods fit for systems thinking? *Ergonomics*, 60:2, 194–205.

Sarmento, H., Anguera, M. T., Pereira, A., & Araújo, D. (2018). Talent identification and development in male football: A systematic review. Sports Medicine, 48:4, 907–931.

Sarmento, H., Clemente, F. M., Araújo, D., Davids, K., McRobert, A., & Figueiredo, A. (2017). What performance analysts need to know about research trends in association football (2012–2016): A systematic review. *Sports Medicine*, 1–38.

Sarmento, H., Marcelino, R., Anguera, M. T., CampaniÇo, J., Matos, N., & LeitÃo, J. C. (2014). Match analysis in football: a systematic review. *Journal of Sports Sciences*, 32:20, 1831–1843.

Smith, M. R., Coutts, A. J., Merlini, M., Deprez, D., Lenoir, M., & Marcora, S. M. (2016). Mental fatigue impairs soccer-specific physical and technical performance. *Medicine and Science in Sports and Exercise*, 48:2, 267–276.

Stanton, N., Salmon, P., & Walker, G. (2018). *Systems Thinking in Practice: Applications of the Event Analysis of Systemic Teamwork Method*. Boca Raton, FL: CRC Press.

Stanton, N., Salmon, P. M., & Rafferty, L. A. (2013). *Human Factors Methods: a Practical Guide for Engineering and Design*. Farmham, UK: Ashgate Publishing, Ltd.

Stewart, R., Stanton, N. A., Harris, D., Baber, C., Salmon, P., Mock, M., Tatlock, K., Wells, L., & Kay, A. (2008). Distributed situation awareness in an airborne warning and control system: Application of novel ergonomics methodology. *Cognition, Technology and Work*, 10:3, 221–229.

Taylor, J. B., Mellalieu, S. D., James, N., & Shearer, D. A. (2008). The influence of match location, quality of opposition, and match status on technical performance in professional association football. *Journal of Sports Sciences*, 26:9, 885–895.

van Duijn, M. A., & Vermunt, J. K. (2006). What is special about social network analysis? *Methodology*, 2:1, 2–6.

Vicente, K. J. (1999). *Cognitive Work Analysis: Toward Safe, Productive, and Healthy Computer-Based Work*. Boca Raton, FL: CRC Press.

Vilar, L., Araújo, D., Davids, K., Correia, V., & Esteves, P. T. (2013). Spatial-temporal constraints on decision-making during shooting performance in the team sport of futsal. *Journal of Sports Sciences*, 31:8, 840–846.

Walker, G. H., Stanton, N. A., Kazi, T. A., Salmon, P. M., & Jenkins, D. P. (2009). Does advanced driver training improve situational awareness? *Applied Ergonomics*, 40:4, 678–687.

Wasserman, S., & Faust, K. (1994). *Social Network Analysis: Methods and Applications* (Vol. 8). Cambridge, UK: Cambridge University Press.

Williams, A. M., Davids, K., Burwitz, L., & Williams, J. G. (1994). Visual search strategies in experienced and inexperienced soccer players. *Research Quarterly for Exercise and Sport*, 65:2, 127–135.

Williams, A. M., & Ericsson, K. A. (2005). Perceptual-cognitive expertise in sport: Some considerations when applying the expert performance approach. *Human Movement Science*, 24:3, 283–307.

Section IV

Systems HFE Applications

14 Environmental Factors Influencing Early Participation in Para-Sport

Bridie Kean and Christina Driver

CONTENTS

14.1　INTRODUCTION

For people with a disability, increasing access to physical activity can lead to enhanced health outcomes (Martin, 2013). Furthermore, improved access to high-level sport can provide greater sporting opportunities within the Paralympic movement. The Paralympic Games is the pinnacle of high-performance para-sport (Gold & Gold, 2007), and were first held in Rome in 1960 with 400 athletes competing. The subsequent growth of the Paralympic Games is evident in current standards of competition and increased competitor numbers, with more than 4,000 athletes competing across 480 events at the London 2012 Paralympic games (Brittain, 2009; Dieffenbach & Statler, 2012). While there has been significant progress in the Paralympic movement, Paralympic athletes face 'a much rockier road to travel to achieve athlete success' than their able-bodied counterparts (Martin, 2015, p. 96). This may be due in part to the barriers experienced by people with disabilities in beginning physical activity, the exclusivity of sports settings, and differing environmental impacts faced by para-athletes compared to their able-bodied counterparts (Martin, 2013; Jaarsma et al., 2014). This chapter will explore factors which impact initial and early sports participation for elite-level wheelchair basketball athletes.

14.2 BARRIERS AND FACILITATORS TO SPORT FOR PEOPLE WITH A DISABILITY

The Physical Activity for People with a Disability Model (PAPDM) explores the impact of environmental factors in sports participation for people with a disability, and therefore such models collectively informed this research (van der Ploeg et al., 2004).

14.2.1 Physical Activity for People with a Disability Model

The PAPDM (van der Ploeg et al., 2004) conceptualises the factors which can influence engagement in physical activity for people with a disability. Acknowledging the role of both personal and environmental factors, the PAPDM identifies elements in the environment which may act as a barrier or facilitator to exercise engagement (van der Ploeg et al., 2004).

The elements the model identifies as impacting athletes include accessible exercise facilities, adaptive equipment, transport, and social influences. The PAPDM acknowledges that these factors can either act as a barrier or a facilitator to physical activity participation depending on the resources available to people with a disability.

Previous research has applied the PAPDM demonstrating that barriers such as inaccessible exercise facilities, inaccessible public transport, and lack of knowledge of gym staff, all affect physical activity participation. Conversely, adequate assistance from others (Buffart et al., 2009), accessible facilities (Buffart et al., 2009; Jaarsma et al., 2014), and adaptive equipment (Jaarsma et al., 2014) were identified as environmental facilitators to physical activity participation (van der Ploeg et al., 2004).

Aligning with the PAPDM, this research adopts the stance that for a person with a disability, their surroundings determine the level of sports participation. It is argued that many of the constraints limiting sports participation for people with a disability are controllable, yet to reduce barriers and increase facilitators it is important to understand what those barriers are within specific sporting contexts. This is an area requiring further exploration, particularly given para-sports systems are complex, and it is therefore difficult to apply a 'one size fits all' approach to developing sports policies which promote success across varied para-sports contexts (Patatas et al., 2018). Consequently, focusing on the intricacies of individual para-sports systems, may be a beneficial starting point.

To investigate how the system and environmental surroundings affect athletes with a disability, the current study explored the perspectives of elite level athletes in one para-sport. Specifically, the purpose of this study was to explore factors which impacted initial and early sports participation for elite-level wheelchair basketball athletes.

14.3   METHOD

The research design used to explore the experiences of wheelchair basketball athletes was a qualitative case study method (Yin, 2013). Underpinning this research was a constructivist perspective, a view that athlete participant insights construct insider knowledge about how the environment can impact athletes.

14.3.1   PARTICIPANTS

Wheelchair Basketball in Australia formed the case study of focus in this research. Participants included nine elite wheelchair basketball athletes. Elite was defined as competing in wheelchair basketball at a state, national, or international level, as guided by the definition of elite in previous studies with para-athletes (Smith et al., 2016). Participants were either recruited at a national junior tournament or contacted directly by the first author, who had been previously engaged in wheelchair basketball programme as an athlete, requesting them to participate in the study. Interviews occurred at training and competition venues, as well as a single interview at a participant's place of work. There were nine interviews included in the study. Three interviews were undertaken in January 2015, four in April 2015 and two occurred later in August 2015 to ensure data saturation. Overall, participants originated from four different Australian states: Queensland, New South Wales, Victoria, and Western Australia. Further demographic information is not provided to protect the anonymity of the players. For this reason, all participants are referred to using a pseudonym.

14.3.2   DESIGN

The interview guide was informed by the PAPDM model and there were questions based around how the environment influenced initial participation in wheelchair basketball (van der Ploeg et al., 2004). The interviews conducted were open-ended to explore the experiences of athletes. Five interviews were conducted at the participant's home training venues and three interviews were conducted at competition venues. For the convenience of the participant, one interview was conducted at a location near a participant's place of employment, following recruitment at the participant's training venue.

14.3.3   DATA ANALYSIS

Thematic analysis is an appropriate technique for identifying people's views and perspectives on a certain issue and was used in this study (Braun et al., 2016). The PAPDM underpinned the study and therefore provided direction in the analysis, which, similarly to the PAPDM, was broadly concerned with how environmental factors influenced participation physical activity and specifically was focused on sports participation in wheelchair basketball.

14.3.4 Data Familiarisation

After each interview, the audio interview was transcribed verbatim. The transcript was then read in full. This guided the multiple stages of data collection, which was completed when it was established that the interview data had achieved saturation. Once data collection and familiarisation of each interview had been completed, the first author coded all the interview data using NVivo software.

14.3.5 Coding and Theme Development

In the first pass of coding, initial codes were developed by applying labels to segments of data which were interpreted as exploring how the environment impacted participants (Braun et al., 2016). After the final interview was coded, the first author returned to the dataset and contrasted the codes. Through this process similar codes were housed under subthemes. A total of 13 subthemes were developed. The final development of themes was undertaken with cross-reference to the research question and the PAPDM to contrast subthemes, which were further collapsed into four overarching themes which describe the factors which influenced initial participation in wheelchair basketball.

14.4   RESULTS

Analysis of interviews led to the development of four themes which describe how the environment impacts sports participation. The themes include: 1) social influences, 2) access to sports programmes, 3) access to wheelchairs, and 4) financial factors as environmental factors which impacted initial sports participation for athletes with a disability.

14.4.1   Social Influences

All but one of the interviewees described how existing wheelchair basketball players were influential in them beginning the sport. For the eight athletes, their introduction to wheelchair basketball and the opportunity to play came from an existing player of the game. This occurred at all different stages of life for the participants in the study.

William described how he 'started quite late'. He spoke about how he 'didn't actually know about wheelchair basketball' and that programmes were run in his hometown of Melbourne. It was an existing player whom he met at a prosthetic clinic that introduced him to the game.

Thomas's introduction to the game was also attributed to an existing player. However, it was earlier on in Thomas's experience of living with a disability: 'I met a guy while I was in the spinal unit who played basketball and then he was on the Gold Coast as well'.

Roger was also introduced to the sport by an existing player who reached out to him. Roger was diagnosed with his impairment in the same year that he started playing the sport. Whilst he was not introduced to wheelchair basketball in rehabilitation he described the process of being contacted by an existing player: 'when I got out of hospital I was looking for a sport and I was in contact with the local radio station, [name] rang up and he was friends with the radio station and got hold of it and that's how I went from there'.

Briony's early experiences and desire to reach a high level of basketball were also influenced by an existing player, after starting to play in a local competition: 'She was talking about college and I was like, wow that would be really cool to expand and then do this'.

Athletes in the study spoke about the role of their social network in their initial participation in wheelchair basketball. There were a range of influential people in the lives of athletes, including parents, siblings, teammates, and coaches.

Four out of the nine athletes stated that their parents often had the role of driving them to training 'my parents, they transported me, so they were quite supportive of any decision I made, just as long as I stayed in sport' (Celeste).

Beyond transporting athletes, parents also took on supportive roles. William described how morale support influenced his drive to improve: 'my family, them seeing me improve motivated them to support me more because I feel like they didn't really think I'd make it but seeing improvements and stuff like that, it opened their eyes and they're helping me'.

For Peter, it was his brother who was influential in his reason for starting basketball. He reflected on the shared love of basketball and how his initial interest in wheelchair basketball had stemmed from him playing the stand-up version of the game with his brother: 'he loved basketball too. So seeing my brother out, we'd play out in the backyard and I'd be on my prosthetic leg and I guess I just kept playing with him outside and its sort of where I grew a passion for it' (Peter).

Whilst Peter had played on a prosthetic leg in the backyard, it was wheelchair basketball which offered an avenue for him to continue the sport he was passionate about on an equal playing field: 'when I played my brother, I couldn't really keep up to him speed wise. He'd slow it down to match my prosthetic leg speed. When I'd get on the wheelchair court I could play the same speed as people. I guess that was a boost for me'.

14.4.2 Access to Sports Programmes

The second identified theme described how the environment influenced the athletes' access to sports programmes. To enable athletes to even begin wheelchair basketball, an accessible programme needs to be already established within a reasonable distance from where the athlete lives. The existence of programmes was deemed a facilitator to sports participation for James, who was initially introduced to the game through a 'come and try' day and then went on to compete in his local competition. James described how his career started 'through a local disabled sports group; it was a "Come and Try Night". There were a couple of different sports; I tried wheelchair basketball and loved it'.

James went onto describe how he was 'reliant on mum and dad to get me to and from training' (James). Whilst sporting programmes were recalled as facilitators to starting basketball, there were also barriers to attending sports programmes. Three of the nine athletes described the distance of attending wheelchair basketball programme prevented them from starting or continuing the sport at a younger age. For example, William, a state-level player, described some of the barriers to initially starting the sport: '[programs] were all the way in the city, so it makes it hard for my parents to take me there' (William).

Angie, a national squad member, also identified that the large distance to access wheelchair basketball was one of the reasons she started the sport late, as she waited until she held a drivers licence: to start out 'mum could only really drop me at the swimming pool around the corner. So yeah, that was probably a big reason why I didn't until about five years after my accident' (Angie).

Distance was also identified as a barrier for some athletes in accessing quality programmes which would advance their progress in wheelchair basketball required to reach the elite level. Phillip, an Australian Paralympic athlete, described how access to sporting facilities alone was not enough, as the quality of competition was also important. Phillip described what it was like growing up in a remote town: 'It's easy to get to everywhere we needed to go, but the level of competition there was pretty basic, if I wanted to play at a higher level, transportation was obviously an issue as far as getting there'.

For this reason, factors that led to being able to attend high-quality sessions was deemed a facilitator to initial sports participation. In Thomas's situation this was also connected to a feeling of independence. He had initially been introduced to wheelchair basketball after rehabilitation. Whilst he had been able to drive before acquiring his disability, Thomas described that driving was a facilitator and it was important for him to have a vehicle modified so that he could transport himself to basketball: 'I didn't want anyone to drive me there, even though dad offered to drive me and whatnot, but I was just waiting until I had my car and then went there'.

14.4.3 Access to Wheelchairs

To play wheelchair basketball, a basketball wheelchair is required. At the early participation level, 'borrowing equipment' (Thomas) was common and often imperative for wheelchair basketball players starting out the sport. Four of the nine players specifically mentioned borrowing a chair in their initial involvement in the sport.

Thomas, an Australian national player, accessed a wheelchair through his wheelchair basketball club who 'had plenty of club chairs'. He acknowledged that borrowing a chair was vital for staying in the sport. He described the process of using wheelchairs available through his wheelchair basketball club:

> [I] had the club chair and I used that one until I got the chair I'm in now. So I used that for a while actually which was really good, like I don't know what I would have done if I had to fork out the money or something like that. Probably wouldn't have pursued it that's for sure.

Prior to his parents buying his chair, Peter, a national squad member, also borrowed a number of wheelchairs throughout his adolescent years. He borrowed wheelchairs off adult teammates who were able to pass on their equipment to him whilst he was still growing:

> From that point on I had older people on the team, and some of had finished playing basketball and would hand their chair down to me, they just sort of passed me the chair. I guess that sort of helped with my growth, because I didn't have to keep buying new chairs through that young teenager age. I was just sort of handed down chairs.

Access to equipment through clubs and loaning chairs from teammates was a common facilitator for wheelchair basketball players, enabling continued participation as well as allowing athletes to trial the sport without investing in a specialised wheelchair.

14.4.4 FINANCIAL FACTORS

The theme of financial factors intersects aspects of access to equipment. The high cost of equipment was deemed one of the challenges in beginning wheelchair basketball. However, this was offset by borrowing a wheelchair to commence the sport, allowing athletes to consider the next step in their pathway. Thomas describes this process by reflecting on his early participation in wheelchair basketball:

> First year or two I used a team mate's chair until I knew that the sport was serious and I wanted to have a crack at it and that gave me enough time to save up. So, yeah, cost of equipment was a big one.

Thomas was also able to receive financial support which facilitated his involvement in sport, and supported him with other expenses such as travel. Thomas reflected on the importance on this financial support as he became more serious about sport and was restricted in his ability to also take on part-time work: 'got a number of grants and scholarships as well from local sporting groups and asked for donations and that kind of stuff, just so that I wouldn't have to work and train and study' (Thomas).

Financial support for wheelchairs was referred to as a facilitator by Angie. Angie's first wheelchair was developed to be suitable for multiple sports as she decided which sporting pathway she would pursue: 'I got a grant from PMH for a tennis and basketball chair, so it was like an all-rounder sort of thing because I had no idea what I wanted to do' (Angie).

14.5 DISCUSSION AND PRACTICAL APPLICATION IN SPORT

The PAPDM is central to understanding how exercise for people with disabilities can be influenced by external factors such as assistance of others, social influence, transportation, and availability of built and natural facilities (van der Ploeg et al., 2004). The analysis revealed how environmental factors outlined in the PAPDM influenced elite athletes in their early participation in wheelchair basketball. In the current study, environmental factors were described within four themes which included: social influences, access to sports programmes, access to wheelchairs, and financial factors. The influence of these factors in initial sports participation are demonstrated in Figure 14.1.

Social influences and access to sports programmes were closely linked as factors that influenced initial participation in wheelchair basketball. A facilitator to beginning wheelchair basketball in the current study was the access to a wheelchair basketball programme, which required the existence of programs, awareness of programmes in existence, and access to such programs. To facilitate awareness of programs, existing players as role models often introduced new players to sports programs, yet the ability to access the programmes was heavily dependent on factors such as parental support, transport and proximity.

From a systems perspective, there needs to be a programme in existence in order for people to be aware of the opportunities in para-sport. Research which has examined barriers and facilitators to physical activity found that limited knowledge about physical activity opportunities were a barrier for young adults with a disability (Buffart et al., 2009). Therefore, increased awareness of para-sport may also lead to development of programmes and adaptations of mainstream sports programmes to be inclusive of people with disabilities. For example, in wheelchair tennis there are two divisions in which athletes complete. The quadriplegic division is for tennis players who impairments that impact both their arms and legs, and the para division is for tennis players who have impairments which impact their lower extremities. In 2019, the finals of the quadriplegic division of the Australian Open was televised in

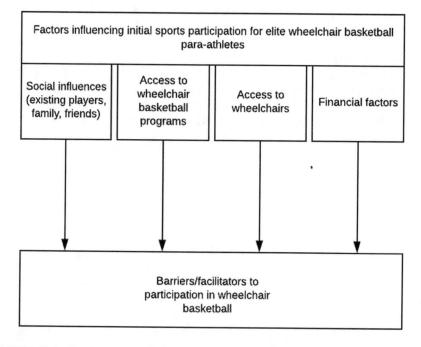

FIGURE 14.1 Barriers and facilitators influencing participation in wheelchair basketball.

2019 on mainstream TV (ABC, 2019). Dylan Alcott, an Australian wheelchair tennis player won the final and in his acceptance speech acknowledged the milestone of broadcasting wheelchair tennis in para-sport, stating the 'match was broadcast into every single TV in Australia. That meant a lot to me and it meant a lot to the 4.5 million people in Australia with a disability' (ABC, 2019). This increased awareness may lead to people with a disability becoming aware of sports opportunities as well as sports programme and policy-makers aiming to increase the programmes and opportunities for people with disabilities to engage in sport. Dylan Alcott is an example of a Paralympic role model who had used his position as an elite athlete to encourage further inclusion in sport and society for people with disabilities.

In the current study, the importance of social influences was highlighted, particularly with existing players introducing new players to the sport. This introduction was also instrumental in increasing awareness; existing players were role models who were fundamental in recruiting and growing wheelchair basketball participation. For example, William's insight showed that he did not know wheelchair basketball competitions were available for him to play in until he met an existing player at a prosthetics appointment. Consequently, lack of awareness often leads to delays in starting sport for some para-athletes in the current study. This can affect the opportunities available to new players. The late entry into sport for para-athletes compared to their able-bodied counterparts, which may also be related to injuries later in life, is one of the reasons why the talent development pathway in para-sport is generally non-linear (Mann et al., 2017).

Considering the multiple factors that may delay participation in para-sport, focusing on increasing awareness may be paramount in promoting sporting opportunities. Increased media coverage of para-sport is a way forward in this regard and the Australian Open 2019 was an example of how this can educate the public about para-sport. In the current study, new players often met role models in person; however, role models in the media are a way to achieve broader exposure. Whilst the Australian Open final was a milestone, further coverage of a variety of para-sports on mainstream TV is recommended. For clubs, it is also recommended that existing programmes encompass building awareness of sports opportunities into their club structures and vision.

Whilst broader exposure increases awareness of para-sport, it is most beneficial if there are accessible programmes for people with a disability to engage in sports. Previous research suggests this may not be the case considering people with disabilities report lower rates of physical activity than their able-bodied counterparts (Martin, 2013; Rimmer et al., 2004). In wheelchair basketball, the current study found there were a range of factors that influenced whether a new player could access sporting programs. The ability to access the programmes was heavily dependent on factors such as parental support, transport, and proximity. This is consistent with existing studies suggesting people with a disability face additional barriers to accessing sport and physical activity compared to their able-bodied counterparts, such as lack of possibilities and facility accessibility (Jaarsma et al., 2014).

Parents play a vital role in early sports participation (Davids & Baker, 2007; Li et al., 2014). In the current study, this was also the case as parental support was influential for both transporting new players and supporting their goals in sport.

Social influences, such as support from parents (Buffart et al., 2009; Shields et al., 2012), and involvement of peers (Shields et al., 2012) facilitated physical activity for people with a disability. In the current study, new players were motivated when their family and parents watched them play and were involved in their development as athletes. This shows that the role of parents extends beyond transporting athletes to programmes which is in a similar finding to that in existing talent development research conducted with able-bodied participants, where the role of the parents are in providing emotional support is heightened in the years of early participation in sport (Keegan et al., 2014; Spray et al., 2014).

Whilst parents were instrumental in facilitating transport, this was challenging for some new players due to their proximity to sports programmes. Furthermore, the quality of training within local proximity was a consideration for athletes as they progressed in their sporting development. Further along the talent development pathway necessitated higher quality training for emerging athletes which was not available in regional settings. This supports previous findings where the requirement of an able-bodied person to transport players was deemed a barrier for people with a spinal cord injury in accessing sport (Stephens et al., 2012).

Once an athlete had access to a programme, additional factors impacting participation were access to specialised equipment and the financial cost involved. This is consistent with a systematic review identifying barriers to physical activity for children with disabilities concluding that lack of programmes and cost were barriers to participation (Shields et al., 2012).

For the participants in this study, being able to loan a wheelchair from a club or teammate was identified as a facilitator, as the high cost of equipment was highlighted as a barrier to participating in the sport. Accordingly, in wheelchair basketball the high cost of equipment may prevent potential athletes from reaching the elite level if it is not feasible for a player to access a specialised wheelchair. The high cost of equipment was also regarded as a barrier to sports participation in previous research conducted with people with a spinal cord injury (Stephens et al., 2012). Furthermore, this supports previous findings in the literature stating that when children without a disability have restricted access to the resources and necessary equipment to participate in a sport they may never realise their complete potential (Bailey & Morley, 2006). Specifically, in the context of this study, the basketball wheelchair is fundamental for an athlete to reach optimal performance, and specialised equipment necessitates progression to perform at the elite level (Cooper & De Luigi, 2014). As technology evolves and wheelchairs become lighter and stronger, the cost of playing wheelchair basketball is likely to remain a barrier at both grassroots and high-performance levels. Therefore, consideration of all levels of participation is paramount.

14.6   PRACTICAL IMPLICATIONS FOR PARA-SPORT

In terms of implications for practice, the role of social influences in increasing access to sports programmes were often interconnected, as are financial factors and access to equipment. Whilst there is limited research in wheelchair basketball, the findings from this case study offer considerations for sports systems. In terms of facilitating involvement in sport, access to sports programmes and local competitions, support-ive social influences, access to sports programmes, the ability to loan or borrow wheelchairs and financial support for the cost involved in para-sport were enablers to sports participation.

To enable greater participation, sporting organisations and policy-makers should increase the availability of equipment for players at a grassroots level. For example, sports clubs and organisations should prioritise buying club chairs for new players to loan and schools could aim to provide sports equipment for students with a disability to enable greater participation in para-sports. Until para-sporting opportunities are socialised into school systems in Australia there will remain reliance on the club based local opportunities to initially start the sport and provide high-performance opportunities.

To facilitate athletes to reach the high-performance level, this study recom-mends that sports systems should consider how emerging athletes will finance the equipment required to play the sport. Leading the way in this space, the Australian Paralympic Committee acknowledged the importance of providing access to equip-ment early in a para-athletes talent development pathway. This was evidenced by a commitment to fund para-sport equipment to the value of $2 million for emerging athletes around Australia. Based on the findings in this study, this type of investment is key to providing a system that enables equitable pathways for potential athletes with a disability to reach the elite level.

Based on the insights provided by those athletes who have experienced grassroots and elite para-sport, further research utilising qualitative methods should be under-taken to incorporate the para-sports voice into policy and direction for para-sports systems. Athletes are vital stakeholders in sport, and similarly to previous sports policy research, the athlete voice should be considered when identifying factors that affect sporting success (De Bosscher et al., 2009).

14.7 LIMITATIONS

There are some limitations to the current study. The study is limited to one para-sport in one nation. To advance knowledge of the factors that influence initial sports participation, further research should be conducted in other para-sports.

Another limitation of the current study was that athletes were recalling their experiences of being elite; they were no longer in the initial participation phase of their talent development pathway. Rather, the athletes had all reached national or international standard and their most recent experiences may be the environmental factors which influence them in their current level of sport, which are presented in case studies published elsewhere (see Kean et al., 2018; Kean et al., 2017). As these insights are from athletes who are recalling their initial participation, to gain further insights into other sports future research conducted with people with a dis-ability who are beginning sports may capture further insights into how the envi-ronment impacts grassroots sports participation. Despite these limitations, this research has contributed the para-athlete voice to better understand how wheel-chair basketball systems can facilitate sports participation and the talent develop-ment pathway.

14.8 CONCLUSION

This research explored how environmental elements impacted initial sports participation for a group of athletes with a disability who reached the elite level. Qualitative interviews were used to gain an in-depth understanding of the experience when athletes began playing wheelchair basketball. The findings indicated there are a number of factors that affected awareness and access to the sporting programmes that began the sporting pathway of elite wheelchair basketball athletes. In terms of awareness, most athletes were introduced to the sport by existing players. There were also influential factors in accessing programs, such as the role of parents in transporting players and being able to access wheelchairs. Recommendations are provided to increase awareness of para-sports programmes, whilst considering how to facilitate better access to these programs.

REFERENCES

ABC. (2019). Dylan Alcott wins fifth consecutive Australian Open quad wheelchair singles title in a row. *ABC News*.

Bailey, R., & Morley, D. (2006). Towards a model of talent development in physical education. *Sport, Education and Society*, 11:3, 211–230. doi: 10.1080/13573320600813366.

Braun, V., Clarke, V., & Weate, P. (2016). Using thematic analysis in sport and exercise research. In: *Routledge Handbook of Qualitative Research in Sport and Exercise* (pp. 191–205).

Brittain, I. (2009). *The Paralympic Games Explained*. Routledge.

Buffart, L. M., Westendorp, T., Van Den Berg-Emons, R. J., Stam, H. J., & Roebroeck, M. E. (2009). Perceived barriers to and facilitators of physical activity in young adults with childhood-onset physical disabilities. *Journal of Rehabilitation Medicine*, 41:11, 881–885. doi: 10.2340/16501977-0420.

Cooper, R. A., & De Luigi, A. J. (2014). Adaptive sports technology and biomechanics: Wheelchairs. *PM&R*, 6:8 Supplement, S31–S39. doi: 10.1016/j.pmrj.2014.05.020.

Davids, K., & Baker, J. (2007). Genes, environment and sport performance: Why the nature-nurture dualism is no longer relevant. Sports Medicine, 37:11, 961–980. doi: 10.2165/00007256-200737110-00004.

De Bosscher, V., De Knop, P., van Bottenburg, M., Shibli, S., & Bingham, J. (2009). Explaining international sporting success: An international comparison of elite sport systems and policies in six countries. *Sport Management Review*, 12:3, 113–136. doi: 10.1016/j.smr.2009.01.001.

Dieffenbach, K. D., & Statler, T. A. (2012). More similar than different: The psychological environment of paralympic sport. *Journal of Sport Psychology in Action*, 3:2, 109–118. doi: 10.1080/21520704.2012.683322.

Gold, J. R., & Gold, M. M. (2007). Access for all: The rise of the Paralympic Games. *Journal of the Royal Society for the Promotion of Health*, 127:3, 133–141. doi: 10.1177/1466424007077348.

Jaarsma, E. A., Dijkstra, P., Geertzen, J., & Dekker, R. (2014). Barriers to and facilitators of sports participation for people with physical disabilities: A systematic review. *Scandinavian Journal of Medicine and Science in Sports*, 24:6, 871–881.

Kean, B., Gray, M., Verdonck, M., Burkett, B., & Oprescu, F. (2017). The impact of the environment on elite wheelchair basketball athletes: a cross-case comparison. *Qualitative Research in Sport, Exercise and Health*, 9:4, 485–498.

Kean, B., Oprescu, F., Gray, M., & Burkett, B. (2018). Commitment to physical activity and health: a case study of a Paralympic Gold medallist. *Disability and Rehabilitation*, 40:17, 2093–2097. doi: 10.1080/09638288.2017.1323234.

Keegan, R. J., Harwood, C. G., Spray, C. M., & Lavallee, D. (2014). A qualitative investigation of the motivational climate in elite sport. *Psychology of Sport and Exercise*, 15:1, 97–107. doi: 10.1016/j.psychsport.2013.10.006.

Li, C., Wang, C. K. J., & Pyun, D. Y. (2014). Talent development environmental factors in sport: A review and taxonomic classification. *Quest*, 66:4, 433–447. doi: 10.1080/00336297.2014.944715.

Mann, D. L., Dehghansai, N., & Baker, J. (2017). Searching for the elusive gift: advances in talent identification in sport. *Current Opinion in Psychology*, 16, 128–133. doi: 10.1016/j.copsyc.2017.04.016.

Martin, J. J. (2013). Benefits and barriers to physical activity for individuals with disabilities: A social-relational model of disability perspective. *Disability and Rehabilitation*, 35:24, 2030–2037. doi: 10.3109/09638288.2013.802377.

Martin, J. J. (2015). Determinants of elite disability sport performance. *Kinesiology Review*, 4(1), 91–98.

Patatas, J. M., De Bosscher, V., & Legg, D. (2018). Understanding para-sport: an analysis of the differences between able-bodied and para-sport from a sport policy perspective. *International Journal of Sport Policy*, 10:2, 235–254. doi: 10.1080/19406940.2017.1359649.

van der Ploeg, H. P., van der Beek, A. J., van der Woude, L. H. V., & van Mechelen, W. (2004). Physical activity for people with a disability: A conceptual model. *Sports Medicine*, 34:10, 639–649. doi: 10.2165/00007256-200434100-00002.

Rimmer, J. H., Riley, B., Wang, E., Rauworth, A., & Jurkowski, J. (2004). Physical activity participation among persons with disabilities: Barriers and facilitators. *American Journal of Preventive Medicine*, 26:5, 419–425.

Shields, N., Synnot, A. J., & Barr, M. (2012). Perceived barriers and facilitators to physical activity for children with disability: a systematic review. *British Journal of Sports Medicine*, 46:14, 989. doi: 10.1136/bjsports-2011-090236.

Smith, B., Bundon, A., & Best, M. (2016). Disability sport and activist identities: A qualitative study of narratives of activism among elite athletes' with impairment. *Psychology of Sport and Exercise*, 26, 139–148. doi: 10.1016/j.psychsport.2016.07.003.

Spray, C. M., Harwood, C. G., & Lavallee, D. E. (2014). A qualitative synthesis of research into social motivational influences across the athletic career span AU – Keegan, R.J. Qualitative Research in Sport, Exercise and Health, 6:4, 537–567. doi: 10.1080/2159676X.2013.857710.

Stephens, C., Neil, R., & Smith, P. (2012). The perceived benefits and barriers of sport in spinal cord injured individuals: a qualitative study. *Disability and Rehabilitation*, 34:24, 2061–2070. doi: 10.3109/09638288.2012.669020.

Yin, R. K. (2013). *Case Study Research: Design and Methods*. New York, NY: SAGE Publications.

15 Round and Round and Up and Down We Go Again

Using Causal Loop Diagrams to Model Football Club Performance

Paul M. Salmon, Scott McLean, Karl Dodd, and Nicholas Stevens

CONTENTS

15.1 INTRODUCTION

Complexity science and systems thinking have a long history of applications within HFE (Dekker, 2011; Leveson, 2004; Rasmussen, 1997; Salmon et al., 2017; Walker et al., 2017, 2017; Wilson, 2014). In short, this approach involves attempting to describe overall work or social systems and then identifying emergent properties that arise when components across the system interact with one another. Whilst this work is most prominent in the areas of safety, risk, and accident causation, increasingly this approach is being applied to understand and optimise system behaviour (e.g. Read et al., 2017).

In sports science, complexity and systems thinking are currently receiving increasing attention and there is a burgeoning set of applications attempting to understand features of complexity in sports performance (Bittencourt et al., 2016; Davids et al., 2013; Balagué et al., 2013; Hulme et al., 2018, 2019; McLean et al., 2017; Salmon & McLean, 2020). The adoption of a 'complex systems approach' has been identified as a critical requirement in areas such as sports injury prevention (Bittencourt et al., 2016; Hulme et al., 2018), sports performance analysis (McLean et al., 2017), coaching (McLean et al., 2019; Soltanzadeh & Mooney, 2016), and sports science generally (Davids et al., 2013; McGarry et al., 2002; Mooney et al., 2017). These arguments are made on the basis that existing reductionist approaches produce outputs which do not provide a comprehensive account of the network of factors which influence athlete, team, club, and sports system performance. To illustrate, in the area of football performance analysis specifically, it is argued that existing reductionist approaches such as player pass completion rates, shots on target, and distances ran do not support the identification of wider systemic factors that may influence team performance (McLean et al., 2017; Salmon & McLean, 2020). As a result, only a limited understanding of individual player performance is achieved, and interventions designed to improve performance may not be as effective as other factors beyond the athlete, team, and match continue to have an influence.

Such arguments are beginning to gain traction in the football literature (McLean et al., 2017, 2019; Salmon & McLean, 2020; Stockl et al., 2017; Vilar et al., 2013). Despite this, there are few examples of where appropriate methods have been used to analyse or model football 'systems' (Salmon & McLean, 2020). Whilst previous research has investigated aspects of complexity, the focus has been on complexity associated with player and team behaviour within matches and training scenarios, such as small-sided games (e.g. Araújo & Davids, 2016; Silva et al., 2016). For example, ecological dynamics has been used to describe and understand some of the emergent properties of football teams (Araújo & Davids, 2016), whereas dynamical systems theory has been used to model patterns of player movement and performance (Davids et al., 2005). Whilst such applications are useful, there are limitations from a complexity and systems thinking science point of view (Salmon & McLean, 2020). Most pertinent with regard to the present chapter is that key aspects of the broader football system have not been considered (Salmon & McLean, 2020). As a result, it is not clear what factors across football systems interact to influence football club performance.

In a recent commentary on the use of complexity and systems thinking in football research, Salmon and McLean (2020) discussed various systems analysis methods which could be used to model football systems with a view to enhancing the knowledge base on the factors that influence football performance. One of the methods identified was Causal Loop Diagrams (CLDs; Sterman, 2000). CLDs have been used extensively to identify and depict the feedback loops underpinning behaviour in a diverse set of areas ranging from obesity (Allender et al., 2015), food security (Purwanto et al., 2019), social media use (Comrie et al., 2019), terrorism (Schoenenberger et al., 2014), and transport (Shepherd, 2014). The aim of this chapter is to present a proof-of-concept CLD of football club performance that depicts the key variables that interact to influence elite football club performance. The intention is to demonstrate the utility of the CLD approach for understanding complexity in sports systems and to encourage further applications in football and other sports settings.

15.2 CAUSAL LOOP DIAGRAMS

CLD is a method that is used to elicit and represent positive and negative feedback loops, the relationships between them, and how they dynamically influence behaviour in a given system. This is based on the notion that all systems comprise interacting networks of positive and negative feedback loops that ultimately influence behaviour (Sterman, 2000). According to Sterman (2000), CLDs are useful for capturing hypotheses about causal dynamics, for eliciting and capturing mental models, and for communicating the feedbacks that may be responsible for a particular behaviour or issue. The CLDs method is typically used to inform the basis and development of systems dynamics models which are used to simulate the behaviour of systems over time (see Sterman, 2000); however, it can also be used independently for systems analysis purposes (e.g. Comrie et al., 2019).

CLDs comprise variables connected by arrows which depict the causal influences between the variables (Sterman, 2000). There are two types of causal loops: positive loops (or reinforcing loops) and negative loops (or balancing loops) (Sterman, 2000). Positive feedback loops are self-reinforcing; for example, as shown in Figure 15.1, as a population of chickens increases, more eggs are laid, which in turn increases the chicken population, which in turn leads to more eggs, and so on (Sterman, 2000). Negative loops are self-correcting. As the chicken population expands, negative loops will act to balance the chicken population. For example, as the population of

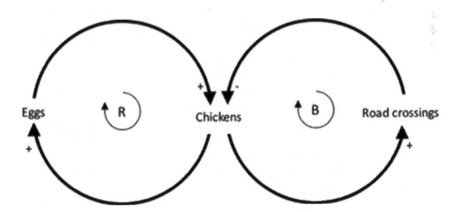

FIGURE 15.1 Simple chicken and egg causal loop diagram (adapted from Sterman, 2000). Positive links (+) refer to links where the effect is in the same direction as the cause. Negative links (–) refer to links where, if the cause increases, the effect decreases, and if the cause decreases, the effect increases. R is used to denote a positive or reinforcing loop. B is used to denote a negative or balancing loop.

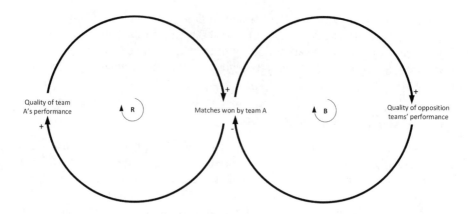

FIGURE 15.2   Simple football CLD. Positive links (+) refer to links where the effect is in the same direction as the cause. Negative links (–) refer to links where, if the cause increases, the effect decreases, and if the cause decreases, the effect increases. R is used to denote a positive or reinforcing loop. B is used to denote a negative or balancing loop.

chickens increases, more chickens will attempt to cross roads, leading to more collisions with vehicles and deaths, and a reduction in the chicken population which acts as a balance to population growth (Sterman, 2000).

To demonstrate this logic in football, a simplistic football CLD is presented in Figure 15.2. The CLD in Figure 15.2 shows how, as the quality of a team A's performance increases, the number of matches won by team A increases concomitantly. As increased wins lead to more successes in league and cup competitions, quality of performance can also increase (for example through increased confidence and the attraction of more talented players and coaching staff through enhanced finances and reputation). However, the CLD also shows that over time opposition teams adapt, either through copying the successful approaches employed by team A, or by improving their playing squad and tactical approach through avenues such as player recruitment and enhancing their coaching and support staff. The subsequent improvement in the quality of other teams' performances then acts as a balance on the successes of team A and they win fewer matches as a result.

The CLD presented in Figure 15.2 provides a simple view on one aspect of football club performance surrounding the idea that success breeds more success. This is characteristic of a well-known systems thinking archetype, known as 'success to the successful', that has been used previously to explain system behaviour (Kim, 1993). It is presented here to demonstrate causal loop logic, but also to highlight specifically where the football science literature is lacking. For example, little is known about what specific variables across football systems interact to influence football club performance. Although some of the dynamics influencing club performance are understood at a high level, as in Figure 15.2, the specific variables which interact to influence performance are not clear. For instance, in relation to Figure 15.2, exactly what variables interact to create opposition team adaptation remains unclear. Likewise, exactly why successful teams continue to be successful in terms of winning league and cup titles, but cannot maintain success for long periods, has not fully been explained, especially in terms of the systemic variables that facilitate this. The aim of the study described in this chapter was to develop a proof-of-concept CLD of football club performance that could be used to explore some of these gaps.

15.3 METHOD

15.3.1 STUDY AIMS

The aim of this study was to develop a CLD to depict the feedback loops that influence football club performance. Whilst we did not focus on a particular football club, the English Premier League was used as the conceptual basis for the study. The boundary for the analysis was established as football club performance within the English Premier League system where the club's primary and realistic goal is to win the league title. The variables identified were limited to only those that could be influenced by the football club itself, rather than some other entity within the English Premier League football system, e.g. governing bodies and the national team.

15.3.2 CLD DEVELOPMENT

CLDs are typically developed using a group model building process (Berard, 2010; Sterman, 2000). Berard (2010) describes two forms of group model building. The first involves the research team constructing models using data derived from appropriate sources such as the peer-reviewed literature and organisational documentation. The second involves engaging relevant Subject Matter Experts (SME) to assist in building the models via avenues such as workshops, survey, or Delphi studies. The involvement of SMEs is particularly important, as studies have demonstrated the importance of involving many participants in the model building process to incorporate multiple stakeholder perspectives and ultimately to enhance model validity (Berard, 2010). In the present study, both processes were utilised, with a draft model being developed initially by the research team, and the model subsequently being refined following SME review.

Initially, the authors developed a generic causal loop structure of football club performance in collaboration with five other researchers who have extensive experience in the application of complex systems modelling techniques in various areas, including sport (Hulme et al., 2018, 2019). A draft CLD was developed and was subsequently reviewed and refined via discussion between the authors and the five researchers. The draft CLD was further refined by the authors based on peer-reviewed literature, publicly available data on football club performance, and other relevant documentation such as media reports and conference articles. The CLD was finally constructed in the system dynamic modelling software Vensim.

The VenSim CLD was then reviewed by an SME who is currently an international level football coach with over ten years' experience of playing at the elite level, which included a period playing in the UK. The SME was asked to review the CLD and to question whether it provided a valid representation of the causal loops underpinning football club performance. The SME was asked specifically to review the variables and relationships to determine whether the variables and relationships included were appropriate and whether they were expressed correctly. Further, the SME was asked to identify any variables and relationships that were not included in the CLD. The model was subsequently updated based on the SMEs feedback. This included modifying the terminology used within the model, adding new variables and relationships, and removing any inappropriate variables and relationships.

15.4 RESULTS

The football club performance CLD is presented in Figure 15.3.

The CLD includes 35 variables that were identified as having an influence on football club performance. Within the model, quality of team performance is directly influenced by nine variables (see Figure 15.4): quality of opposition, player fitness levels, quality of training, quality of tactics, quality of squad, player contentment, the media (positive and negative), and quality of fitness. The greater the quality of performance is, the more wins the club has, and as the number of wins increase success in league and cup competitions increases also. Successes lead to increases in the club's finances through more prize money, sponsorship, broadcasting revenue, and ticket and merchandise sales. This enables the club to attract and recruit a higher calibre of players and coaching and support staff, which leads to improved quality of performance through the variables described above and shown in Figure 15.4. This dynamic is representative of the success to the successful archetype described earlier.

The model shows how the adaptation of opposition teams acts as a balance to the success to the successful archetype. Here the successes of one club lead to improvements in the quality of opposition teams' performances, which acts to reduce the number of wins of the initially successful club. Although not included specifically in the model, it is likely that opposing club adaptation is driven by many of the variables that drive the successes of the initial club. These include modifications to club strategy and management, enhancing the quality of the playing squad, coaching and support staff, and improving training and facilities.

The model also identifies a number of highly influential variables that ostensibly play a critical role in football club performance. For example, within the model the variables with the most outgoing links are 'Quality of club strategy & management' (eight outgoing links) and 'Club finances' (seven outgoing links). Quality of club strategy & management (see Figure 15.5) has a direct influence on important variables such as club finances, quality of coaches, scouting staff, and performance staff, quality of player recruitment, and quality of facilities and the club's academy. Likewise, club finances play a key role in various critical aspects of club functioning, such as player recruitment, salaries, and outgoing transfers.

The CLD model also includes variables relating to player health and well-being. For example, the model shows how, as the number of games increases due to involvement in more competitions, load and accumulated fatigue increases, which in turn can lead to an increase in injuries. As injuries increase, player contentment is adversely impacted, which can negatively impact performances. As with the adaptation of opposition teams, this acts as a balance on the success to the successful archetype within the model, as increased games brought about by successes serve to create injuries which in turn impact player contentment.

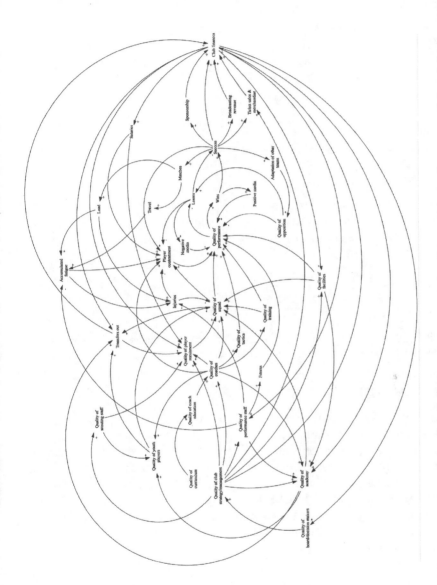

FIGURE 15.3 Football club performance causal loop diagram. Note, for clarity of presentation we have removed the reinforcing and balancing loop notations and loop descriptors.

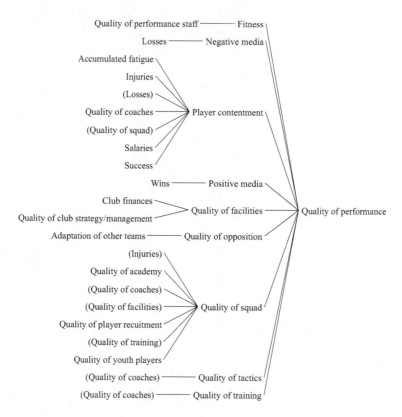

FIGURE 15.4   Variables influencing the quality of team performance (parentheses indicate that the variable appears elsewhere in the causal tree).

15.5   DISCUSSION

The aim of the study described in this chapter was to develop a proof-of-concept CLD of football club performance that depicts the key variables that interact to influence football club performance. The impetus for the study was the growing number of calls to apply complex systems modelling approaches in football, and sport more generally (Salmon & McLean, 2020). Whilst our primary intention was to demonstrate the utility of the CLD method for analysing the behaviour of sports systems, the CLD provides a number of implications for optimising football club performance. These are discussed below along with pertinent implications for football and sports research and practice.

15.5.1   SUCCESS TO THE SUCCESSFUL, BUT ONLY FOR SO LONG

One of the central set of loops within the model aligns with the already well known 'success to the successful' archetype. This archetype explains how, as one entity becomes successful, they are allocated more resources, which in turn helps to create more success (Kim, 1993; Senge, 1990). Conversely, unsuccessful entities are

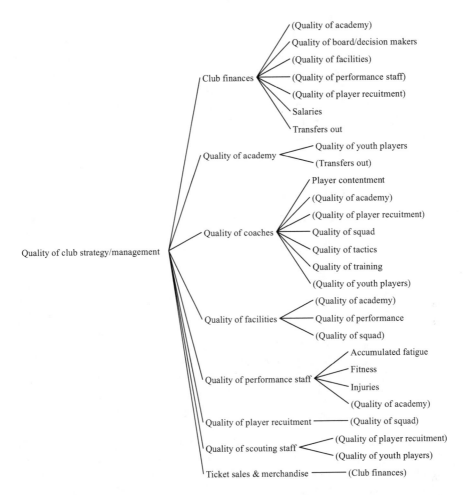

FIGURE 15.5 Variables influenced by the quality of club strategy and management (parentheses indicate that the variable appears elsewhere in the causal tree).

allocated fewer resources and thus find it even more difficult to achieve success. This can be seen in the CLD as improvements in team performance generate success, which in turn leads to increased financial resources, which enables clubs to improve their playing squad and coaching and support staff, which in turn potentially generates more success. The model indicates, however, that the strength of the success to the successful archetype diminishes over time and is eventually disrupted by an 'adapting opposition' loop. This ensures that opposing teams adapt in various ways to improve their own performance and generate their own successes, which in turn potentially diminishes the success of the initial team and ensures that adapting teams begin to experience some success and enter into success to the successful archetype. This dynamic is one that is commonly seen across professional football leagues whereby one club enjoys a period of sustained success and is then replaced by another club, and so on. For example, in the English Premier League, although a set of clubs have dominated in terms of league titles, no club has ever won the league title in more than three consecutive seasons. Manchester United won the league title three seasons in a row between 1998 and 2001 and 2006 and 2009; however, on both occasions were unable to win a fourth league title in a row. Current league champions Manchester City have won the league title two seasons in a row; however, at the time of writing are second in the league table after 24 matches, 13 points behind leaders Liverpool. The CLD suggests that this lack of prolonged success can be explained by the interaction between the success to the successful and adapting opposition loop.

15.5.2   THE LIMITS TO SUCCESS

As well as the adapting opposition archetype, other limits to success were identified in the model. This is in line with another commonly used archetype knowns as the 'Limits to success' archetype whereby improved performance and associated successes are slowed down and ultimately ended by natural or emergent limits within the system (Kim, 1993; Senge, 1990). The CLD suggests that, in addition to adapting opposition teams there are various other potential limits to success, including player fatigue, injuries, and reduced player contentment. Paradoxically, these variables are adversely impacted by success, as more games are played due to involvement in more competitions. Similar to the adaptation of opposition teams, reduced player contentment through fatigue and injury acts to diminish performance and success. Another variable which the model suggests can act as a limit to success is outgoing transfers. Again, conversely, these can be influenced through success, whereby squad and player quality increases leading to an increased interest in players from other clubs.

Encouragingly, the CLD suggests that most of the variables that act as limits to success can be managed by clubs. For example, strategies designed to manage player health and well-being and contentment could be adopted to ensure that players remain fit and content throughout periods of success.

15.5.3   THE IMPORTANCE OF CLUB STRATEGY AND MANAGEMENT

The CLD clearly shows the importance of football club strategy and higher-level management in ensuring optimum performance. Within the model, club strategy and management had the most outgoing connections to other variables, suggesting that it is the most prominent in terms of influence on other variables in the model. Whilst this finding is perhaps not surprising, it does highlight the importance of ensuring club strategy considers multiple aspects of football club performance, rather than only selected variables such as financial management and coach and player recruitment. Moreover, this finding again highlights the importance of considering factors beyond players and coaches when attempting to understand and optimise performance. It is therefore recommended that further research be undertaken to explore the relationship between club strategy and higher-level management and team and player performance.

15.5.4 Implications for Football Practice

Based on the insights discussed above, the model has a number of potential implications for football club management. First, the model confirms that elite football clubs are complex systems, and that performance can only be optimised by considering a broad set of variables and their interrelations. Whilst researchers and practitioners may be aware of many of the variables influencing system performance, they are often unaware of how these factors interact dynamically and so may not consider the interactions and potential knock-on effects when implementing strategies designed to improve performance. The CLD suggests that, whilst quick fixes that focus on individual components (e.g. coaches and players) may have an initial positive effect, they will not be sustained over time as there are various other components influencing club performance (Dekker, 2011; McLean et al., 2019). Rather, the CLD suggests that performance will be better optimised by seeking to modify key leverage points. One such point of leverage indicted by the model is club strategy and management. Second, the success to the successful and limits to success archetypes identified within the model provide two interrelated implications. Namely, that clubs should attempt to optimise when they appear to be already optimised (i.e. are enjoying sustained success), and that clubs should also plan for periods of reduced success. As the model suggests that top clubs will likely enjoy periods of success followed by periods of non-success, it is suggested that efforts to optimise future performance should be made during periods of success so to ensure that unsuccessful periods are as short as possible. Managing each of the variables identified in the model provides a useful indication of how this optimisation could occur. Leicester City football club may represent an example of failing to do this, as they won the Premier League title in 2015, but finished 12th in the following season, and at the time of writing have not finished in the top eight since 2015. Further, clubs should also acknowledge and embrace unsuccessful periods and view them as transitions to successful periods. Rather than make knee jerk interventions such as appointing a new coach, clubs should instead ensure that optimal adaptations are made. A third and final implication of the model is that football clubs should attempt to optimise players' health and well-being outside of purely physical condition and injuries. Player contentment was identified as a limit to success and was shown in the model to be influenced by a number of variables, many of which are under the control of the football club.

15.5.5 Study Limitations and Areas for Further Research

The present proof-of-concept study had three limitations. First, the use of only one elite-level football SME used may impact the quality, validity, and generalisability of the CLD. Further validation using a larger sample of SMEs may strengthen the CLD. This could involve a Delphi study incorporating elite level football SMEs with specific knowledge of the English Premier League. Second, the CLD was developed using the English Premier League as a conceptual base. This could impact the global generalisability of the CLD; however, it was necessary to ensure that the analysis was achievable given finite project resources. It is the authors' opinion that the similarities between the English Premier League and other countries ensure that the findings are generalisable to other football league systems worldwide. A follow-up study could involve adapting the model to specific contexts where similarities may not be present, for example, football league systems in lower-middle-income countries. A third and final limitation of the model was the analysis boundary, which was set to consider only variables that football clubs could have some influence over. As such, important wider systemic variables may have been overlooked. For example, additional variables that could ostensibly influence football club performance include those relating to the relevant national football association and international team and also broader societal issues such as the economy and unemployment. Further applications of the CLD method to consider a broader set of variables are encouraged. A logical next step would be to develop a CLD model of the English football system.

15.5.6 FUTURE RESEARCH DIRECTIONS

The aim of this chapter was to demonstrate the potential utility of CLD as a method for identifying and depicting the wider systemic influences on sports performance. As such, further applications of the CLD method in football and other sporting contexts are encouraged. Whilst these may include match and club systems, it is our view that applications in which broader sports systems are modelled would be useful. Whilst sports performance is one area that can benefit from such modelling applications, there are a various other areas which could also be explored such as doping and the use of performance-enhancing drugs, injury causation and prevention (Hulme et al., 2019), player transfer systems, corruption, sports governance, and technology insertion.

A second pertinent area of further research involves using CLDs to support the development and application of quantitative computational models (Hulme et al., 2018, 2019). System dynamics is one computational modelling method that is used to simulate the behaviour and dynamics of complex systems over time (Sterman, 2000). System dynamics models are developed based on an initial CLD of the system or issue in question. Although system dynamics is acknowledged to be a powerful approach for understanding system behaviour and for simulating the impacts of policy change, to date there have been few applications in the sporting context, and even fewer in football. Development of system dynamics models for football systems is therefore recommended as an important area for further research (Salmon & McLean, 2020). This will enable researchers and practitioners to better understand the feedback loops that influence performance, and also offers a mechanism for evaluating the likely impacts of different interventions designed to optimise performance. Moreover, the development and application of system dynamics in other sporting contexts is recommended (Hulme et al., 2019).

15.6 CONCLUSION

This chapter presented a proof-of-concept CLD of football club performance. It concluded that football club performance is influenced by a range of variables relating to players (e.g. skill, fitness level, personal lives), coaches and support staff (e.g. skill, experience), coaching and training programmes, the club (e.g. strategy, finances, transfers, marketing, culture, academy), sponsors, other clubs, and supporters to name only a few. Moreover, the dynamics of football club performance suggest that even the most successful clubs will experience periods of success followed by unsuccessful periods based on the capacity of opposition teams to adapt and optimise their own practices and performance and also limits to success such as player fatigue, injuries, and contentment. Various implications were identified, such as the requirement for clubs to continue to manage and optimise many variables during periods of success, and also the need to effectively manage player health and well-being. It is hoped that this chapter has demonstrated the potential utility of CLDs as a method for modelling sports systems, and that sports researchers and practitioners will apply the method in their own context.

REFERENCES

Allender, S., Owen, B., Kuhlberg, J., Lowe, J., Nagorcka-Smith, P., Whelan, J., & Bell, C. (2015). A community based systems diagram of obesity causes. *PloS one*, 10(7), e0129683. https://doi.org/10.1371/journal.pone.0129683

Araújo, D., & Davids, K. (2016). Team synergies in sport: theory and measures. *Frontiers Psychology*, 7, Article ID 1449.

Balague, N., Torrents, C., Hristovski, R. *et al.* (2013). Overview of complex systems in sport. *J Syst Sci Complex* 26, 4–13. https://doi.org/10.1007/s11424-013-2285-0

Berard, C. (2010). Group model building using system dynamics: An analysis of methodological frameworks. *Journal of Business Research*, 8:1, 13–24.

Bittencourt, N. F., Meeuwisse, W. H., Mendonça, L. D., Nettel-Aguiree, A., Ocarino, J. M., & Fonseca, S. T. (2016). Complex systems approach for sports injuries: moving from risk factor identification to injury pattern recognition-narrative review and new concept. *British Journal of Sports Medicine*, 50, 1309–1314.

Comrie, E. L., Burns, C., Coulson, A. B. Quigley, J., & Quigley, K. F. (2019). Rationalising the use of Twitter by official organisations during risk events: Operationalising the social amplification of risk framework through causal loop diagrams. *European Journal of Operational Research*, 272:2, 792–801.

Davids, K., Araújo, D. and Shuttleworth, R. 2005. Applications of dynamical systems theory to football. In *Science and Football* V, Edited by: Reilly, T., Cabri, J. and Araújo, D. 537–550. London, UK: Routledge.

Davids, K., Hristovski, R., Araujo, D., Serre, N. B., Button, C., & Passos, P. (2013). *Complex Systems in Sports*. Routledge Research in Sport and Exercise Science.

Dekker, S. (2011). *Drift into Failure: from Hunting Broken Components to Understanding Complex Systems*. Aldershot, UK: Ashgate.

Hulme, A., Thompson, J., Nielsen, R. O., Lane, B., McLean, S., & Salmon P. M. (2019). Computational methods to model complex systems in sports injury research: agent-based modelling and system dynamics modelling. *British Journal of Sports Medicine*, 53:24, 1507–1510.

Hulme, A., Thompson, J., Nielsen, R. O., Read, G. J. M., & Salmon, P. M. (2018). Formalising the complex systems approach: Using agent-based modelling to simulate sports injury aetiology and prevention. *British Journal of Sports Medicine*, 53, 560–569.

Kim, D. H. (1993). *Systems Archetypes I: Diagnosing Systemic Issues and Designing High-Leverage Interventions*. Waltham, MA: Pegasus Communications.

Leveson, N. G. (2004). A new accident model for engineering safer systems. *Safety Science*, 42:4, 237–270.

McGarry, T., Anderson, D. I., Wallace, S., Hughes, M., & Franks, I. M. (2002). Sport competition as a dynamical self-organising system. *Journal of Sports Science*, 20, 771–781.

McLean, S., Read, G. J. M., Hulme, A., Dodd, K., Gorman, A., Solomon, C., & Salmon, P. M. (2019). Beyond the tip of the iceberg: using systems archetypes to understand common and recurring issues in sports coaching. *Frontiers in Sports and Active Living*, 1, 49.

Mclean, S., Soloman, C., Gorman, A., & Salmon, P. M. (2017). What's in a game? A systems approach to enhancing performance analysis in football. *Plos One*, 1–15.

Mooney, M., Charlton, P. C., Soltanzadeh, S., & Drew, M. K. (2017). Who 'owns' the injury or illness? Who 'owns' performance? Applying systems thinking to integrate health and performance in elite sport. *British Journal of Sports Medicine*, 51, 1054–1055.

Purwanto, A., Susnik, J., Suryadi, F. X., de Fraiture, C. (2019. Using group model building to develop a causal loop mapping of the water-energy-food security nexus in Karawang Regency, *Journal of Cleaner Production* 240:118170. DOI: 10.1016/j.jclepro.2019.118170

Rasmussen, J. (1997). Risk management in a dynamic society: A modelling problem. *Safety Science*, 27:2/3, 183–213.

Read, G. J. M., Beanland, V., Lenne, M. G., Stanton, N. A., & Salmon, P. M. (2017). Integrating human factors methods and systems thinking for transport analysis and design. Boca Raton, FL: CRC Press.

Salmon, P. M., & McLean, S. (2020). Complexity in the beautiful game: implications for football research and practice. *Science and Medicine in Football*, 4:2, 162–167.

Salmon, P. M., Walker, G. H., Read, G. J. M., Goode, N. & Stanton, N. A. (2017). Fitting methods to paradigms: are ergonomics methods fit for systems thinking? *Ergonomics*, 60:2, 194–205.

Schoenenberger, L., Schenker-Wicki, A., Beck, M. (2014). Analysing terrorism from a systems perspective. *Perspectives on Terrorism*, 8:1, 16–36.

Senge, P. (1990). *The Fifth Discipline. The Art and Practice of Learning Organization.* New York, NY: Doupleday Currence.

Shepherd, S. P. (2014). A review of system dynamics models applied in transportation. *Transportmetrica B: Transport Dynamics*, 2:2, 83–105.

Silva, P., Vilar, L., Davids, K., Aruajo, D., & Garganta, J. (2016). Sports teams as complex adaptive systems: manipulating player numbers shapes behaviours during football small sided games. *SpringerPlus*, 5, 191. doi: 10.1186/s40064-016-1813-5.

Soltanzadeh, S., & Mooney, M. (2016). Systems thinking and team performance analysis. *International Sport Coaching Journal*, 3(2), 184–191.

Sterman, J. D. (2000). *Business Dynamics: Systems Thinking and Modeling for a Complex World.* Boston, MA: Irwin McGraw-Hill.

Stockl, M., Pluck, D., & Lames, M. (2017). Modelling games sports as complex systems – application of recurrence analysis to golf and soccer. *Mathematical and Computer Modelling of Dynamical Systems Methods, Tools and Applications in Engineering and Related Sciences*, 23:4, 399–415.

Vilar, L., Araújo, D., Davids, K., & Bar-Yam, Y. (2013). Science of winning soccer: Emergent pattern-forming dynamics in association football. *Journal of Systems Science and Complexity* 26:1, 73–84.

Walker, G. H., Salmon, P. M., Bedinger, M., & Stanton, N. A. (2017). Quantum ergonomics: shifting the paradigm of the systems agenda. *Ergonomics*, 60:2, 157–166.

Wilson, J. (2014). Fundamentals of systems ergonomics/human factors. *Applied Ergonomics*, 45:1, 5–13.

16 Performance Pathways in the Sport of Dressage
A Systems Ergonomics Approach

Elise Berber, Vanessa Beanland, Amanda
Clacy, and Gemma J. M. Read

CONTENTS

16.1 INTRODUCTION

Dressage is one of three Olympic equestrian disciplines, together with eventing and show jumping (Barrey et al., 2002). The objective of dressage is to harmoniously develop an intimate unity between human and non-human which combines balance, suppleness and power with a good horse displaying both elegance and expression (Dyson, 2000; Hawson et al., 2010). An Olympic dressage test comprises of 32–36 movements which consist of specific gymnastic exercises that horse and rider must perform within a 60 x 20 metre sized arena. Examples of these movements include half-passes, extended trot, piaffe, passage, canter pirouettes, and flying changes every stride. Seven judges mark each movement out of ten based on the technical accuracy of the movements and fulfilment of the training scale (rhythm, suppleness, contact, impulsion, straightness, and collection) (Fédération Equestre Internationale [FEI], 2007; Hawson et al., 2010). Olympic dressage consists of three rounds across three non-consecutive days. All nominated competitors compete in the team competition and in the Grand Prix test. The top 25 high scoring combinations go on to compete in the Grand Prix Special. Team medals are calculated from these two rounds (Hawson et al., 2010). The top 18 scoring combinations qualify to compete in the Grand Prix Freestyle event where individual medals are determined (Waltemeyer, 2016).

The current world record for the highest competitive level of dressage, the Grand Prix Freestyle, is 94.3% held by British rider Charlotte Dujardin (British Dressage, 2014), whereas Australia's record held by Kristy Oatley is 78.1% (Equestrian Life, 2018). Although dressage was first included in the Olympics at Stockholm in 1912, Australia did not send any representatives until 1984, and has sent a full team on only four occasions. Even when Australia has sent a team, the team has not placed high enough to qualify for team medals, and only one combination has ever been in contention for an individual medal (Equestrian Australia, n.d.a.). This is in stark contrast to Australia's Olympic record in eventing, which boasts multiple team and individual gold medals. These statistics suggest that the Australian dressage performance pathway could be better developed and improved.

There is currently no literature on sports performance pathways within dressage to inform improvements to the current situation of Australia's limited intentional success at the elite level. The following sections therefore will discuss sports performance pathways in general including all sports where sports performance pathways have been the focus of the research.

Talented combination	A talented combination includes both a trained horse and rider performing together.
National futures squad	A squad which is dedicated to identifying talented horses between the ages of 5–9 years of age who show potential to develop and perform at the world stage.
National elite squad	A squad which is dedicated to identifying horse and rider combination who are consistently scoring above 67% at Grand Prix level at international level events.
State level squad	Squad which are dedicated to identifying horse and rider combination who are consistently scoring from medium level to Grand Prix level above the minimum qualifying score.
Dressage Team	A team of four horse and rider combinations who were the highest-scoring talented combinations at a series of international Grand Prix events.
Squad selection events	Events that are regional level and above which have squad selectors in attendance.
Squad selectors	Individuals recognised to be selectors. Usually these individuals are A level (Grand Prix level) qualified judges.
FEI	International federation for equestrian sports.
Half Pass	A lateral movement where the horse moves forward and sideways simultaneously and is bent around the rider's leg in the direction of travel.
Extended Trot	A movement which requires the horse to length its frame and stride whilst increasing the suspension between each footfall.

Piaffe	A highly collected trot where the horse moves in place or nearly in place.
Passage	A highly elevated trot.
Canter Pirouette	A movement in the canter where the horse bends through its body in the direction of travel for six to eight strides whilst turning on a small diameter circle.
Flying Changes	A movement where the horse remains in canter but changes the canter lead during a moment of suspension when all four hooves are off the ground.
Australian Sports Commission	Government statutory agency responsible for distributing funds and providing strategic guidance and leadership for sporting activity in Australia.
Sports performance pathway	The pathway that outlines an athlete's journey from amateur level to elite level competition.

16.1.1 What Are Sports Performance Pathways?

Sports performance pathways (SPPs) outline an athlete's journey from the initial stages of talent identification through to elite level performance. Clear SPPs are essential to provide opportunities for talented athletes to be recognised and identified, give direction in training and competition, facilitate mental and physical skill acquisition, and develop self-confidence and performance motivation (Cripps et al., 2015; Kirk & Gorely, 2000; Martindale et al., 2007; Smith, 2005). SPPs also contribute to successful retention of athletes within a sport and improve the overall level of competition by enabling a greater number of athletes to achieve elite status. Community identity can also be developed through sport, which often serves as a central point for community engagement, pride, and achievements (Richards & May, 2017). Thus, performance pathways have positive impacts for individuals, sports, and the wider community.

Australia, like many other countries, supports two broad levels of sports participation: community and elite (Department of Health, 2013). The SPP is an important aspect of sport as it allows athletes to transition from recreational club level sport to elite competition (Lloyd et al., 2015). The community SPP is a learning environment that fosters the development and realisation of athletic potential while also creating a link between mass participation and specialisation in a sport (Brouwers et al., 2015; Vaeyens et al., 2008). Research suggests successful community-based SPPs facilitate the attraction, retention, and nurturing of athletes within a sporting system (Brouwers et al., 2015; Sotiriadou et al., 2008, 2016).

Research to date has focused on the ability of SPPs to provide information for goal setting, planning, development and recognition, as well as identifying key stakeholders in the pathway process. For example, Brouwers et al. (2015) identified national tennis associations as key stakeholders who initiate and support the programmes delivered by community clubs and coaches that promote the attraction and retention of athletes into the sport. The nurturing aspects of community sports include guiding and supporting players transitioning from junior level to senior levels of competition by providing them with opportunities to develop their sporting potential (Cripps et al., 2015). Similarly, Kirk, and Gorely (2000) suggest SPPs provide support by identifying the steps (e.g. moving through age groups or being selected into state and national teams) needed to make this transition. Martindale et al., (2007) further argue SPPs permit decision-making regarding the possibilities for progression in sport, as options can be viewed clearly and logically. This provides realistic expectations, standards and goals (Martindale et al., 2007). SPPs also facilitate planning where attempts to progress have been unsuccessful, pointing athletes to additional other possibilities for remaining in the sport (Kirk & Gorely, 2000). This could involve remaining at the community level but competing at a higher level than previously, or exploring possibilities connected to the sport through a different role such as coaching.

Many SPPs models have been developed; however, they are reductionist in nature, focusing mainly on individual traits such as skill level (Côté et al., 2009; Gagné, 2004; Vaeyers et al., 2008) or on policy and the athletic environment (Henriksen et al., 2010a, 2010b, 2011). Accordingly, no model has critically considered how all of the factors interact to optimise talent development.

16.1.2 Sports Systems

Systems thinking approaches and methods in the sports science context are increasingly being applied to understand and optimise entire sports systems (Clacy et al., 2017; Hulme & Finch, 2015; Neville & Salmon, 2015; Salmon et al., 2017). From the perspective of SPPs, a systems approach enables exploration of how the entire system of sport facilitates or hinders the ability of individuals to move through the pathway. This is because a systems perspective considers the impact of many different systemic and latent factors (e.g. political, regulatory, socio-cultural determinants) which influence an athlete's chance of reaching the elite level.

One systems-based framework that has been previously applied in sporting and outdoor recreation contexts (Salmon et al., 2017) is Rasmussen's (1997) Risk Management Framework (RMF). The RMF is domain generic and theorises that systems are comprised of a hierarchy of actors whose decisions and actions interact to shape the overall performance of systems. At the top of the hierarchy sits the government and legislative level, whereas the bottom levels are reserved for actors' activities and work processes (Rasmussen, 1997). Based on the RMF, the Systems Theoretic Accident Model and Processes (STAMP; Leveson, 2004) method is an updated systems approach that offers the opportunity to address the gaps evident within the SPP literature. STAMP posits that systems are controlled using adaptive feedback mechanisms, whereby actors and organisations within the system hierarchy impose controls on actors and organisations at the level below in order to constrain their behaviour. Further, information regarding the appropriateness and effectiveness of controls is fed back up the hierarchy to inform improvements to control mechanisms (Leveson, 2004). Although STAMP was developed to understand how safety risks are controlled within complex engineering systems (such as aerospace), it lends itself well to adaptation given its focus on system hierarchies, including describing and identifying points of weakness within any system (e.g. points of inadequate control or feedback). A fundamental assumption of STAMP is that dysfunctional outcomes can occur even when all actors within the system are working in accordance with the rules. It is therefore the system itself that is producing failure through its weaknesses, rather than individual actors or organisations. This makes STAMP highly applicable to the study of any system that is functioning sub-optimally despite the best efforts of actors within that system.

Recently, Hulme and colleagues (2017) applied STAMP to injury prevention in recreational distance running. The application of STAMP allowed for the description and examination of the recreational running injury system and identification of practical, system-wide opportunities to implement sustainable interventions for running injury prevention. Through the application of a systems approach to the issue, this study demonstrated the need for new safety controls and controllers, therefore pointing to areas of the system which could be the focus of future interventions.

Using Rasmussen's (1997) RMF to map previous SPP research shows that research to date has focused predominantly on the top 'government' level and/or the bottom two levels, comprising the activity environment and the individual athletes (Gagne, 2004; Gowthorp et al., 2016; Henriksen et al., 2011; Kirk & Gorely, 2006; Llyod et al., 2015; Sotiriadou et al., 2016; Vaeyers et al., 2008). SPP research has failed to consider the influence of actors and organisations at the management, company, and association levels. Actors at these levels are concerned with implementation of policy, company standards, and industry policy and therefore could play a key role in SPPs due to their ability to influence opportunities, development, and selection processes. Therefore, applying STAMP will allow for the identification of who is involved in the dressage system, their role and their influence in the system, and the feedback mechanisms in place. Providing a model which includes all interacting factors that can impact progression through the performance pathway without isolating any particular factor and examining it separately.

16.1.3   THE NEED FOR STRENGTHENING PERFORMANCE PATHWAYS IN DRESSAGE

The current Australian method for developing pathways in dressage is based on the Australian Sports Commission's (2012) Winning Edge scheme. Talented horse and rider combinations are identified and move through the pathway using squads at State, National, and Olympic. To be considered for the Olympic team, the horse and rider combination must be a member of a national high-performance squad (Equestrian Australia, 2016). There are several national and state squads (see Figure 16.1) which recognise and reward combinations at different levels. For example, selection to the National Elite Squad theoretically means combinations are well-established at Grand Prix level and are seeking selection for the next major championship (Olympic or World Equestrian Games). The purpose of National Futures Squad is to identify promising combinations and help them to develop correctly, enabling future contention for a place on the Australian team (Equestrian Australia, n.d.b). To be selected for a national squad, a horse and rider combination must be a member of a state-level squad (Equestrian Australia, n.d.c).

Traditional methods for identifying talented horse and rider combinations have relied upon the performance of the horse and rider at certain events. For example, to qualify for the squad, combinations must attain a minimum of four qualifying scores at national or regional level competitions or at identified squad selection events which are outlined at the beginning of each competition season. A panel of squad selectors

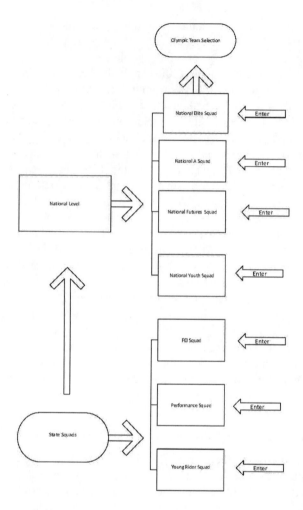

FIGURE 16.1   Australian dressage pathway.

consider each combination's scores across these events and the highest scoring combinations are named on the squad. As horse and rider combinations move through the levels of dressage, they are eligible to be named on the squad relative to the level they are now competing at but only if they continue to gain the qualifying scores required for that level.

Currently there is little formal knowledge or evidence to demonstrate the effectiveness of this SPP. As equestrian sport is one of the few sports offering lifelong elite-level participation opportunities, with many Olympians over the age of 40 and some competing into their 60s (Dumbell et al., 2018), Equestrian Australia recognises the need to 'develop a fully integrated National High Performance Pathway from talent identification through to elite competitor' (Equestrian Australia, n.d.d). The previous Olympic results for dressage, however, indicate a need to examine the current dressage performance pathway system as a basis to explore opportunities for improvement. Within the equestrian discipline of eventing Australia is competing on par with top countries, such as the UK and Germany, but in dressage is not performing to the same standard. Current Australian dressage results may not be fostering national identity or development of the sport in general.

16.1.4 AIM OF THE STUDY

Given the nonexistence of SPP research in the sport of dressage, the current study was designed to consider all the components of Australian dressage and their interactions, in contrast with the traditional SPP research approach of reducing the system into components (e.g. coach, athlete, etc.) and attempting to improve the performance of each individually. An adapted one-round Delphi study (Linstone & Turoff, 1975) was employed to refine and validate a preliminary STAMP model of the Australian dressage performance pathway. The Delphi was administered using interviews. The aim was to gather feedback from experts on the appropriateness of the model, refine the model, and reach consensus that its contents were valid. The application of STAMP to SPPs, and dressage, in particular, is a first of its kind analysis and it was expected that STAMP would offer the opportunity to comprehensively model the SPP of dressage, including all actors involved and the various constraints imposed throughout the system.

16.2 METHOD

16.2.1 PARTICIPANTS

Participants were 11 subject matter experts (SMEs) with expertise in equestrian sport ($n = 8$; average 25.8 years' experience), Australian Government ($n = 1$), and/or human factors and systems thinking ($n = 3$; average 7.7 years' experience). One participant reported expertise in both equestrian sport and human factors; all others had expertise in a single domain. Participants were identified through the researcher team's professional networks and by reviewing relevant websites (Equestrian Australia, State Equestrian Branches, and State Dressage Branches). Participants were recruited via email. All participants provided written, informed consent and no incentive or compensation was provided. Ethical aspects of the research were approved by the University of the Sunshine Coast Human Research Ethics Committee (S/17/1059).

Participants from the Australian equestrian industry included riders, coaches, judges, event organisers, squad selectors, and committee members. The most common role was rider ($n = 6$), with most equestrian-focused participants holding multiple roles. Most dressage SMEs had competed at an international level, and several had judged and coached at the national and local levels. All human factors experts were full time researchers.

16.2.2 MATERIALS

The draft STAMP model was developed by the research team based on a review of documentation (articles, policies and websites) sourced from Equestrian Australia, dressage state branches, Australian Sports Commission, and government websites. The research team comprised of three human factors experts with experience applying RMF and STAMP to multiple domains (Clacy et al., 2016; Hulme et al., 2017; Salmon et al., 2016, 2019; Read et al., 2019). The draft model was further validated via interviews with 11 SMEs. An *actor* was defined as any person, organisation, or government agency with an ability to influence progress through the dressage performance pathway. Elements of the environment were also included (i.e. the crowd, the surface) as these elements have the ability to affect the performance of talented combinations during competition. *Control mechanisms* were defined as any action, decision, or policy implemented by a higher level actor that can potentially limit actions in lower levels. *Feedback mechanisms* were defined as any ways in which actors at lower levels of the dressage system reported (or could report) to the levels above on the effectiveness of the controls.

The STAMP model included five levels. Level 1 included Australian legislators and international governing bodies (e.g. International Olympic Committee); Level 2 included government agencies, sporting associations and funding and research organisations (e.g. Equestrian Australia); Level 3 outlined general service and pathway providers (e.g. Equestrian state branches such as Equestrian Queensland); Level 4 included immediate pathway influencers (e.g. coaches); and Level 5 involved the rider, horse, equipment, and riding environment (e.g. horse and rider combination, riding surface and weather). The draft STAMP model presented to participants is shown in Figure 16.2.

An interview schedule was developed to support a structured interview process with the participants. The schedule included demographic questions to gather information on location of residence, their experience, expertise, and qualifications in either human factors or equestrian industry and questions to gain feedback on the draft STAMP model. Thirty-nine interview questions were formulated based on the content of the draft STAMP model. Questions were broken into four sections: actors, direct controls, indirect controls, direct feedback loops, and indirect feedback loops. Questions included: Do you agree with the actors, control structures, and feedback mechanism? Are there any actors, control structures, and feedback mechanism you believe are missing from the model? Do you disagree with any actors, control structures, and feedback mechanism in the model?

16.2.3 PROCEDURE

Participants were interviewed individually either in person ($n = 2$) or via telephone ($n = 9$). Each interview took 30–45 minutes. Participants were given a research participation information sheet at the beginning of the interview and asked to provide written informed consent and demographic information before commencing the interview. For phone interviews, ethics documentation, and the draft STAMP model were emailed prior to the interview.

During the interview, participants first answered the demographic questions and were then prompted by the interviewer to review the draft STAMP model. Responses were recorded using audio recordings and later were transcribed verbatim.

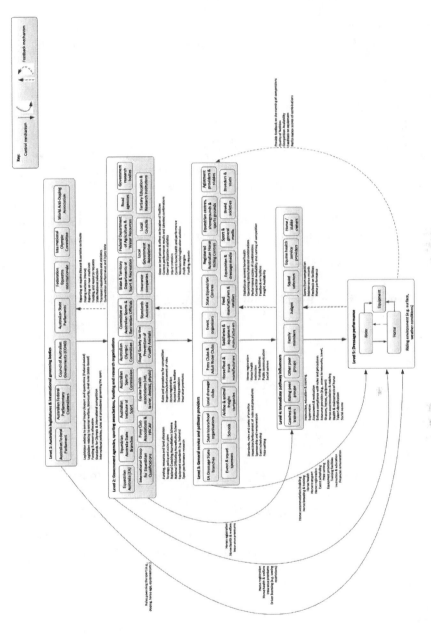

FIGURE 16.2 Draft dressage performance pathway STAMP model given to participants during interviews.

16.2.4 DATA ANALYSIS

Data analysis involved first calculating the percentage of participant agreement with the different aspects of the draft model (i.e. the actors, control mechanisms, feedback mechanisms). Next, the proposed changes to the model were analysed by three members of the research team to determine how the feedback should be incorporated into the model. The model was revised and finalised based on participant feedback.

16.3 RESULTS

There was overall a high level of agreement across the whole model. As shown in Table 16.1, the participants in the study agreed that the actors included in the draft STAMP model were correct however some ($n = 6$) suggested additions. Suggested actor additions included: Minister of Sport (Level 1); international welfare organisations (Level 1); Australian Dressage Committee (Level 2); event staff, officials, and stewards (Level 4); human and equine health professionals (Level 4); and quality of facility/surface (Level 5).

Participants mostly agreed that the controls included in the draft model were correct (see Table 16.2), however, there were some ($n = 10$) suggested additions.

TABLE 16.1
Level of Agreement with Actors Represented in the STAMP Model

Level	Agree	Disagree	Unsure
1	100%	0%	0%
2	100%	0%	0%
3	100%	0%	0%
4	100%	0%	0%
5	100%	0%	0%

TABLE 16.2
Level of Agreement with Control Mechanisms Represented in the STAMP Model

Level	Agree	Disagree	Unsure
1 →2	91%	0%	9%
2→3	100%	0%	0%
3→4	100%	0%	0%
4→5	91%	0%	9%
1→5	100%	0%	0%
2→5	91%	9%	0%
2→4	100%	0%	0%
3→5	100%	0%	0%
Average	97%	1%	2%

The changes included the addition of biosecurity measures, strategy, policy, and action plans, standards for squad selection; role models and social support; standards for international competition; accreditation for vets, coaches and judges; event scheduling and hosting; and rider membership and registration.

Majority of participants agreed that the feedback mechanisms included in the draft model were correct (see Table 16.3), however, there were suggested amendments ($n = 10$). Suggested changes to feedback mechanisms included removing state selection (Level 2 to 1), adding a feedback loop from Level 5 to 3, and adding several additional feedback mechanisms: doping statistics/information; feedback regarding horse purchased; strategy and policy; transport of horse; quality of facilities; complaints and event participation; rider registration; injury (horse and rider); and equine disease information and statistics. Figure 16.3 displays the final STAMP model with changes highlighted in bold italic text.

16.4 DISCUSSION

The current study aimed to apply a systems approach to identify the current dressage performance pathways in Australia. The STAMP analysis produced a comprehensive model of actors, controls, and feedback mechanisms. Participants indicated a high level of agreement with the draft STAMP model and recommended minor changes which were incorporated into the final model.

By employing STAMP, the current study extended previous research on SPPs, which has focused on individual athletes and their immediate environment. It has also provided a novel contribution by studying dressage SPPs for the first time. The STAMP model comprehensively identifies all of the actors and organisations who influence the dressage SPP, and the control and feedback mechanisms in dressage that may affect progression through pathways. The actors identified within the model were confirmed by SMEs to have a shared responsibility and play an active role in the development and/or facilitation of the dressage performance pathway. This is in line with contemporary safety theory which posits that safety and performance are the responsibility of all actors within the system (Leveson, 2004; Rasmussen, 1997). Actors such as Equestrian Australia, Dressage State Branches, coaches, FEI,

---

TABLE 16.3
Level of Agreement with Feedback Mechanisms Represented in the STAMP Model

Level	Agree	Disagree	Unsure
5→4	100%	0%	0%
4→3	100%	0%	0%
3→2	82%	0%	18%
2→1	82%	0%	18%
5→3	100%	0%	0%
Average	93%	0%	7%

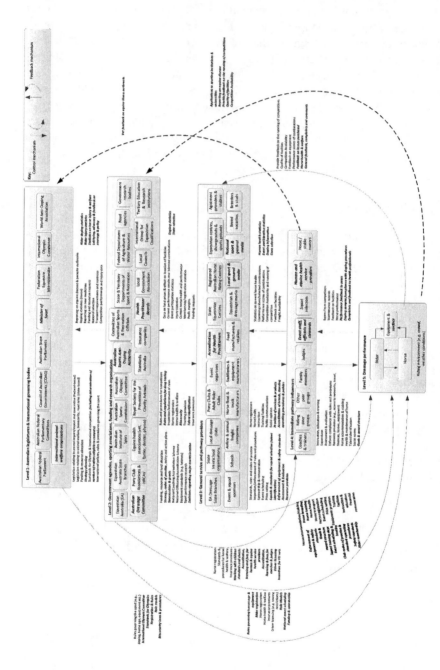

FIGURE 16.3   Final dressage performance pathways STAMP model. Amendments are shown in bold and italics.

and squad selectors have an obvious role in the system. However, other actors such as road agencies, Standards Australia, media, and tertiary research groups were also considered to influence the performance pathway for dressage. These actors are involved in setting up the conditions that allow riders to access competitions, equestrian research and dressage news, all of which facilitate progression through performance pathways.

The final STAMP model outlined the large number of constraints on actors at the lower levels of the dressage system, such as riders. The control mechanisms in the dressage performance pathway represent the expected relationships for establishing the rules and procedures for a high-performance pathway. For example, standards for Olympic and international competition (set by FEI and International Olympic Committee at Level 1) are controls imposed on Equestrian Australia (Level 2), its state branches (Level 2) and dressage state branches (Level 3), which in turn influence the rider's ability to compete at events (Level 5). A further example is government funding and resource allocation. Government agencies such as the Australian Sports Commission (Level 2) exert a political-based control on Equestrian Australia and its sub-branches to fulfil their duties such as organising coaching development, rider development and national events, which influence the rider indirectly.

Just as higher-level actors place constraints on those below, actors and organisations also provide feedback on the effectiveness and functioning of the system to levels above (Leveson, 2004). The feedback mechanisms represented in the STAMP model include the expected and potential communication between the levels, particularly in relation to how actors at lower levels experience the constraints imposed upon them, and whether these controls are functioning effectively. For example, riders can give feedback to both local clubs and event organisers regarding the running of competitions (Level 5 to 3). Equestrian Australia and state branches may also give feedback regarding current performance of combinations and membership statistics to the federal government to inform funding allocation requests.

Beyond the specific focus of this study being performance pathways, this model could also be used as a general model of dressage as it outlines all the actors, controls, and feedback loops involved in the whole dressage system. Utilising it as a model of the dressage system as a whole could provide insights into how the dressage system is performing overall and could focus its attention on all aspects of the system instead of focusing purely on its ability to produce elite level combinations.

Although the STAMP model identified actors, controls, and feedback mechanisms that would have been overlooked by methods focusing on individual athletes, it still has some limitations. For example, the region in which a rider lives was suggested as a potential influencing factor, as it influences a rider's access to quality competitions, coaching, and services. However, this was not able to be captured within the STAMP structure. Therefore, future research could build on this study by applying other systems-based methods, such as Cognitive Work Analysis (Vicente, 1999) or Hierarchical Task Analysis (Annett, 2004), to gain a more complete picture of the factors influencing access to performance pathways.

A further theme that emerged was actors *within* levels were suggested to give feedback and impose constraints on one another. For example, within Level 2, Equestrian Australia is technically higher than, and imposes constraints on, its state branches. State branches also provide feedback to Equestrian Australia, but within the model both have similar roles and are below the FEI (Level 1) and above dressage state branches and local clubs (Level 3), so therefore were placed within the same level even though there is some level of control/feedback occurring. This was a limitation of applying STAMP to a whole system, whereby it is challenging to represent control, feedback loops and relationships existing between actors who operate at the same level of the hierarchy (Salmon et al., 2018). Future research should consider ways in which within level relationships can be better represented.

A third emerging theme was that rider pathways, which we focused on here, are closely linked and interdependent with coach and judge pathways. For example, poor coaching pathways will decrease the number of high-standard coaches for riders, limiting rider opportunities for development. Again, pathway crossover cannot be fully represented by STAMP but could be explored using other methods.

16.5   PRACTICAL APPLICATION TO SPORT

The implications of the validated model are not only applicable to dressage specifically but can be applied across a range of different sports. For example, Hulme et al.'s (2017) STAMP model of running injury highlighted a range of actors that were also involved in the dressage performance pathway. This included the Federal Parliament, state sporting branches, local clubs, anti-doping agencies, the Australian Sports Commission, and coaches. Not only is it possible to now see which individuals and organisations are likely to be involved in the majority of sporting performance pathways but also the control and feedback mechanisms which could also be generic across sports. For example, the International Olympic Committee create the standards for Olympic and international competition for all sports, the Australian Sports Commission is involved in funding and resource allocation as well as strategy and planning for all sports within Australia, and coaches are universally responsible for instruction, education, training, and supervision. The types of influence these actors have on the dressage pathway therefore can be transferred across to other sports and their influence is likely to be similar.

What makes dressage, and equestrian in general, unique as a sport is the addition of an animal, the horse. This means the pathway also includes animal-related actors and complex issues around facilities and land use. These are slightly different to other sports as it is even harder to get access to facilities not only for training and competition purposes but also for the general care and maintenance of the horse due to regulations and limited land availability to build equestrian centres. Another difference is the interaction the rider has with the horse. Equestrian performance is not solely a property of the human athlete as it would be for a runner; performance also rests upon the horse's physical and psychological strength to perform; performance is highly dependent on the horse and rider combination. There is also greater regulation and cost around international travel with a horse, which affects the pathway to a greater extent than if it were a human team. The current study therefore sheds light on phenomena that impact all sports, but also supports the notion that individual sports have unique features that may not fit within a one-size-fits-all approach.

The validated model can also be used to evaluate the effectiveness of the dressage performance pathway. For example, using STAMP allowed for both existing and potential feedback mechanisms to be identified. Potential feedback mechanisms included: injury statistics, state squad selection and general feedback, complaints, and comments. This enhances the practical applications of this research as actors such as Equestrian Australia can use the model to improve feedback between levels of the system to increase performance. One option could be better communication on websites and social media; for example, explicitly including feedback avenues for athletes to contact Equestrian Australia. Therefore, interventions can be created based on the findings of the current STAMP model. Other sports could also use this finding and strategy to ensure they are communicating effectively with their athletes.

16.6 CONCLUSION

Due to the lack of a formally evaluated SPP and current underperformance in dressage on the world stage, there was a need for a systems approach to shed new light on talent development in Australian dressage. Taking a systems approach allowed the current study to diverge from previous research in that it provides a systemic view of the SPP and how it can limit athletes' abilities to achieve success and recognition. This research therefore contributes to the literature regarding factors influencing SPPs in that other sports may be able to adapt the model to guide understanding of their SPP. It can also be used to describe complexity of a pathway and highlight areas for improvement. For example, where there are limited feedback mechanisms or inflexible control mechanisms. This can then inform development of interventions to increase the effectiveness of the system. Ultimately, this can improve SPPs across a range of sports to provide opportunities for more talented individuals to be recognised and developed.

REFERENCES

Annett, J. (2004). Hierarchical task analysis. In: D. Diaper & N. A. Stanton (eds.), *The Handbook of Task Analysis for Human-Computer Interaction* (pp. 67–82). Mahwah, NJ: Lawrence Erlbaum.

Australian Sports Commission. (2012). *Australia's Winning Edge*. Canberra, Australia: Australian Sports Commission. Retrieved from: http://www.ausport.gov.au/__data/as sets/pdf_file/0011/509852/Australias_Winning_Edge.pdf.

Barrey, E., Desliens, F., Poirel, D., Biau, S., Lemaire, S., Rivero, J.-L. L., & Langlois, B. (2002). Early evaluation of dressage ability in different breeds. *Equine Veterinary Journal*, 34, 319–324. doi: 10.1111/j.2042-3306.2002.tb05440.x.

British Dressage. (2014). *Oh What a Night!* Retrieved from: http://www.britishdressage.co.uk /news/show/2229-oh-what-a-night.

Brouwers, J., Sotiriadou, P., & De Bosscher, V. (2015). An examination of the stakeholders and elite athlete development pathways in tennis. *European Sport Management Quarterly*, 15:4, 454–477. doi: 10.1080/16184742.2015.1067239.

Clacy, A., Goode, N., Sharman, R., Lovell, G. P., & Salmon, P. M. (2017). A knock to the system: A new sociotechnical systems approach to sport-related concussion. *Journal of Sports Sciences*, 35:22, 2232–2239. doi: 10.1080/02640414.2016.1265140.

Côté, J., Lidor, R., & Hackfort, D. (2009). ISSP position stand: To sample or to specialize? Seven postulates about youth sport activities that lead to continued participation and elite performance. *International Journal of Sport and Exercise Psychology*, 7:1, 7–17. doi: 10.1080/1612197X.2009.9671889.

Cripps, A. J., Hopper, L. S., Joyce, C., & Veale, J. (2015). Pathway efficiency and relative age in the Australian Football League talent pathway. *Talent Development and Excellence*, 7:1, 3–11.

Department of Health. (2013). *Australian Sport: Emerging Challenges, New Directions.* Canberra, Australia: Australian Government. Retrieved from: https://www.health.gov.au/internet/main/publishing.nsf/Content/aust_sport_emerg/$file/australian_sport_emerg.pdf.

Dumbell, L. C. Rowe, L. & Douglas, J. L. (2018) Demographic profiling of British Olympic equestrian athletes in the twenty-first century. *Sport in Society*, 21:9, 1337–1350. doi: 10.1080/17430437.2017.1388786.

Dyson, S. (2002). Lameness and poor performance in the sport horse: Dressage, show jumping and horse trials. *Journal of Equine Veterinary Science*, 22, 145–150. doi: 10.1016/S0737-0806(02)70139-1.

Equestrian Australia. (2016). *2016 Australian Olympic Team. Equestrian Australia Nomination Criteria: Dressage.* Retrieved from: http://www.equestrian.org.au/sites/default/files/2016%20Australian%20Olympic%20Team%20Nomination%20Criteria_Equestrian%20Dressage.pdf.

Equestrian Australia. (2017). *Equestrian Brings More Than $1 Billion to the Economy.* Retrieved from: http://www.equestrian.org.au/news/equestrian-brings-more-1billion-economy.

Equestrian Australia. (n.d.a). Australia's Equestrian Olympic Record. Retrieved from: http://www.equestrian.org.au/sites/default/files/Australian%20Equestrian%20Record.pdf.

Equestrian Australia. (n.d.b). *High Performance Program: Dressage High Performance Squad Criteria.* Retrieved from: http://www.equestrian.org.au/sites/default/files/Dressage%20High%20Performance%20Squad%20Criteria.pdf.

Equestrian Australia. (n.d.c). *National High Performance Program: National Dressage Squad Selection Criteria.* Retrieved from: http://www.equestrian.org.au/sites/default/files/Dressage%20Squad%20Selection%20policy.pdf.

Equestrian Australia. (n.d.d). *Strategic Priorities.* Retrieved from: http://www.equestrian.org.au/Strategic-Priorities.

Equestrian Life. (2018). *Kristy Oatley Sets Australian Grand Prix Freestyle Record in Denmark.* Retrieved from: http://www.equestrianlife.com.au/articles/Kristy-Oatley-sets-Australian-Grand-Prix-Freestyle-record-in-Denmark.

Fédération Equestre Internationale. (2007). FEI Dressage Handbook – *Guidelines for Judging.* Lausannce, Switzerland: Fédération Equestre Internationale.

Gagné, F. (2004). Transforming gifts into talents: The DMGT as a developmental theory. *High Ability Studies*, 15:2, 119–147. doi: 10.1080/1359813042000314682.

Gowthorp, L., Toohey, K., & Skinner, J. (2016). Government involvement in high performance sport: an Australian national sporting organisation perspective. *International Journal of Sport Policy and Politics*, 9, 153–171. doi:10.1080/19406940.2016.1220404.

Hawson, L. A., McLean, A. N., & McGreevy, P. D. (2010). Variability of scores in the 2008 Olympic dressage competition and implications for horse training and welfare. *Journal of Veterinary Behavior: Clinical Applications and Research*, 5:4, 170–176. doi: 10.1016/j.jveb.2009.12.010.

Henriksen, K., Stambulova, N., & Roessler, K. K. (2010a). Holistic approach to athletic talent development environments: A successful sailing milieu. *Psychology of Sport and Exercise*, 11:3, 212–222. doi: 10.1016/j.psychsport.2009.10.005.

Henriksen, K., Stambulova, N., & Roessler, K. K. (2010b). Successful talent development in track and field: Considering the role of environment. *Scandinavian Journal of Medicine and Science in Sports*, 20(s2) Supplement 2, 122–132. doi: 10.1111/j.1600-0838.2010.01187.x.

Henriksen, K., Stambulova, N., & Roessler, K. K. (2011). Riding the wave of an expert: A successful talent development environment in kayaking. *The Sport Psychologist*, 25:3, 341–362. doi: 10.1123/tsp.25.3.341.

Hulme, A., & Finch, C. F. (2015). From monocausality to systems thinking: A complementary and alternative conceptual approach for better understanding the development and prevention of sports injury. *Injury Epidemiology*, 2:31, 1–12. doi: 10.1186/s40621-015-0064-1.

Hulme, A., Salmon, P. M., Nielsen R. O., Read, G. J. M., & Finch, C. F. (2017). From control to causation: Validating a 'complex systems model' of running-related injury development and prevention. *Applied Ergonomics*, 65, 345–354. doi: 10.1016/j.apergo.2017.07.005.

Hulme, A., Thompson, J., Plant, K. L., Read, G. J., Mclean, S., Clacy, A., & Salmon, P. M. (2018). Applying systems ergonomics methods in sport: A systematic review. *Applied Ergonomics*.

Kirk, D., & Gorely, T. (2000). Challenging thinking about the relationship between school physical education and sport performance. *European Physical Education Review*, 6:2, 119–134. doi: 10.1177/1356336X000062002.

Leveson, N. (2004). A new accident model for engineering safer systems. *Safety Science*, 42:4, 237–270. doi: 10.1016/S0925-7535(03)00047-X.

Linstone, H. A., & Turoff, M. (1975). *The Delphi Method: Techniques and Applications.* Reading, MA: Addison-Wesley.

Lloyd, R. S., Oliver, J. L., Faigenbaum, A. D., Howard, R., De Ste Croix, M. B., Williams, C. A., Best, T. M., Alvar, B. A., Micheli, L. J., Thomas, D. P., Hatfield, D. L., Cronin, J. B., & Myer, G. D. (2015). Long-term athletic development – Part 1: A pathway for all youth. *The Journal of Strength and Conditioning Research*, 29:5, 1439–1450. doi: 10.1519/JSC.0000000000000756.

Martindale, R. J. J., Collins, D., & Abraham, A. (2007). Effective talent development: The elite coach perspective in UK sport. *Journal of Applied Sport Psychology*, 19:2, 187–206. doi: 10.1080/10413200701188944.

Neville, T. J., & Salmon, P. M. (2015). Seeing officiating as a sociotechnical system – the case for applying distributed situation awareness to officials in sport. *Lecture Notes in Computer Science*, 9174, 164–175.

Rasmussen, J. (1997). Risk management in a dynamic society: A modelling problem. *Safety Science*, 27:2–3, 183–213. doi: 10.1016/S0925-7535(97)00052-0.

Read, G. J., Beanland, V., Stanton, N. A., Grant, E., Stevens, N., Lenné, M. G., Thomas, M., Mulvihill, C. M., Walker, G. H., & Salmon, P. M. (2019). From interfaces to infrastructure: extending ecological interface design to re-design rail level crossings. *Cognition, Technology and Work*, 1–19.

Richards, R., & May, C. (2017). *Sport for Community Development.* Retrieved from: https://www.clearinghouseforsport.gov.au/knowledge_base/organised_sport/value_of_sport/sport_for_community_development.

Salmon, P. M., Dallat, C., & Clacy, A. (2017). It's not all about the bike: Distributed situation awareness and teamwork in elite women's cycling teams. In: R. Charles & J. Wilkinson (eds.), *Contemporary Ergonomics and Human Factors 2017* (pp. 240–248). Loughborough, UK: Chartered Institute of Ergonomics and Human Factors.

Salmon, P. M., Read, G. J. M., Beanland, V., Thompson, J., Filtness, A. J., Hulme, A., McClure, R., & Johnston, I. (2019). Bad behaviour or societal failure? Perceptions of the factors contributing to drivers' engagement in the fatal five driving behaviours. *Applied Ergonomics*, 74, 162–171.

Salmon, P. M., Read, G. J. M., & Stevens, N. J. (2016). Who is in control of road safety? A STAMP control structure analysis of the road transport system in Queensland, Australia. *Accident Analysis and Prevention*, 96, 140–151. doi: 10.1016/j.aap.2016.05.025.

Salmon, P. M., Read, G. J. M., Walker, G. H., Goode, N., Grant, E., Dallat, C., Carden, T., Naweed, A., & Stanton, N. A. (2018). STAMP goes EAST: Integrating systems ergonomics methods for the analysis of railway level crossing safety management. *Safety Science*, 110, 31–46. doi: 10.1016/j.ssci.2018.02.014

Smith, A. C. (2005). Junior sport participation programs in Australia. *Youth Studies Australia*, 24, 54–59.

Sotiriadou, K., Shilbury, D., & Quick, S. (2008). The attraction, retention/transition, and nurturing process of sport development: Some Australian evidence. *Journal of Sport Management*, 22:3, 247–272. doi: 10.1123/jsm.22.3.247.

Sotiriadou, P., Brouwers, J., & De Bosscher, V. (2016). High performance development pathways. In: E. Sherry, N. Schulenkorf & P. Phillips (eds.), *Managing Sport Development: An International Approach* (pp. 63–76). New York, NY: Routledge.

Vaeyens, R., Lenoir, M., Williams, A. M., & Philippaerts, R. M. (2008). Talent identification and development programmes in sport: Current models and future directions. Sports Medicine, 38:9, 703–714. doi: 10.2165/00007256-200838090-00001.

Vicente, K. J. (1999). *Cognitive Work Analysis: Toward Safe, Productive, and Healthy Computer-Based Work*. Boca Raton, FL: CRC Press.

17 Network Analysis of the Goals Scored at the 2018 Football World Cup

Scott McLean, Paul M. Salmon, and Orito Forsyth

CONTENTS

17.1 INTRODUCTION

Network analysis (NA) is a method borne out of mathematical and graph theory, and is commonly used in the social sciences with a strong focus on systemic analysis (Wäsche et al., 2017). Network analysis is used to understand the relationships (communications, transactions) between system components (people, tasks), as well as to identify the prominent entities within systems via quantifiable metrics (Lusher et al., 2010; Wäsche et al., 2017). In systems ergonomics research, network analysis is used on its own (Houghton et al., 2006) or within other methods, such as Event Analysis of Systemic Teamwork (EAST) (Stanton et al., 2013). Although network analysis is relatively new to sport, there are an increasing number of applications, specifically in football, which are aimed at understanding team functioning via the interactions of team members (Sarmento et al., 2017). Recent applications of network analysis in football have analysed entire match passing networks (Clemente et al., 2015; Grund, 2012), goal scoring passing networks (GSPN) (McLean et al., 2017a, b), intra-team communication (McLean et al., 2018), unsuccessful passing networks (McLean & Salmon, 2019), and for passing analysis of different playing formations (Clemente et al., 2015; McLean et al., 2018). This chapter will demonstrate the use of NA for system analysis in sport, by analysing the goals scored at the 2018 football World Cup held in Russia.

When using network analysis for passing connectivity in sport, the players are seen as a series of nodes within a network that are connected by passes (Grund, 2012). An advantage of using network analysis over traditional notational analysis methods (individual frequency and percentages) for passing is the consideration of the relational perspective and interdependencies of the entire team (Gonçalves et al., 2017). Network analysis moves beyond traditional passing analyses in football by understanding the relationships between team members rather than the isolated individual contributions (Wäsche et al., 2017). The outputs derived from passing network analyses provides a more detailed analysis, such as the level of team connectivity, and the key players involved in connecting the team.

Two previous applications of network analysis to the GSPN from major football tournaments, i.e. the 2016 EUROs and COPA (McLean et al., 2017a, 2017b), have shown that the GSPN were characterised by few players, minimal reciprocity, fast duration, and involving minimal pitch locations. Although these two continental tournaments are major footballing events, the World Cup is the pinnacle of football competition and involves teams from all continents. As such, analysing all of the GSPN from the World Cup will provide additional information on the mechanisms of GSPN at the ultimate level of competition.

To extend upon previous investigations of GSPN, the aim of the current chapter was to conduct a mixed-methods analysis by combining network analyses, notational analysis, and shot and goal efficiency measures to provide a highly detailed analysis of the World Cup GSPN. Notational analysis measurements, such as duration of networks, and pitch locations involved in the GSPN, have provided useful information and complemented the use of network analysis in previous research (McLean et al., 2017b). Furthermore, efficiency measures have been used to determine the number of passes leading to shots, and the amount of shots taken to score goals, however, efficiency measures have not been used alongside network analysis to attempt to describe the mechanisms of GSPN and will provide an additional and detailed level of analysis to describe efficiency of possession.

17.2 METHODS

17.2.1 Study Design

All goals scored ($n = 169$) in the 2018 World Cup tournament held in Russia were analysed from television footage. Fourteen GSPN were removed from the analysis, which included 12 own goals, and two GSPN that could not be determined from the television coverage, providing a total of 155 GSPN. The analysis included comparisons of the GSPN between the group and knock out stages, and between the 16 teams that were successful (progressed from group stage) and unsuccessful (knocked out of the tournament at the group stage). The comparison between the successful and unsuccessful teams was conducted from the goals scored in the group stage only, as each team at the tournament played three matches in the group stage of the tournament.

Additional analyses were performed to determine the shot efficiency and goal efficiency as a function of match outcome (win, draw, loss) for all matches, and between the teams finishing the tournament placed 1–8 and 9–16. Teams ranked 17–32 were not included in the shot and goal efficiency analysis because these teams, in general, scored a minimal number of goals. The pitch locations involved in the GSPN networks were determined by tracking and coding the progression of the GSPN through the pitch zones (Figure 17.2). The goal-scoring strike to the goal was included as the final connection within the GSPN.

Cohens Kappa was used for the inter-rater reliability, one coder analysed all 155 GSPN, and a second coder analysed 35 GSPN (22.5%). There was a high level of agreement between coders for network edges (1.000), players involved in the GSPN (0.971), network duration (0.907), and the pitch zones involved in the GSPN (0.914).

17.2.2 Procedure

17.2.2.1 Network Analysis

The network analysis involved viewing the GSPN and then constructing a social network matrix (Table 17.1) and network diagram (Figure 17.1) comprising the nodes (players) in the network and the connections (passes) between them (Clemente et al., 2015) (Figure 17.1). The GSPN were directional (e.g. player A passes to player B) and weighted where the passes between the players have a weight assigned to them

TABLE 17.1

Network Matrix for an Example GSPN Comprising Players 2–7–8–11–8–9 and Goal

	1	2	3	4	5	6	7	8	9	10	11	goal
1												
2							1					
3												
4												
5												
6												
7								1				
8									1		1	
9												1
10												
11								1				
goal												

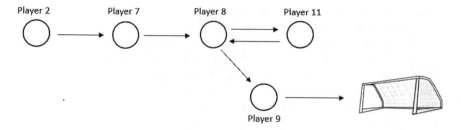

FIGURE 17.1   Network diagram for an example GSPN 2–7–8–11–8–9–goal.

(e.g. total number of passes connecting player A and player B). A GSPN commenced when the goal-scoring team gained possession and was completed when the ball crossed the goal line (McLean et al., 2017b).

17.2.2.2   Density

Density provides a measure of the level of connectivity of the team members in the passing network. Density is expressed as a value between 0 and 1, with 0 representing a network with no passing connections between team members, and 1 representing a network in which every team member is connected by passes to every other team member within the network (McLean et al., 2017a).

17.2.2.3   Cohesion

Cohesion is defined as the number of reciprocal connections in the network divided by the maximum number of possible connections. In this context, cohesion gives an indication of how often a player was involved within a network. For example, a reciprocal pass occurs when player A passes to player B who then passes back to player A.

17.2.2.4   Socio-Metric Status

The socio-metric status (SMS) of a network is a measure of the level of contribution made by each node (player) to the network (McLean et al., 2017a). In the context of football passing networks, a network with a higher SMS indicates that more players are repeatedly involved in the passing network in terms of giving and receiving the ball.

17.2.2.5   Notational Analysis

Notational analysis methods were used for total network connections, network duration, and the number of pitch zones involved in the networks. The total connections of a GSPN were the total number of passes between players and included the goal-scoring strike. The duration of the networks was the time (seconds) from when the network commenced to when the ball crossed the goal line. The pitch locations involved in the progress of the GSPN were recorded. The pitch zones used in the current study are presented in Figure 17.2.

17.2.2.6   Efficiency (Shot and Goal)

To calculate shot efficiency, the total number of completed passes in a match was divided by the shots taken in each match for each team, which represents the number

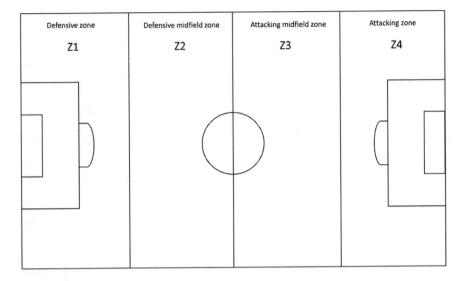

FIGURE 17.2   Pitch zones. Direction of attack is from left to right.

of passes completed per shot. For goal efficiency, the shot efficiency was divided by the number of goals scored by each team in each match, essentially providing a goal per passes per shot value. The data for completed passes, shots, and goals scored was sourced from the official FIFA World Cup 2018 website.

17.2.3   STATISTICAL ANALYSIS

For the GSPN, independent samples t-tests were used to determine differences between the group and knock out stages, the successful and unsuccessful teams for all network analysis and notational analysis metrics, and for the comparison of goal efficiency between the teams ranked 1–8 and teams ranked 9–16. The level of significance was set at $p < .05$.

One-way analysis of variance (ANOVA) tests were performed to determine the shot efficiency and goal efficiency as a function of match outcome (win, draw, loss). The level of significance was set at $p < .05$. Post hoc pairwise comparisons were conducted when the probability between groups was $p < .05$ using the least significant difference (LSD) test.

Where t-tests and ANOVAs were performed, effect size (Cohens d) was calculated to indicate the magnitude of the difference between the variables. Effect size categories were defined as small (0–.2), moderate (.2–.5), and large (>.5) (Cohen, 1988). Median and inter-quartile range (IQR) values were included in the analysis due to the large range of values within variables.

17.3   RESULTS

There were no significant differences ($p < .05$) between the group and knock out stages, or between the successful and unsuccessful teams from the group stage for network connections, density, cohesion, SMS, and duration (Tables 17.2 17.3 and 17.3). There were moderate effect sizes for network duration ($d = .283$) in the group versus knock out stage, and for network density ($d = .200$) in the successful versus unsuccessful comparison (Table 17.3). For all goals scored, 88% of the GSPN ranged from 1–6 network connections. The distribution of the GSPN connections and goal frequency is presented in Figure 17.3.

For shot efficiency, there was no significant ($p > .05$) main effect as a function of match outcome (Figure 17.4). There were moderate effect sizes between matches won and drawn ($d = .236$), and for matches won and lost ($d = .281$), and a low effect size between matches drawn and lost ($d = .087$). The teams ranked 1–8 at the tournament had a significantly ($p = .025$) lower shot efficiency with a large effect size ($d = 1.26$) compared to teams ranked 9–16 at the tournament (Figure 17.5).

For goal efficiency, there was a significant ($p = .022$) main effect for match outcome (Figure 17.4). Pairwise comparisons indicate that in matches won there was a significantly lower goal efficiency compared to matches lost ($p = .010$). There was a moderate effect size between matches won and drawn ($d = .484$), a large effect size for matches won and lost ($d = .636$), and a moderate effect size between matches drawn and lost ($d = .248$). The teams ranked 1–8 in the tournament had a significantly

---

TABLE 17.2

Network Analysis and Notational Analysis Metrics for the GSPN

		Connections	Density	Cohesion	SMS	Duration (s)
Overall	Mean ± SD	3.5 ± 3.6	.031 ±.055	.00 ±.01	.05 ±.06	8.1 ± 10.7
	Median	2	.015	.00	.058	3.5
	IQR	2	.028	.00	.049	2.5
	Max	27	.454	.09	.438	64
Groups	Mean ± SD	3.3 ± 3.7	.031 ±.062	.004 ±.011	.047 ±.062	7.3 ± 9.7
	Median	2	.015	0	.03	3
	IQR	3	.023	0	.049	9
	Max	27	.454	.09	.44	56
KO	Mean ± SD	4.0 ± 3.5	.028 ±.024	.005 ±.010	.057 ±.058	10.5 ± 12.7
	Median	3	.022	0	.041	6
	IQR	4	.023	0	.049	11
	Max	16	.106	.045	.256	64
Successful	Mean ± SD	3.4 ± 3.6	.035 ±.072	.004 ±.012	.047 ±.059	7.1 ± 8.8
	Median	2	.015	0	.025	4.5
	IQR	3.3	.023	0	.049	10.3
	Max	27	.454	.090	.438	56
Unsuccessful	Mean ± SD	3.2 ± 4.1	.024 ±.029	.004 ±.011	.046 ±.069	7.8 ± 11.7
	Median	2	.015	.011	.025	2
	IQR	2	.016	0	.033	9.5
	Max	19	.136	.045	.305	54

Successful (78 goals) and unsuccessful (35 goals) teams taken from group stage only. Successful (progressed out of the group stage)

TABLE 17.3

Probability (*P*) and Effect Size (Cohens *d*) of Variables

Group vs Knock Out Stages			Successful vs Unsuccessful		
Variable	*P*	Cohen's *d*	Variable	*P*	Cohen's *d*
Network density	.763	.063	Network density	.402	.200
Cohesion	.751	.095	Cohesion	.921	.000
SMS	.387	.166	SMS	.951	.015
Connections	.347	.194	Connections	.790	.051
Network duration	.099	.283	Network duration	.728	.067

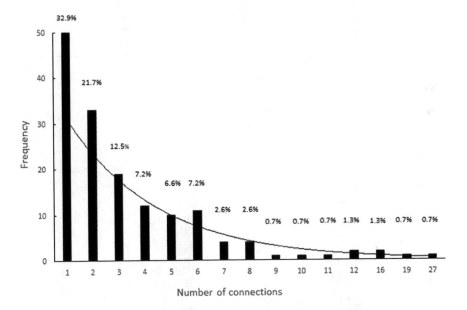

FIGURE 17.3 Distribution of the network connections for the frequency of goal scored.

($p = .001$) lower goal efficiency with a large effect size ($d = 2.06$) compared to teams ranked 9–16 in the tournament (Figure 17.5). Table 17.4 presents the shot and goal efficiency for each team which participated at the 2018 World Cup, and Figure 17.6 presents the pitch locations involved in all of the GSPN.

17.4 DISCUSSION

The purpose of this chapter was to present the use of NA as a method for understanding the network characteristics of the GSPN at the 2018 World Cup. In order to conduct a highly detailed analysis, NA was used together with notational analysis, shot and goal efficiency measures, and pitch location to investigate the mechanisms of the goals scored at the 2018 World Cup.

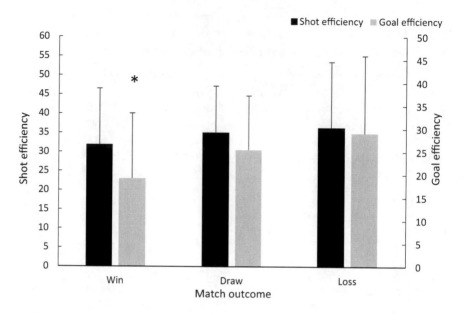

FIGURE 17.4 Shot efficiency and goal efficiency as a function of match outcome. Data are mean ± SD. * denotes a significant difference ($p < .05$) to matches lost.

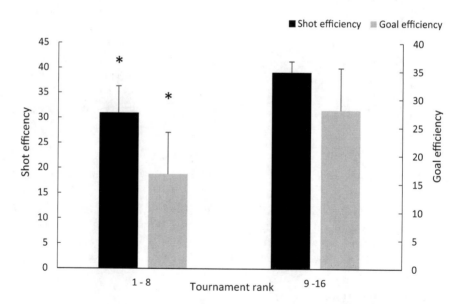

FIGURE 17.5 Shot efficiency and goal efficiency of the teams that ranked 1–8 compared to the teams ranked 9-16 at the 2018 World Cup. Data are mean ± SD. * denotes a significantly ($p < .05$) lower efficiency.

TABLE 17.4

Shot Efficiency and Goal Efficiency for Each Team at the 2018 World Cup

Country	Position	Shot efficiency	Goal efficiency	Position	Country	Shot efficiency	Goal efficiency
France	Winner (n = 7)	32.3 ± 11.9	15.5 ± 7.3	Eliminated group stage (n = 3)	Tunisia	35.6 ± 9.1	25.2 ± 16.8
Croatia	Runner up (n = 7)	28.9 ± 7.0	15.6 ± 5.3		Panama	37.3 ± 11.1	34.2 ± 13.8
Belgium	Semi-finalist (n = 7)	34.6 ± 14.6	15.3 ± 10.3		Senegal	26.6 ± 7.1	13.1 ± 5.0
England		35.2 ± 7.1	20.4 ± 11.1		Costa Rica	35.2 ± 14.7	10.1 ± –
Russia	Quarter finalist (n = 5)	38.9 ± 33.9	13.9 ± 13.7		Serbia	29.7 ± 8.6	26.3 ± 8.8
Sweden		22.3 ± 10.9	16.1 ± 8.7		Australia	49.6 ± 29.0	57.3 ± 36.4
Brazil		25.8 ± 10.6	17.1 ± 6.1		Iceland	18.9 ± 4.3	17.5 ± 4.9
Uruguay		29.5 ± 7.0	19.7 ± 13.8		Saudi Arabia	54.7 ± 24.7	13.4 ± –
Colombia	Final 16 (n = 4)	46.8 ± 28.8	41.4 ± 34.0		Peru	48.8 ± 33.5	42.4 ± –
Japan		42.6 ± 9.7	20.5 ± 5.5		Morocco	28.5 ± 6.0	17.3 ± –
Mexico		25.4 ± 6.4	17.9 ± 1.9		Iran	32.3 ± 11.19	32.3 ± 11.19
Denmark		35.4 ± 8.4	32.3 ± 6.9		Egypt	35.8 ± 3.1	35.8 ± 3.1
Argentina		44.5 ± 14.9	24.2 ± 6.2		Poland	40.1 ± 7.2	41.5 ± 9.6
Spain		45.8 ± 7.1	31.0 ± 12.5		South Korea	29.4 ± 11.7	12.4 ± 6.2
Swiss		38.7 ± 16.6	31.2 ± 25.7		Germany	28.3 ± 9.9	19.8 ± –
Portugal		33.2 ± 7.6	26.6 ± 10.7		Nigeria	23.0 ± 2.2	16.6 ± 5.8

Data are mean ± SD. – Denotes no SD as only scored in one game (i.e. one value). Shot efficiency = passes completed / shots. Goal efficiency = shot efficiency / goals scored. Number of matches played displayed in parenthesis.

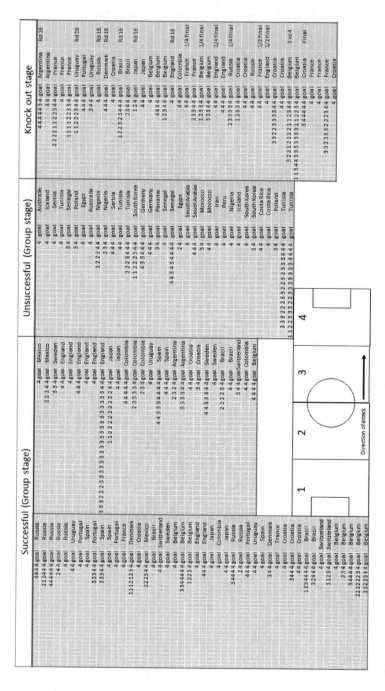

FIGURE 17.6 Pitch zones involved in each GSPN. Column A represents the GSPN of the Successful teams in the group stage, column B represents the GSPN of the unsuccessful teams in the group stage, and column C represents the GSPN from the knock-out stages. Each line in the columns represents the pitch zones involved in the GSPN, where connections were made. For example, in column A the first goal scored in the World Cup, Russia's GSPN involved connections from 4–4–4–goal.

17.4.1 OVERALL TOURNAMENT ANALYSIS

The overall tournament GSPN had low values for density, cohesion, and socio-metric status, with an average of 3.5 connections (passes) with a duration of 8.1 s, involving less than two pitch zones. These results indicate that the GSPN at the tournament were rapid and direct attacks containing loosely connected networks with low reciprocity of passes. Similar results have been found in previous research using NA to analyse the GSPN of the 2016 European championships (EURO), and the 2016 COPA America (COPA) (McLean et al., 2017a, 2017b). The current World Cup analysis, together with the 2016 EURO and COPA analyses, have investigated the GSPN of the three biggest football tournaments in the world using the most recent editions, with results showing that GSPN are similar across each tournament. The GSPN results of the three tournaments are characterised by rapid and direct attacks containing loosely connected networks with low reciprocity of passes. Furthermore, a look at historical GSPNs provides a similar pattern to the abovementioned studies, albeit using different methods of analysis. For example, similar results were found from 3,213 matches between 1953 and 1968 (Reep & Benjamin, 1968), and from the 1990 and 1994 World Cups (Hughes & Franks, 2005). In these studies, approximately 80% of goals were scored from four or fewer passes. This suggests that, despite the apparent evolution of football playing styles over the past decades, the method in which goals are scored has stayed largely stable. The distribution of the network connections (Figure 17.3) in the current study highlights the large number of goals scored from short networks, where 88% of goals were scored with six connections or less.

17.4.2 COMPARATIVE ANALYSIS

There were no differences in the measured variables between the group and knock out stages, or for the successful and unsuccessful teams, and the current results suggest that GSPN characteristics were consistent irrespective of tournament stage, or the success of teams. However, there was a moderate effect size for the decreased network duration in the group stage compared to the knock out stage of the 2018 World Cup. In comparison to the 2016 EUROs group stage, the GSPN network duration of the 2018 World Cup was approximately 4 s shorter (i.e. 7.3 s compared to 11.7 s) at the EUROs. A potential explanation for the 4 s difference in GSPN duration may be the high number of penalties awarded during the group stage of the World Cup. For instance, there were 22 penalties scored at the World Cup, 18 of which were scored in the group stage. As the scored penalties were coded as 1 s in duration, this presents a potential explanation for the reduced GSPN duration in World Cup group stage compared to the EUROs.

An interesting and potential influence on the current results is that the 2018 World Cup was the first occasion the Video Assistant Referee (VAR) had been used at a major tournament. Although we did not measure the effectiveness of the VAR, it is assumed from the reduction in VAR awarded penalties in the knock out stages of the tournament that the players adapted to having the VAR available during matches. Having the VAR available meant that referees were able to seek clarification on decisions relating to goal scoring actions, as such fouls and the subsequent penalties that were awarded at the 2018 World Cup may not have been awarded in previous tournaments because there was no VAR. This indicates that players were potentially more cautious when defending inside the penalty box in the knock out stages of the tournament, hence the reduced penalties and subsequent increase in the average GSPN duration in the knock out stages.

> The number of penalties awarded in the group stage significantly increased from previous editions, partly due to the presence of VAR.
>
> **(FIFA 2018 World Cup technical report)**

The moderate effect found for increased density in the successful compared to the unsuccessful teams suggests that the successful teams were more skilled at retaining possession and involving more players in the GSPN. Teams that reached the later stages, compared to initial stages, of the 2014 World Cup had higher values for network density for entire match passing networks, which indicates increased team connectivity (Clemente et al., 2015). Although the current analyses investigated the GSPN only, the increased connectivity previously measured in successful compared to unsuccessful teams (Clemente et al., 2015) may have been represented in the current GSPN of successful teams.

17.4.3 Shot and Goal Efficiency

The shot and goal efficiency analyses provide an assessment of how effectively teams used their possessions to create shots, and to score goals. For the match outcome analysis, the results show that winning teams were more efficient at creating shots at goal, and at scoring goals relative to the number of completed passes, compared to the drawing and losing teams (Figure 17.4). To investigate the efficiency measures of highly successful teams, the current analysis compared the efficiency measures of the teams that progressed to the knock out stages by dividing the top 16 teams into two groups 1–8 and 9–16 based on their final tournament rankings. The teams ranked 1–8 had significantly lower shot and goal efficiency values (i.e. more effective), compared to the teams ranked 9–16 in the tournament (Figure 17.5). This suggests that the highly successful teams make better use of their possession in terms of creating shots and scoring goals, relative to the number of completed passes. The top three ranked teams from the tournament France, Croatia, and Belgium each had low-efficiency values, which indicates effective use of possessions to create and score goals and this may be one potential component for their success at the tournament (Table 17.5). An, example of a team that had poor efficiency values was Australia (Table 17.5), which indicates that they did not create chances or score goals very often relative to the amount of completed passes.

The use of the efficiency measures in this chapter may also provide an insight into the success of teams who were not highly ranked prior to the World Cup but exceeded many expectations, such as Sweden and Russia. This may be explained in some part by looking at their goal efficiency values, which were comparable to those of France, Croatia, and Belgium (Table 17.5). Irrespective of the amount of possession they had, they were able to use the possession efficiently to score goals, and subsequently experience success in the tournament.

The current efficiency results support expert opinions that there appears to be a developing trend in modern football that is characterised by being organised defensively whilst relinquishing the majority of possession but then having the ability to launch the efficient, rapid, loosely connected GSPN. A good example of a victim of this trend is Spain who typically have excellent players in highly connected passing networks, a style that has bought them much success. However, as shown in Table 17.5 their efficiency values were high, (i.e. not effective), and their results were sub-par at this and the previous World Cup. It seems that teams now understand that being efficient offensively and organised defensively is perhaps a successful formula, compared to having a high volume of possession that is inefficient at creating and scoring goals.

17.4.4 PITCH ZONES

The context of pitch location of where actions are performed is important when analysing performance (McLean et al., 2017b). At the 2016 EUROs approximately 50% of the GSPN originated in Zone 4, meaning that half of the GSPN originated in the quarter of the pitch closest to the goal. Similarly, in professional Norwegian football more goals were scored by regaining possession in the final third (closest to goal) of the pitch, compared to the middle and first thirds of the pitch (Tenga et al., 2010). An additional analysis of the 2016 European and Copa America championships, showed that 63% of the GSPN at the European championships, and 85% of the GSPN at the Copa America commenced by regaining possession in the opposition half (McLean et al., 2017a). In the current study, each GSPN from the World Cup is shown (Figure 17.6), and highlights the minimal number of pitch zones used for the majority of GSPN, and to demonstrate the relatively low number of long networks involving multiple pitch zones. In this analysis, there were less than two pitch zones involved in the GSPN, and moderate effect size for between the group and knock out stages. The fewer pitch zones involved in the GSPN in the group stage may also

TABLE 17.5
Pitch Zones Utilised in the GSPN

	Mean ± SD	P	Cohens d
Group	1.72 ± 0.95	.113	.268
Knock out	2.02 ± 1.26		
Successful	1.68 ± 0.91	.531	.133
Unsuccessful	1.81 ± 1.04		
Overall	1.81 ± 1.05		

be explained by the high number of penalties in the group stage, as a penalty is performed in Zone 4 only.

17.4.5 PRACTICAL IMPLICATIONS

This chapter has presented a detailed analysis of the GSPN of the 2018 World Cup using a mixed-methods approach. The current chapter has implications for researchers, match analysts, and coaches. NA provides an understanding for the structure of the connections between players, the notational analyses allows some context of the location of specific actions, and the efficiency measures provide a measure of the usefulness of passing connectivity. The current analysis has also shown that when combing NA with other analysis methods, a highly descriptive assessment of performance can be achieved. Coaches could design training practices that include the efficient use of attacking plays by incorporating the findings of this chapter.

REFERENCES

Clemente, F. M., Martins, F. M. L., Kalamaras, D., Wong, D. P., & Mendes, R. S. (2015). General network analysis of national soccer teams in FIFA World Cup 2014. *International Journal of Performance Analysis in Sport*, 15:1, 80–96.
Cohen, J. (1988). *Statistical Power Analysis for the Behavioral Sciences*. 2nd edition. Hillsdale, MI: Erlbaum Associates.
Gonçalves, B., Coutinho, D., Santos, S., Lago-Penas, C., Jiménez, S., & Sampaio, J. (2017). Exploring team passing networks and player movement dynamics in Youth Association Football. *PLoS One*, 12:1, e0171156.
Grund, T. U. (2012). Network structure and team performance: The case of English Premier League soccer teams. *Social Networks*, 34:4, 682–690.
Houghton, R. J., Baber, C., McMaster, R., Stanton, N. A., Salmon, P., Stewart, R., & Walker, G. (2006). Command and control in emergency services operations: a social network analysis. *Ergonomics*, 49:12–13, 1204–1225.
Hughes, M., & Franks, I. (2005). Analysis of passing sequences, shots and goals in soccer. *Journal of Sports Sciences*, 23:5, 509–514.
Lusher, D., Robins, G., & Kremer, P. (2010). The application of social network analysis to team sports. *Measurement in Physical Education and Exercise Science*, 14:4, 211–224.
McLean, S., & Salmon, P. M. (2019). The weakest link: a novel use of network analysis for the broken passing links in football. *Science and Medicine in Football*, 1–4.
McLean, S., Salmon, P. M., Gorman, A. D., Dodd, K., & Solomon, C. (2018). Integrating communication and passing networks in football using social network analysis. *Science and Medicine in Football*, 1–7.
McLean, S., Salmon, P. M., Gorman, A. D., Naughton, M., & Solomon, C. (2017a). Do intercontinental playing styles exist? Using social network analysis to compare goals from the 2016 EURO and COPA football tournaments knock-out stages. *Theoretical Issues in Ergonomics Science*. doi:10.1080/1463922X.2017.1290158.
McLean, S., Salmon, P. M., Gorman, A. D., Stevens, N. J., & Solomon, C. (2017b). A social network analysis of the goal scoring passing networks of the 2016 European Football Championships. *Human Movement Science*, 57, 400–408.
McLean, S., Salmon, P. M., Gorman, A. D., Wickham, J., Berber, E., & Solomon, C. (2018). The effect of playing formation on the passing network characteristics of a professional football team. *Human Movement*, 2018:5, 14–22.

Reep, C., & Benjamin, B. (1968). Skill and chance in association football. *Journal of the Royal Statistical Society. Series A (General)*, 131:4, 581–585.

Sarmento, H., Clemente, F. M., Araújo, D., Davids, K., McRobert, A., & Figueiredo, A. (2017). What performance analysts need to know about research trends in Association Football (2012–2016): A systematic review. *Sports Medicine*, 1–38.

Stanton, N., Salmon, P. M., & Rafferty, L. A. (2013). *Human Factors Methods: A Practical Guide for Engineering and Design*. Farnham, UK: Ashgate Publishing Ltd.

Tenga, A., Holme, I., Ronglan, L. T., & Bahr, R. (2010). Effect of playing tactics on goal scoring in Norwegian professional soccer. *Journal of Sports Sciences*, 28:3, 237–244.

Wäsche H. G., Woll A., Brandes U. 2017. Social network analysis in sport research: an emerging paradigm. *European Journal for Sport and Society*, 14(2):138–165.

18 Preventing, Identifying, and Treating Concussion

A Systems Approach to Concussion Management in Community Sport

Amanda Clacy and Glenn Holmes

CONTENTS

18.1   INTRODUCTION

Concussion is a type of mild traumatic brain injury and refers to a sudden disruption to cognition resulting from a forceful acceleration and/or deceleration of the brain caused by a blunt force trauma or blow to the body that transmits a serious inertial force to the brain. Concussive injuries that occur in the context of sport (i.e. sport-related concussion [SRC]) are complex, both in terms of factors that influence injury risk as well as factors that determine injury management. A multitude of risk factors for SRC have been identified, such as playing position (i.e. offensive versus defensive; Gardner et al., 2014), fitness and training (Giza et al., 2013), athlete age (Harmon et al., 2013), and previous concussion history (Guskiewicz et al., 2013; Harmon et al., 2013). Notably, these risk factors are influenced by actors at all levels of the system (e.g. governing bodies decide playing separations, medical experts assess concussion history, coaches dictate training duration). Despite this, current concussion management strategies to address these risk factors do not adequately consider the interactions between elements of the system; rather they aim to address single known risk factors or elements of those factors in isolation, known as reductionism.

To illustrate, individual characteristics such as body mass index have been shown to increase the risk of concussion in community rugby union (Giza et al., 2013). In demonstration of a reductionist approach to risk management, strategies to address this have included models of weight-based grading to replace traditional age banding have been implemented (Patton et al., 2016). Without considering the wider contextual systems factors that influence concussion risk, this kind of reductionist strategy (i.e. it only focusses on one element of the issue in isolation) not only has limited effectiveness in reducing concussion risk, it also overlooks potential ramifications of this singular change. For example, by enforcing weight-based team grading negative physical and psychological effects have been shown to occur, such as dieting and dehydration to 'make weight' and negative body image (Patton et al., 2016). Consequentially, players are engaging in sport while undernourished or dehydrated, which may then reduce their performance and attention to the game, actually making them at higher risk of sustaining injuries such as concussion. Alternatively, it may lead young players to withdraw from participation, removing them from opportunities to remain physically active and socially engaged, placing them at risk of experiencing other pathologies (e.g. mental illness, obesity). This type of reductionist approach demonstrates a lack of consideration both for the complexity of SRC, as well as a lack of understanding of the wider systemic factors that influence SRC management. This failure to take into account that the management of SRC as multifaceted is a key gap in the literature. Analysis at the wider systems perspective enables the development of interventions that address SRC management limitations in the context of the system in which the injury occurs.

It is widely accepted that safety science is shifting away from individual, reductionist approaches to injury prevention and management, and instead recognising systemic influences on the occurrence of accidents and injuries (e.g. Dekker, 2016; Rasmussen, 1997; Salmon et al., 2014, 2016). The systems, or human factors, approach to safety science instead posits that injuries have multiple interrelated contributory factors that are created by the decisions and actions of people across all levels of a particular system, from the physical environment to the overarching governing bodies (Rasmussen, 1997). Rather than focus analyses and interventions on those injured, this approach instead attempts to understand the network of systemic contributory factors involved and develop interventions targeted at system reform rather than individual behaviour change. With this shift towards a systems approach to safety management, both research and industry bodies have become increasingly concerned with developing more system-based injury management strategies. While most current sport-related concussion prevention and management strategies focus on only the first component (e.g. rule changes and use of protective equipment in sport), the application of systems analysis methods to better understand and manage injury provides the opportunity to revise concussion management strategies to ensure they are context-specific and reflect the needs of the end-users.

Community sport is typically delivered through a network of local clubs that is administered in a hierarchical manner in international, national, state/provincial, and regional structures. Unfortunately, the paucity of concussion management knowledge in the different levels of sporting participation has been consistently demonstrated (e.g. Hollis et al., 2012; King et al., 2014; McAllister-Deitrick et al., 2014). Attitudes and behaviours of physicians, trainers, coaches, parents, and athletes on the identification and management of concussion lack both consistency and accuracy (see Donaldson et al., 2016 for review). As a result, applications of systems theory to sports performance and safety are emerging (see Bittencourt et al., 2016; Dallat et al., 2015; Hulme et al., 2017a; Salmon et al., 2010, 2014). In relation to SRC, a systems theory perspective considers the concussive event and subsequent management as emergent properties arising from the interactions between actors and factors across a systems hierarchy. These factors include policy and guidelines, rules and regulations, culture, training, and equipment (Clacy et al., 2013, 2017; Leveson, 2004; Salmon et al., 2016). The aim of the following chapter is to demonstrate the significant contribution a systems approach can have to understanding and improving SRC management in community sport by investigating the actors and factors that exist within the system.

18.2 MANAGING CONCUSSION

To effectively manage SRC a comprehensive knowledge of the physical complexities of injury and an understanding of the tools and processes required to manage injury is necessary. Sports organisations are increasingly active in their movements towards improving the safety of their players; however, many of these prevention measures have been reactive and reductionist. An example of a this has been the development and introduction of helmets (made from hard materials) and headgear (soft shell caps with fixed firm foam padding), which have been shown to have an inconsistent impact on SRC incidence and/or severity. For example, in rugby union (rugby), Marshall et al. (2005) and McIntosh et al., (2009) indicated that headgear has no significant effect on reducing concussion incidence in either junior or senior levels of rugby competition. In stark contrast, Hollis et al. (2009) and Kemp et al. (2008) found that the use of headgear in amateur and professional rugby decreased athletes' concussion incidence. Research exploring the efficacy of helmets and headgear for SRC prevention has not only been contentious, but has also highlighted systemic considerations to injury prevention that have been overlooked. Head protection policy to date has failed to take into account the wider systematic implications of changes, such as culture and sports psychology. Indeed, research has shown that the use of protective equipment, such as headgear and body-guards, may lead players to feel more protected and therefore play harder, through a concept known as 'risk compensation' or 'risk homeostasis'. In a study of adolescent rugby union players undertaken by Finch et al. (2001), players with no prior head or neck injury were more likely to report that they felt safer wearing protective headgear compared with those with a prior injury. Further, of the players wearing protective headgear, 67% reported that they played with more confidence when they wore the headgear. Finch et al., (2001) concluded that 'players report that they are more confident and able to tackle harder if they wear headgear, suggesting that a belief in its protective capabilities may influence behaviour'. Additionally, headgear reduces pain from scalp injuries and lacerations sustained while playing sport. By reducing the perceived risk of head injury, players may become accustomed to a playing style in which contact to the head was no longer avoided or feared. Consequently, this could lead to an increase in the frequency of total collisions to the head and potentially increase the total number of concussions as well. By reducing the issue of concussion management down to one strategy, with minimal consideration for systemic contributing factors such as player beliefs and behaviours, sporting culture, skills coaching, and supervision, the introduction of headgear not only fails to mitigate concussion risk, but also present the opportunity for greater injury to potentially occur.

Similarly, the introduction of changes to game rules and enforcement strategies again overlook the complex nature of SRC. For example, the introduction of the blue card in rugby union; the introduction of the 'concussion sub' in Australian football and cricket; and strict bans on head high tackles and impacts in ice hockey and various codes of football. Contrary to this direction taken by the governing bodies of each sport, there is scant evidence to suggest rule changes have a significant impact on the reduction of injuries such as concussion (Fraas & Burchiel, 2016). In fact, the number of SRCs in the National Hockey League (NHL) actually *increased* after the impact bans such as 'Rule 48' into effect even though the intended purpose of this intervention was explicitly to reduce the incidence of SRC (Donaldson et al., 2013). The effect was not due to players ignoring the new ruling (i.e. avoiding head hits), rather the majority of SRCs were due to hits that failed to draw a penalty (Donaldson et al., 2013). The most dangerous hits in the NHL are actually body checks (64.2% of concussions), which became the preferred impact method once head hits were outlawed (Donaldson et al., 2013).

This reactive rule change demonstrates how imperative it is to assess all of the factors that influence injury in context and with consideration for emergent risk that may occur when only one factor in isolation is addressed. This is not to say that rule changes are not appropriate in some instances, indeed consideration of rule changes to reduce head injury incidence or severity is thought to be appropriate where a clear-cut mechanism is identified (McCrory et al., 2017). Problems occur in injury management when changes to how an environment is governed or run fail to consider actor roles and behaviours, the system's capacity for translation of information to stakeholders at all levels of the system, and the variations in sporting contexts that influence the decisions and actions of actors.

The factors relevant to SRC management span identification, diagnosis, treatment, recovery, and return to play. The sideline identification of SRC has been an ongoing problem across multiple sports and levels of competition (i.e. junior and elite sport alike). Unfortunately, SRC is often only assessed if there is an obvious event that is likely to cause an SRC (Clacy et al., 2017). Given the dynamic and fast-paced nature of contact team sports, identification by incident observation is problematic, especially in junior and amateur leagues of sport which rarely have the same degree of supervision or recording. In fact, more than half of SRC's go undiagnosed (Roberts et al., 2017), due to both the subtle presentation of the injury and inadequate sideline identification and reporting (Littleton & Guskiewicz, 2013; McCrory et al., 2017).

Several factors have been found to contribute to SRC under-reporting. First, a lack of player education regarding concussion and its potential severity has been shown to lead to non-reporting behaviour (McAllister-Deitrick et al., 2014; McCrea et al., 2004). In junior and amateur sport, SRC communication and education is potentially limited due to the more recreational approach to participation. This was shown by Dawson et al. (2017) who looked at reporting behaviours in community Australian football and found that underestimating the consequences of concussion was the most frequent factor influencing a player's non-report decision. It is also important to note, however, that research has shown that up to 60% of athletes who have received formal concussion education fail to report suspected SRCs (Chrisman et al., 2013; McCrea et al., 2004; Sye et al., 2006; Torres et al., 2013). As with the use of protective gear, the issues surrounding the identification and sideline reporting of concussion are complex. Emerging from inconsistent education and communication in conjunction with extended stand-down rulings (i.e. mandatory removal from play), under-reporting, or failing to report an SRC has been an ongoing issue in SRC management.

In response to the growing need to manage SRC, many sporting bodies adopted one-size-fits-all stand-down rulings (e.g. junior athletes have a mandatory three-week removal from play). This guidance, however, requires efforts from multiple actors to ensure it is consistently implemented, enforced, and maintained; however, research has shown that the evidence to support any specific type or duration of return-to-play protocol is sparse (Burke et al., 2015). Thus, it is not surprising that there is resistance to this ruling from both athletes, who want to continue to play their sport, and coaches, who are trying to field a full team. A player may risk possible career advancement opportunities if they report a concussive injury and are withdrawn from play, thus may prioritise their sporting career over their own health (Clacy et al., 2013; Kroshus et al., 2014). The social and cultural factors that influence reporting behaviours should also be considered. For example, an athlete may perceive some degree of pressure from teammates, spectators, or family to remain active in the game and prevent them from reporting a suspected concussion (Kroshus et al., 2015). Concussed athletes who experience pressure from multiple sources (e.g. coaches, teammates, spectators, parents) have been found to be significantly more likely to continue playing than athletes who had not perceived this pressure (Kroshus et al., 2015). In addition, Caron et al. (2013) found that former professional ice hockey players often hid concussive symptoms from teammates and coaches, acting in ways they perceived to be normative for their masculine sports culture.

If a player does report the perceived presence of concussion, or an impact is observed, the assessment of an injured player is generally conducted by a medically trained individual. However, community sport often relies on volunteer or trainee medical staff; consequently fully trained medics are not always present at training and game events (Borich et al., 2013; Cohen et al., 2009; Makdissi et al., 2013). Thus, the responsibility for determining whether to remove an athlete from play falls equally on coaches, parents, players, and officials (Cohen et al., 2009). This presents a potential problem, as inconsistencies in SRC management responsibilities have been identified in community sport (Clacy et al., 2017). Additionally, the extent to which concussion management tools are implemented in community sports to identify a suspected concussion is largely unknown (Clacy et al., 2017; Finch et al., 2013; Hollis et al., 2012). From a systems perspective, this limitation demonstrates issues in the vertical integration of SRC management guidelines into the context of grassroots sport.

Fundamental to SRC management is the effective dissemination and consistent implementation of evidence-based practice (Finch et al., 2013; Hollis et al., 2012). Unfortunately, the paucity of concussion management knowledge in the different levels of sporting participation has been consistently demonstrated (e.g. Hollis et al., 2012; King et al., 2014; McAllister-Deitrick et al., 2014). This failure is the result of a breakdown in the promulgation of information downwards through the system. Inadequate controls at higher levels of the system permit the behaviour of the lower levels in the immediate environment of SRC management. This inability to deliver policy and practices in a coordinated way represents a need for a consideration of the sporting system as a whole.

Overall, a number of factors influencing SRC management have been identified in the literature. It is important to note that, to date, these factors have been viewed independently and in isolation of the system in which they occur. When considered using a systems approach, however, it can be seen that the current issues undermining the management of SRC both occur across multiple levels, and are influenced by a network of actors operating at multiple levels throughout the system (see Table 18.1).

Given the complexities involved in the delivery of sport across multiple grades of participation, the linear approaches that have been taken in existing literature limit the effectiveness of the current SRC guidelines as they fail to consider wider system factors that influence injury management behaviours. The key challenges currently faced in SRC management need to be looked at in context, especially in settings with limited resources (i.e. community sport). By applying complex systems methodology, however, the community sports context can be better conceptualised and

TABLE 18.1

Summary of Identified Factors Impacting SRC Management Divided According to Emergent Level Using Rasmussen's (1997) Risk Management Framework

Governing Bodies	Concussion guideline choice (Identification, recovery, RTP) is scarcely known at lower levels, and enforcement is minimal.
	Rule changes that do not consider the system-wide implications.
Regulatory Bodies and Associations, Administration, and Training	Allied medical knowledge in SRC is lacking.
	Promulgation of information from above (to mitigate paucity of knowledge below) is poor.
Club Level, Community Groups, Applied Medical	Access to trained medics at games is limited (Lack of resources to implement concussion management plan).
	Paucity of knowledge is widely demonstrated through the system.
Coaching and Direct Athlete Access	Actor role multiplicity causes conflicts resulting in mismanagement.
Player Level	Non-report behaviour (due to not wanting to be removed from play, social pressure, lack of knowledge of the risks).

appropriate consideration can be given to the actors, factors, and interactions that influence SRC management.

The aims of the following chapter are to demonstrate the utility of a systems approach to injury management in the sporting context, thus providing an alternative to the reductionists approaches that have been traditionally applied. Recent applications of leading systems analysis methods to injury management in organised sport will be applied and discussed. Lastly, considerations for future research in this context will be presented, along with opportunities for future applications of systems analysis methods.

18.3   APPLYING HUMAN FACTORS AND ERGONOMICS METHODS

To address some of the ongoing issues undermining SRC management, a growing programme of research has applied human factors methodologies with the aim of better understanding the intricacies of the community sports context that influence injury management behaviours. The application of these methodologies offers the opportunity to identify and model previously unidentified or omitted phenomenon, such as missing vertical integration of key information. The application of human factors methods to issues and phenomena occurring in the sporting context offers a long-awaited alternative to the typical cause-and-effect approaches to injury prevention and management, which are often centralised on actor training and behaviour.

Traditional approaches to injury management often fail to capture the automated and/or cognitive processes that actors undertake to accommodate contextual limitations or restrictions within the system (Naikar et al., 1999; Vicente, 1999). Of particular note is the capacity of human factors methodologies to identify intricate details about the workings of a specific system. For example, traditional research methods would have difficulty in identifying the uniqueness of the community sporting context and how it operates as a sub-system of the wider sporting context.

Furthermore, a linear perspective on injury prevention and management is often unable to accommodate for adaptive behaviours or organisation drift, which often occurs in dynamic environments or environments that are somewhat peripheral to overarching or governing bodies. The context of community sport exemplifies both of these types of environments.

18.3.1   RASMUSSEN'S RISK MANAGEMENT FRAMEWORK

The initial study in this programme of research utilised Rasmussen's Risk Management Framework (RMF; Rasmussen, 1997) to examine how sport-related concussion is currently managed (i.e. prevented, identified, and treated) in community rugby union (rugby) in Australia (Clacy et al., 2017). As with many community sports, rugby is delivered through a network of local clubs that it is administered in a hierarchical manner in international, national, state/provincial, and regional structures. By taking a systems approach to SRC management, it can be said that injuries such as SRC have multiple interrelated contributory factors that are created by the decisions and actions of people across all levels of a particular system, up to and including those working in government sectors (Rasmussen, 1997). The

RMF therefore offers a sound methodology to identify the role-specific concussion management knowledge of actors from each level of the rugby system; specifically, which actors in the community rugby system are currently involved in concussion management, and their perceived roles and responsibilities.

This study relied on a strong collaborative relationship with local community sporting teams who informed the identification of key stakeholders who are involved in the management of SRC, which were later mapped using RMF (Figure 18.1). To identify both the actual behaviours that actors undertake as well as their involvement in these behaviours, actors from across the system were then asked about their perceived roles and responsibilities in the management of SRC (Clacy et al., 2017). By taking this broader systems approach using RMF, this study was able to identify unique context-specific issues that are potentially undermining SRC management, such as role multiplicity and incongruencies in the perceptions actors have regarding their roles and responsibilities in injury management. The study was also able to identify that many of the current and proposed strategies to SRC management require action from actors throughout the entire sporting system (2017).

Adding to this, Dawson et al. (2017) also applied Rasmussen's (1997) RMF to investigate the interacting factors that influence decision making around SRC management in community sport. In line with the framework, concussion and its management were considered as emergent properties arising from the interactions between factors across the system's hierarchy including policy and guidelines, rules and regulations, culture, training, equipment, etc. Clacy et al., 2017). To explore how these interactions impact decision-making, Dawson et al. (2017) interviewed actors from across the community Australian football system using the Critical Decision Method (CDM; Klein, 2015). CDM has been used extensively to examine decision making in real-world scenarios and experiences across a range of areas, including sport (Salmon et al., 2010). CDM is underpinned by the well-researched and applied concepts of Naturalistic Decision Making (NDM; Klein et al., 1993) and the Recognition-Primed Decision (RPD) model (Klein et al., 1986).

While NDM has applications in understanding how people make decisions in environments familiar to them, RPD describes how people use their experience in the form of a repertoire of patterns (Klein et al., 1986). The patterns highlight the most relevant cues, provide expectancies, identify plausible goals, and suggest typical types of reactions in that type of situation. Therefore when people need to make a decision they can quickly match the situation to the patterns they have learned (Klein, 2015). In the context of managing SRC, application of the RDP model may explain why some coaches respond rapidly to a player receiving a head impact, while others may 'play on' – depending on the previous experiences they have encountered with concussion and their present goals (e.g. player safety versus game results). CDM further builds on these concepts by offering a structured approach to elicit specific, detailed information about the important cues, choice points, options, and action plans and the role of experience in judgement and decision-making.

By integrating Rasmussen's RMF along with CDM, Dawson et al. (2017) was able to identify not only the decisions made by individual actors, but those made by actors from across the system. This provided the opportunity for researchers to explore how key stakeholders with extensive experience in community sport to respond to

FIGURE 18.1 Application of Rasmussen's RMF, illustrating the hierarchical organisation of the community rugby union system (adapted from Clacy et al., 2017).

Governing bodies	International Rugby Union Board (IRB)	Australian Sports Commission	Australian Rugby Union (ARU)	Queensland Sport and Recreation	National Children's Injury Prevention Council
Regulatory and influential bodies	Regional Rugby Union Bodies	Queensland Rugby Union	Public and Private Schools	Sports Medicine Professionals	Sports and Public Media	Sports Celebrities (i.e., role models)
Local clubs and school teams	Local Team Management	Public and Private School Team Managers	Regional Representatives Team Management	Club Members (e.g. Volunteers)	Professional Club Members	Selectors
Direct supervisors	Coaches	Skills Coach/Trainer	Player Guardians	Match Officials (e.g., time keepers, referees)	Medical Aides	Physical Education Teachers
Immediate game/training participants	Players	Opposition	Team Captain	Spectators/Crowd		
Equipment and environment	Groundsmen	Field conditions	Headgear	Mouthguards	Bodyguards/Padding	Post Guards

critical situations (i.e. head injury). Using this rich information, the findings from Dawson et al. (2017) illustrated the complexities of SRC management in concussion sports, including the importance of considering context-specific factors such as the consistency of actor training and education, role multiplicity, resource limitations, and community culture.

The application of Rasmussen's RMF in these studies highlighted the utility of systems theory over previous reductionist methodologies. By not only identifying the factors that influence SRC management, these studies also critically explored how a variety of known and unknown factors that influence SRC management can differ between stakeholders and levels of the system in the unique context of community sport. It is important to note, that this methodology would be appropriate to apply in any community sport that is governed hierarchically to investigate the factors contributing to injuries and their management that would not be otherwise discernible using traditional linear techniques.

A notable limitation of Dawson et al. (2017) was their focus on the experiences of key stakeholders in the community sporting context. Although these findings offered key insights into the issues surrounding concussion management in the community context, minimal consideration was given to the controlling influences from governing bodies that direct SRC management. Systems-Theoretic Accident Model and Processes (STAMP; Leveson, 2004) and Cognitive Work Analysis (CWA; Vicente, 1999) are two salient human factors research methods that offer an opportunity to investigate these influences, and how they interact to shape SRC management behaviours in community-level sport.

18.3.2 Systems-Theoretic Accident Model and Processes (STAMP)

Building on Rasmussen's RMF, Leveson (2004) argues that safety is an emergent property resulting from interactions among humans, physical system components, and the environment that lead to the violation of controls. The behaviour and actions of people within the system are shaped by controls in a hierarchical control structure, where each level of the structure enforces the required constraints on the behaviour of the components at the next lower level. These hierarchies of operation, like those in which organised sports are administrated, are characterised by control processes operating at the interfaces between different levels of influence, ranging from international governance to the immediate environment (Holmes, Clacy, & Salmon, 2019). This integration also operates in reverse through feedback mechanisms which translate information from the lower levels to the higher levels of the system. This upward feedback should ultimately be used in the development and revision of managerial, regulatory, and government decisions (Leveson, 2004). In systems and control theory, every control developed by the upper levels must contain a model of the process it is controlling (Young & Leveson, 2014). This model is used to determine what control actions are necessary, with consideration for the nuances of the level at which the control is being imposed. According to the STAMP model, inappropriate or ineffective controls are not the result of human failure, rather they stem from inconsistencies or inaccuracies in the controlling level's models of the levels below (Figure 18.2).

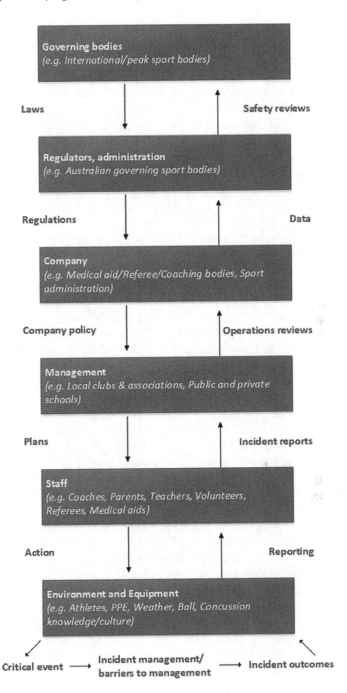

FIGURE 18.2 Example STAMP model of the community sport system, including controls filtering downward on the left-hand side of the diagram, and system feedback filtering upward on the right side of the diagram. (Adapted with permission from Holmes et al., 2019.)

As mentioned previously, ongoing concerns have been raised in both literature and practice regarding the consistency of stakeholder knowledge of the SRC management guidelines throughout all levels of competition (i.e. elite through to community grade sport; Finch et al., 2013; Hollis et al., 2012). The application of STAMP provides the opportunity to better investigate the factors that may be impeding accurate and relevant knowledge transfer throughout the wider system, in terms of the dynamic interactions of actors and the nature of the relationships that exist between them (Holmes, Clacy, & Salmon, 2019). By applying STAMP, not only can the key stakeholders that present both direct and indirect influences on SRC management be identified, but the levels of influence at which they operate can also be modelled. Additionally, STAMP can be used to identify the controls and feedback mechanisms that determine the management practices, as well as any discrete relationships that may be shaping stakeholder knowledge, attitudes, and behaviours within the context of a given system.

Using this method, insufficient, inaccurate, or contradictory components of the system and the processes within it can be identified, such as superfluous actors, unclear role communication, or broken feedback. With a more comprehensive understanding of the factors impacting SRC management behaviours throughout the system in its entirety, controls can be implemented to increase the consistency and appropriateness of safety management practices at all levels of competition. Further, knowledge of the systemic impact that any proposed alterations to management practices may have can be identified through the application of STAMP. With this level of detail, context-specific contingencies can also be developed to maximise the effectiveness of SRC management guidance and practice and minimise nuanced oversights.

Previous research in sports injury prevention has demonstrated the utility of applying STAMP to identify the system-wide actors and control structures that influence injury management (e.g. Holmes et al., 2017; Hulme et al., 2017a, b). Both Hulme et al. (2017a) and Holmes et al. (2017) drew on subject matter input to shape and refine STAMP models of the community running and rugby union context, respectively. Through this process the final models produced were an accurate representation of each context and were able to reflect the nuances of the system with a level of structured detail that would be otherwise unachievable. Hulme et al. (2017a) were able to identify the complex interactions and systemic factors that were influencing running-related injury; looking further beyond the immediate injury environment. Holmes et al. (2017) were able to represent many of the contextual influences that impede current SRC management strategies, including actor role multiplicity, broken communication loops, and some of the cultural pressures that shape player reporting behaviours. Both of these studies were able to approach injury prevention, and the factors that contribute to its success or failure, with specific consideration for the sporting context. The application of STAMP offers many opportunities for researchers and practitioners to understand and manage sports injury in a contextually relevant and meaningful way.

18.3.3 Cognitive Work Analysis (CWA)

Cognitive Work Analysis (CWA; Vicente, 1999) is another human factors method that has been successfully applied in the sports context (e.g. McLean et al., 2019;

McLean et al., 2017). The overarching aim of CWA is to support actors in responding to unanticipated or changing work environments, through identifying the physical, procedural, social, and competency-based boundaries that shape their normal behaviour and decision-making (Vicente, 1999). CWA consists of five domains of analysis: Work Domain Analysis (WDA), Control Task Analysis (ConTA), Strategies Analysis, Social Organisation and Cooperation Analysis (SOCA), and Worker Competencies Analysis (WCA). In the context of SRC management in community sport, the application of each of these systems methodologies may offer much-needed insight into the factors that may be impeding the effectiveness of concussion management strategies in the community sports context.

The application of WDA encourages a deeper level of understanding of the SRC process through the identification of the identifies the high-level purposes, values and priorities, functions, and physical resources of a system. Indeed, the application of WDA has the potential to identify redundant, underutilised, or stressed elements of the community sports system that may be impeding effective SRC management. For example, inconsistent education in community sport has led to an over-reliance on medical staff to prevent, identify, and treat SRC (Clacy et al., 2017). As shown in Figure 18.3, application of CWA would offer the opportunity to gain a systemic understanding of why this element of the system may be being stressed, the decisions that are made by actors that proliferate and/or mitigate it, and how other elements within the system may be better aligned to relieve this pressure and improve SRC management practices.

Control Task Analysis is another domain of CWA that has applicable utility in improving SRC management for the community sports context. Control tasks can be identified in the context of injury management situations where the boundaries between the activities of actors might be different under various situations, such as between training, competition, and recreational sports participation. In the context of community SRC management, application of ConTA would allow researchers to understand the collaborative decision making of the actors associated with managing SRC from identification through to rehabilitation and return to play. Through this analysis method, all of the combinations of management actions (control tasks) can be visually represented to identify the patterns of activity and workload, and provide estimates of spare capacity for additional or shared responsibility (Naikar et al., 2003). By better understanding the current actions and decisions and how they are shared across actors and environments, opportunities for systems-based process revisions would be identified. This would be especially valuable in the community context, wherein the impact of role multiplicity and unclear injury management responsibility is known to reduce the effectiveness of current SRC management efforts (Clacy et al., 2016, 2017). In addition, analysis methods such as Strategies Analysis and SOCA can be used to further identify how different actors from within the community sports system undertake different tasks pertaining to SRC management and how injury prevention tasks are coordinated and distributed. With this level of detail, context-specific management practices could be developed that consider the restraints of the system and propose an alternative process that draws on the unique expertise of the actors within the system.

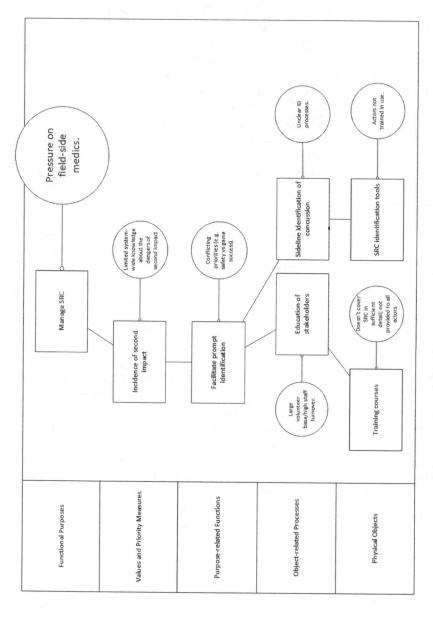

FIGURE 18.3   This model presents an example of part of an optimal SRC management structure.

A key tenet of each component of the CWA framework is the focus on drawing on the expertise and experiences of end-users to inform the analysis of the system. As mentioned previously, most community sports organisations administered by volunteer staff and coaches, who are ultimately the intended end-users or implementers of safety policies and programmes (Donaldson & Finch, 2012; Donaldson et al., 2012; Skille, 2008). Research has shown that policies, such as injury management protocols, need to be developed in collaboration with end-users. According to diffusion of innovations theory, the rate of adoption of an innovation depends more on the end-users' subjective perception of the relative advantage, compatibility, complexity, trialability, and observability of an innovation than it does on the objective evidence of the innovation's efficacy (Rogers, 2004). Therefore, SRC management strategies that have been informed by the staff and coaches who operate within the community sporting system are more likely to improve club engagement and implementation of the management protocols and maintain them consistently. The application of CWA, therefore, offers a systematic model to guide consultations with end-users and inform the revision of SRC management for optimal engagement of the community sports context.

18.4 PRACTICAL APPLICATIONS IN SPORT

Although governed by the same overarching bodies, injury management practices in community sport often have a dramatically different presentation when compared with those of the elite levels. The application of human factors methods, such as those discussed above, have wide-ranging utility in not only better understanding the current injury management environment in community sport, but also present opportunities to identify and develop systemic level changes to improve management practices in this context. At a foundation level, Rasmussen's (1997) RMF offers a unique opportunity to contextually consider the actors and risk factors within a system, and how they interact to influence injury incidence and management. In any sporting code in community sport, this level of systemic detail provides a practical structure to guide the development of context-specific processes and procedures. Building on this framework, the STAMP methodology allows for the exploration of the controls that shape culture and behaviours around injury prevention. This is especially important in the context of organised sport, as the majority of processes and procedures are developed by peak sporting bodies and modelled off the elite sports context, with minimal consideration for the contextual limitations of community sport. To illustrate, some of the current SRC guidelines suggest that all players should receive baseline cognitive testing at the beginning of the season as a way of more clearly identifying concussion based on deviations from 'normal' functioning. In the community context, however, this is rarely feasible due to limitations in resources, expertise, training, and role clarity. By modelling this process using STAMP, it may be seen that that feedback on the limited capacity for community sport to undertake baseline testing is not being communicated upward in the system to help revise these guidelines and/or develop community-level resources to support its implementation. This vertical integration is essential for the effective control of any environment, and STAMP presents the modality to identify and manage this.

While CWA has only recently been applied to a sporting context, it offers possibly the most potential in terms of understanding and remodelling SRC management in terms of system goals, tasks, actor decision-making, and coordinated action. The integration of end-user input into unpacking and understanding the goals and decisions surrounding SRC management in community sport is essential for not only assessing the current state of practice, but also for the redevelopment of context-specific injury management strategies. CWA offers a structured approach to dissecting organisational safety culture, and provides methods for redesign based on the outputs. Indeed Read et al. (2018) developed a toolkit to support the translation of CWA outputs into design concepts using both obvious and non-obvious findings from each of the CWA model components. Although yet to be applied in a community sports context, the CWA design toolkit (CWA-DT; Read et al., 2018) presents a practical methodology to support system-level process design.

18.5   FUTURE RESEARCH DIRECTIONS

Given the current limitations of SRC management in community sport, future research should endeavour to incorporate a systems approach to both understanding injury management, as well as management optimisation. By applying human factors methods such as those discussed above, future injury prevention strategies would more accurately reflect the needs of the entire system. Specifically, more research is needed to better understand the decision-making capacity of actors in the community sports context, in terms of SRC management. This research should also aim to explore the known flexibilities within this context that may be impacting effective concussion management, including identification of cultural influences and organisational drift. Future research would benefit from applying methods such as CWA and its associated design toolkit to achieve these aims while also offering practical solutions for the optimisation of SRC management for the community sports context.

18.6   CONCLUSIONS

The collective findings of the systems methods described here have begun to offer a more in-depth understanding of the contextual factors that influence SRC management in community grade sport. The application of different human factors methodologies to the sporting context expands what is currently known in injury management by providing more holistic methods to guide injury prevention research. It acknowledges that the decisions and actions of different actors, as well as the controls and feedback mechanisms between each level of the sporting system, interact and shape concussion management behaviours in the community context. This insight highlights some of the factors that may be contributing to and perpetuating the known issues in SRC management (e.g. role confusion, under-reporting). SRC management is complex and relies on the actions and behaviours of stakeholders from across the sporting system, including the community sporting system that sits within it. As such, future research should aim to utilise human factors methodologies to ensure that SRC management remains responsive to the dynamic needs of the sporting system.

REFERENCES

Bittencourt, N. F. N., Meeuwisse, W. H., Mendonca, L. D., Nettel-Aguirre, A., Ocarino, J. M., & Fonseca, S. T. (2016). Complex systems approach for sports injuries: moving from risk factor identification to injury pattern recognition-narrative review and new concept. *Br J Sports Med*, 50:21, 1309–1314. doi:10.1136/bjsports-2015-095850.

Borich, M. R., Cheung, K. L., Jones, P., Khramova, V., Gavrailoff, L., Boyd, L. A., & Virji-Babul, N. (2013). Concussion: current concepts in diagnosis and management. *J Neurol Phys Ther*, 37:3, 133–139. doi:10.1097/NPT.0b013e31829f7460.

Burke, M. J., Fralick, M., Nejatbakhsh, N., Tartaglia, M. C., & Tator, C. H. (2015). In search of evidence-based treatment for concussion: characteristics of current clinical trials. *Brain Inj*, 29:3, 300–305. doi:10.3109/02699052.2014.974673.

Caron, J. G., Bloom, G. A., Johnston, K. M., & Sabiston, C. M. (2013). Effects of multiple concussions on retired national hockey league players. *J Sport Exerc Psychol*, 35:2, 168–179.

Chrisman, S. P., Quitiquit, C., & Rivara, F. P. (2013). Qualitative study of barriers to concussive symptom reporting in high school athletics. *J Adolesc Health*, 52:3, 330–335.e333. doi:10.1016/j.jadohealth.2012.10.271.

Clacy, A., Goode, N., Sharman, R., Lovell, G. P., & Salmon, P. M. (2017). A knock to the system: A new sociotechnical systems approach to sport-related concussion. *J Sports Sci*, 35:22, 2232–2239. doi:10.1080/02640414.2016.1265140.

Clacy, A., Sharman, R., & Lovell, G. (2013). Return-to-play confusion: considerations for sport-related concussion. *J Bioeth Inq*, 10:1, 127–128. doi:10.1007/s11673-012-9421-8.

Cohen, J. S., Gioia, G., Atabaki, S., & Teach, S. J. (2009). Sports-related concussions in pediatrics. *Curr Opin Pediatr*, 21:3, 288–293. doi:10.1097/MOP.0b013e32832b1195.

Dallat, C., Salmon, P. M., & Goode, N. (2015). All about the teacher, the rain and the backpack: The lack of a systems approach to risk assessment in school outdoor education programs. *Procedia Manuf*, 3, 1157–1164.

Dawson, K., Salmon, P. M., Read, G. J., Neville, T., Goode, N., & Clacy, A. (2017). Removing concussed players from the field: The factors influencing decision making around concussion identification and management in Australian Rules Football. Paper presented at the Proceedings of the 13th International Conference on Naturalistic Decision Making.

Dekker, S. (2016). *Drift into Failure: From Hunting Broken Components to Understanding Complex Systems*. Ashgate, Surrey: CRC Press.

Donaldson, A., & Finch, C. F. (2012). Planning for implementation and translation: seek first to understand the end-users' perspectives. *Br J Sports Med*, 46:5, 306–307. doi:10.1136/bjsports-2011-090461.

Donaldson, A., Leggett, S., & Finch, C. F. (2012). Sports policy development and implementation in context: researching and understanding the perceptions of community end-users. *Int Rev Sociol Sport*, 47:6, 743–760. doi:10.1177/1012690211422009.

Donaldson, A., Newton, J., McCrory, P., White, P., Davis, G., Makdissi, M., & Finch, C. F. (2016). Translating guidelines for the diagnosis and management of sports-related concussion into practice. *Am J Lifestyle Med*, 10:2, 120–135. doi:10.1177/1559827614538751.

Donaldson, L., Asbridge, M., & Cusimano, M. D. (2013). Bodychecking rules and concussion in elite hockey. *PLoS One*, 8:7, e69122. doi:10.1371/journal.pone.0069122.

Finch, C. F., McCrory, P., Ewing, M. T., & Sullivan, S. J. (2013). Concussion guidelines need to move from only expert content to also include implementation and dissemination strategies. *Br J Sports Med*, 47:1, 12–14. doi:10.1136/bjsports-2012-091796.

Finch, C. F., McIntosh, A. S., & McCrory, P. (2001). What do under 15 year old schoolboy rugby union players think about protective headgear? *Br J Sports Med*, 35:2, 89–94.

Fraas, M. R., & Burchiel, J. (2016). A systematic review of education programmes to prevent concussion in rugby union. *Eur J Sport Sci*, 16:8, 1212–1218. doi:10.1080/17461391.2016.1170207.

Gardner, A. J., Iverson, G. L., Williams, W. H., Baker, S., & Stanwell, P. (2014). A systematic review and meta-analysis of concussion in rugby union. *Sports Med*, 44:12, 1717–1731. doi:10.1007/s40279-014-0233-3.

Giza, C. C., Kutcher, J. S., Ashwal, S., Barth, J., Getchius, T. S., Gioia, G. A., Gronseth, G.S., Guskiewicz, K., Mandel, S., Manley, G., & McKeag, D.B., Zafonte, R. (2013). Summary of evidence-based guideline update: evaluation and management of concussion in sports: report of the Guideline Development Subcommittee of the American Academy of Neurology. *Neurology*, 80:24, 2250–2257. doi:10.1212/WNL.0b013e31828d57dd.

Guskiewicz, K. M., Register-Mihalik, J., McCrory, P., McCrea, M., Johnston, K., Makdissi, M., Dvořák J., Davis G., & Meeuwisse, W. (2013). Evidence-based approach to revising the SCAT2: introducing the SCAT3. *Br J Sports Med*, 47:5, 289–293. doi:10.1136/bjsports-2013-092225.

Harmon, K. G., Drezner, J. A., Gammons, M., Guskiewicz, K. M., Halstead, M., Herring, S. A., Kutcher J., Pana A., Putukian M., & Roberts, W. O. (2013). American Medical Society for Sports Medicine position statement: concussion in sport. *Br J Sports Med*, 47:1, 15–26. doi:10.1136/bjsports-2012-091941.

Hollis, S. J., Stevenson, M. R., McIntosh, A. S., Shores, E. A., Collins, M. W., & Taylor, C. B. (2009). Incidence, risk, and protective factors of mild traumatic brain injury in a cohort of Australian nonprofessional male rugby players. *Am J Sports Med*, 37:12, 2328–2333. doi:10.1177/0363546509341032.

Hollis, S. J., Stevenson, M. R., McIntosh, A. S., Shores, E. A., & Finch, C. F. (2012). Compliance with return-to-play regulations following concussion in Australian schoolboy and community rugby union players. *Br J Sports Med*, 46:10, 735–740. doi:10.1136/bjsm.2011.085332.

Holmes, G., Clacy, A., & Salmon, P. M. (2017). *A Systems Theory Approach to the Management of Sports-Related Concussion* (Unpublished Honours thesis). University of the Sunshine Coast, Australia.

Holmes, G., Clacy, A., & Salmon, P. M. (2019). Sports-related concussion management as a control problem: using STAMP to examine concussion management in community rugby. *Ergonomics*, 62(11), 1485–1494. doi:10.1080/00140139.2019.1654134.

Hulme, A., Salmon, P. M., Nielsen, R. O., Read, G. J. M., & Finch, C. F. (2017a). Closing Pandora's Box: adapting a systems ergonomics methodology for better understanding the ecological complexity underpinning the development and prevention of running-related injury. *Theor Issues Ergon Sci*, 18:4, 338–359.

Hulme, A., Salmon, P. M., Nielsen, R. O., Read, G. J. M., & Finch, C. F. (2017b). From control to causation: Validating a 'complex systems model' of running-related injury development and prevention. *Appl Ergon*, 65, 345–354. doi:10.1016/j.apergo.2017.07.005.

Kemp, S. P., Hudson, Z., Brooks, J. H., & Fuller, C. W. (2008). The epidemiology of head injuries in English professional rugby union. *Clin J Sport Med*, 18:3, 227–234. doi:10.1097/JSM.0b013e31816a1c9a.

King, D., Gissane, C., Brughelli, M., Hume, P. A., & Harawira, J. (2014). Sport-related concussions in New Zealand: a review of 10 years of Accident Compensation Corporation moderate to severe claims and costs. *J Sci Med Sport*, 17:3, 250–255. doi:10.1016/j.jsams.2013.05.0071

Klein, G. (2015). Reflections on applications of naturalistic decision making. *J Occup Organ Psychol*, 88:2, 382–386.

Klein, G. A., Calderwood, R., & Clinton-Cirocco, A. (1986). Rapid decision making on the fire ground. Paper presented at the Proceedings of the Human Factors Society Annual Meeting.

Klein, G. A., Orasanu, J. E., Calderwood, R. E., & Zsambok, C. E. (1993). Decision making in action: Models and methods. Paper presented at the This book is an outcome of a workshop held in Dayton, OH, September, 25–27, 1989.

Kroshus, E., Baugh, C. M., Daneshvar, D. H., & Viswanath, K. (2014). Understanding concussion reporting using a model based on the theory of planned behavior. *J Adolesc Health*, 54:3, 269–274. e262.

Kroshus, E., Garnett, B., Hawrilenko, M., Baugh, C. M., & Calzo, J. P. (2015). Concussion under-reporting and pressure from coaches, teammates, fans, and parents. *Soc Sci Med*, 134, 66–75. doi:10.1016/j.socscimed.2015.04.011.

Leveson, N. (2004). A new accident model for engineering safer systems. *Saf Sci*, 42:4, 237–270. doi:10.1016/S0925-7535(03)00047-X.

Littleton, A., & Guskiewicz, K. (2013). Current concepts in sport concussion management: a multifaceted approach. *J Sport Health Sci*, 2:4, 227–235.

Makdissi, M., Davis, G., Jordan, B., Patricios, J., Purcell, L., & Putukian, M. (2013). Revisiting the modifiers: how should the evaluation and management of acute concussions differ in specific groups? *Br J Sports Med*, 47:5, 314–320. doi:10.1136/bjsports-2013-092256.

Marshall, S. W., Loomis, D. P., Waller, A. E., Chalmers, D. J., Bird, Y. N., Quarrie, K. L., & Feehan, M. (2005). Evaluation of protective equipment for prevention of injuries in rugby union. *Int J Epidemiol*, 34:1, 113–118. doi:10.1093/ije/dyh346.

McAllister-Deitrick, J., Covassin, T., & Gould, D. R. (2014). Sport-related concussion knowledge among youth football players. *Athl Train Sports Health Care*, 6:6, 280–284.

McCrea, M., Hammeke, T., Olsen, G., Leo, P., & Guskiewicz, K. (2004). Unreported concussion in high school football players: implications for prevention. *Clin J Sport Med*, 14:1, 13–17.

McCrory, P., Meeuwisse, W., Dvorak, J., Aubry, M., Bailes, J., Broglio, S., Cantu, R. C., Cassidy, D., Echemendia, R. J., Castellani, R. J., & Davis, G. A. (2017). Consensus statement on concussion in sport-the 5(th) international conference on concussion in sport held in Berlin, October 2016. *Br J Sports Med*, 51:11, 838–847. doi:10.1136/bjsports-2017-097699.

McIntosh, A. S., McCrory, P., Finch, C. F., Best, J. P., Chalmers, D. J., & Wolfe, R. (2009). Does padded headgear prevent head injury in rugby union football? *Med Sci Sports Exerc*, 41:2, 306–313. doi:10.1249/MSS.0b013e3181864bee.

McLean, S., Hulme, A., Mooney, M., Read, G. J. M., Bedford, A., & Salmon, P. M. (2019). A systems approach to performance analysis in women's netball: Using work domain analysis to model elite netball performance. *Front Psychol*, 10, 201. doi:10.3389/fpsyg.2019.00201.

McLean, S., Salmon, P. M., Gorman, A. D., Read, G. J., & Solomon, C. (2017). What's in a game? A systems approach to enhancing performance analysis in football. *PLoS One*, 12:2, e0172565. doi:10.1371/journal.pone.0172565.

Naikar, N., Pearce, B., Drumm, D., & Sanderson, P. M. (2003). Designing teams for first-of-a-kind, complex systems using the initial phases of cognitive work analysis: case study. *Hum Factors*, 45:2, 202–217. doi:10.1518/hfes.45.2.202.27236.

Naikar, N., Sanderson, P. M., & Lintern, G. (1999). Work domain analysis for identification of training needs and training-system design. Paper presented at the Proceedings of the Human Factors and Ergonomics Society Annual Meeting.

Patton, D. A., McIntosh, A. S., & Denny, G. (2016). A review of the anthropometric characteristics, grading and dispensation of junior and youth rugby union players in Australia. *Sports Med*, 46:8, 1067–1081. doi:10.1007/s40279-016-0481-5.

Rasmussen, J. (1997). Risk management in a dynamic society: a modelling problem. *Saf Sci*, 27:2–3, 183–213.

Read, G. J., Salmon, P. M., Goode, N., Lenné, M. G. (2018). A sociotechnical design toolkit for bridging the gap between systems-based analyses and system design. *Hum Factors Ergon Manuf* 28:6, 327–341.

Roberts, S. P., Trewartha, G., England, M., Goodison, W., & Stokes, K. A. (2017). Concussions and head injuries in English community rugby union match play. *Am J Sports Med*, 45:2, 480–487. doi:10.1177/0363546516668296.

Rogers, E. M. (2004). A prospective and retrospective look at the diffusion model. *J Health Commun*, 9:Suppl 1, 13–19. doi:10.1080/10810730490271449.

Salmon, P., Williamson, A., Lenne, M., Mitsopoulos-Rubens, E., & Rudin-Brown, C. M. (2010). Systems-based accident analysis in the led outdoor activity domain: application and evaluation of a risk management framework. *Ergonomics*, 53:8, 927–939. doi:10.1080/00140139.2010.489966.

Salmon, P. M., Goode, N., Lenne, M. G., Finch, C. F., & Cassell, E. (2014). Injury causation in the great outdoors: A systems analysis of led outdoor activity injury incidents. *Accid Anal Prev*, 63, 111–120. doi:10.1016/j.aap.2013.10.019.

Salmon, P. M., Read, G. J. M., & Stevens, N. J. (2016). Who is in control of road safety? A STAMP control structure analysis of the road transport system in Queensland, Australia. *Accid Anal Prev*, 96, 140–151. doi:10.1016/j.aap.2016.05.025.

Skille, E. Å. (2008). Understanding sport clubs as sport policy implementers: a theoretical framework for the analysis of the implementation of central sport policy through local and voluntary sport organizations. *Int Rev Sociol Sport*, 43:2, 181–200.

Sye, G., Sullivan, S. J., & McCrory, P. (2006). High school rugby players' understanding of concussion and return to play guidelines. *Br J Sports Med*, 40:12, 1003–1005. doi:10.1136/bjsm.2005.020511.

Torres, D. M., Galetta, K. M., Phillips, H. W., Dziemianowicz, E. M., Wilson, J. A., Dorman, E. S., Laudano, E., Galetta, S.L., & Balcer, L. J. (2013). Sports-related concussion: Anonymous survey of a collegiate cohort. *Neurol Clin Pract*, 3:4, 279–287. doi:10.1212/CPJ.0b013e3182a1ba22.

Vicente, K. J. (1999). *Cognitive Work Analysis: Toward Safe, Productive, and Healthy Computer-Based Work*. Mahwah, NJ: CRC Press.

Young, W., & Leveson, N. G. (2014). An integrated approach to safety and security based on systems theory. *Commun ACM*, 57:2, 31–35.

19 Using Computational Modelling for Sports Injury Prevention

Agent-Based Modelling and System Dynamics Modelling

Adam Hulme, Jason Thompson, Rasmus Nielsen, Gemma J. M. Read, Scott McLean, Ben R. Lane, and Paul M. Salmon

CONTENTS

19.1 INTRODUCTION TO COMPUTATIONAL MODELLING

19.1.1 What Is Computational Modelling?

Computational modelling involves the use of computers to simulate the behaviour of complex systems using mathematics, physics, and computer science (Luke & Stamatakis, 2012; NIBIB, 2016; Cassidy et al., 2019). A computational model is typically developed by a group of people (e.g. researchers, practitioners, policy-makers) who want to identify and better understand the mechanisms that give rise to a certain problem or outcome of interest (GOS, 2018). Alternatively, computational modelling might be used to forecast the probability that a future event or series of events will occur, as is often the case in the science of climate change. For example, it is not possible to know exactly what the Earth's climate will be like in another century from now. With the aid of computational modelling, however, scientists can make predictions about what future global temperatures might be like based on historical data and behaviour, as well as observable interactions that take place among numerous different atmospheric factors. Methodologically, computational models are programmed with different types of inputs or variables that directly correspond to the system under analysis. By manipulating the value of these variables, computer simulations can be run hundreds or thousands of times if required to theoretically evaluate the short or long-term effects that are associated with a range of potential outcomes (Marshall, 2017; Badham et al., 2018; Davis et al., 2019). Computational modelling is a virtual laboratory that affords users the ability to modify different options and test for a multitude of hypothetical scenarios.

19.1.2 Why Use Computational Modelling?

The world has become increasingly complex and interconnected. New scientific discoveries including advances in healthcare, the evolution of robotics and generalised forms of automation, and the proliferation of tightly coupled information systems have changed the way humans go about their everyday lives (Holman et al., 2019). Notwithstanding the many positives that come with such progress, it is more challenging than ever to understand and maintain the functioning of the systems that surround us. This includes the ability to anticipate both desirable and adverse events long before they happen. Computational modelling could, for example, be used to evaluate whether the introduction of future sports technologies will lead to unintended outcomes across athletic populations of interest. Will automated systems that monitor sports workloads and provide feedback to athletes, such as smart devices and wearable sensors linked to artificial intelligence-based software, actually help to improve performances and reduce injury risk (Claudino et al., 2019), or will there be any unforeseen issues and considerations that bring into question their intended value and utility? Asking these types of questions and aiming to answer them with computational modelling is a viable way forwards due to the growing number of interactions occurring between human and non-human factors within complex sports systems (see Chapter 2).

For these reasons, among others, computational modelling can be useful to test the likely effect associated with the implementation of new technologies, products, services, and interventions when direct observation or experimentation are not feasible or ethical options.

BOX 19.1: COOKING UP A COMPUTATIONAL MODEL

One of the better examples available that captures the essence of why computational modelling can be most useful is offered by The National Institute of Biomedical Imaging and Bioengineering (2016). Imagine there is a maximum of 20 different ingredients that could go into making the perfect cake, and we want to know how each of those ingredients contribute to the final product. One option is to physically bake multiple different cakes and systematically leave out a single ingredient each time (NIBIB, 2016). Unfortunately, this process is time consuming and resource intensive – not to mention the mess that would be created in the kitchen! A better option is to instantiate a computational model with suitable data, such as what the ingredients are and how they interact based on empirical (known) observation. Consequently, a computer simulation would be able to model how each cake could turn out under a predefined set of conditions. Once a relatively simple working model has been established and verified, the next step might be to vary the amount of each ingredient until the desired result is achieved. What complex set of ingredients are required to produce a cake that is rich, soft, sticky, and/or firm? (NIBIB, 2016). As increasingly more rules and inputs are specified, possibly into the thousands depending on the goal of the modelling exercise, the more it becomes necessary to use computational modelling to not only make sense of complex factor interactions, but also circumvent ethical, financial, and logistical constraints that would otherwise need to be accounted for. By way of comparison, computational modelling is used in the contexts of public policy and healthcare, business and economics, engineering and science, industry and manufacturing, and urban planning and design (Calder et al., 2018). The rationale and basis for using computational modelling in these diverse areas can be highly unique to a specific problem of interest.

19.1.3 Purposes of Computational Modelling

According to a recent United Kingdom's Government Office for Science (2018) report, 'Computational Modelling: Technological Futures', there are five general purposes for using computational modelling: (i) prediction; (ii) explanation; (iii) theoretical understanding; (iv) illustration; and (v) analogy (Calder et al., 2018). Although it is beyond the scope of this chapter to provide a detailed account of each purpose from a methodological standpoint (i.e. model parameterisation, verification, validation), a general summary with examples is provided in Table 19.1.

TABLE 19.1

Purposes of Computational Modelling with Examples

Modelling purpose	Description	Example
Predictive modelling	Predictive modelling anticipates unknown data, including the forecasting of behaviours and outcomes. Access to a large quantity of extant data can increase the degree of accuracy when wanting to predict future trends and patterns. The outcome(s) are to be clearly defined and well justified, and a given model should first be tested before it is applied in a predictive capacity.	Forecasting climate change dynamics or predicting fluctuating prices in economic systems are examples to this end. In a sports setting, computational modelling might aim to make predictions about future outcomes (e.g. goals conceded, championship wins, points scored) based on past performances, or attempt to predict the likelihood of sports injury occurrence based on the interplay between many different biologic and behavioural factors. Predictions can be made through simple or advanced forms of computational analyses (Hernán, Hsu, & Healy, 2019). Quantifying the association between functional lower limb biomechanics and sports injury risk can be achieved with the use of more traditional statistical approaches such as 'simple' regression modelling and correlation analyses. Alternatively, sophisticated pattern recognition methods, such as supervised machine learning algorithms that use hundreds or thousands of variables, are representative of a more advanced computational modelling approach (Hernán, Hsu, & Healy, 2019).
Explanatory modelling	Explanatory modelling conceptualises the underlying mechanisms and complex processes that give rise to observed phenomena. Here, it is useful to explore and understand why things happen, even if the full extent of a given problem cannot be predicted.	An example of this type of modelling might involve exploring the political and sociocultural factors and processes that lead to population segregation or racial disparity. On a much smaller scale, explanatory modelling has elucidated the mechanisms underpinning the self-organisation of ant colonies in relation to nest selection and construction (Pratt et al., 2005). In the context of teamwork performance in sports, researchers have used a computational model to simulate how individual-level dynamics in a cycling peloton give rise to overall group-level behaviours (Martins Ratamero, 2015). The study by Ratamero (2015) involved assigning a limited number of simple rules to each agent in the model, such as the cohesive and separating forces between each cyclist, as a way to study and observe the mechanisms by which pelotons self-organise to reduce energy expenditure.

(Continued)

TABLE 19.1 (CONTINUED)
Purposes of Computational Modelling with Examples

Modelling purpose	Description	Example
Theoretical modelling	Theoretical modelling asks the 'what if' questions and can be used to evaluate hypotheticals, some of which might require a best- and worst-case scenario. This type of modelling approach offers a great deal of scope and freedom to test whatever it is the user desires.	A civil engineer might want to know whether the structural integrity of a building is enhanced if a different type of material is used during a simulated earthquake. Alternatively, sports scientists with an interest in public health might evaluate the potential effects of changing a policy within a community sports setting. That is, do more or less club-level athletes make better training and health choices? This chapter will later describe in more detail the application of a theoretical agent-based simulation in the context of recreational distance running injury.
Illustrative modelling	Illustrative modelling involves communicating the dynamics of a (usually known) problem so that people can see how factors, structures, and systems behave and interact. This type of modelling is a useful way to supplement explanations of *how* something works and can make clear the consequences of a given action.	How does an aeroplane stay in the sky? Which areas of the human brain light up when an opioid is administered? How do musculoskeletal structures absorb and dissipate load when exposed to a sports workload stimulus?
Analogy modelling	Analogy modelling allows users to immerse themselves in the simulation exercise. Analogy models can facilitate learning, allowing people to make different choices and evaluate the effects of their decisions and actions. Outcomes are predefined, and so it is simply a matter of playing around with the model to see what happens under a given set of circumstances.	Simulation-based video games are the archetypal example of analogy modelling as they require the user to manipulate different aspects of the model to experience what it would be like to make similar decisions and actions in real, complex settings. An example would be running and managing a sports organisation.

Adapted from United Kingdom's Government Office for Science (2018) and the article by Calder et al. (2018) with the use of sports examples.

19.2   COMPUTATIONAL MODELLING IN SPORTS INJURY PREVENTION RESEARCH

The main characteristics of complex systems as described from the perspective of sports injury prevention research are described in Chapter 2 of this book. These characteristics and the corresponding sports injury examples provided justify the use of two computational approaches, namely Agent-Based Modelling (ABM) and System Dynamics (SD) modelling.

19.2.1   AGENT-BASED MODELLING (ABM) AND SYSTEM DYNAMICS (SD) MODELLING

Simple forms of static modelling (e.g. spreadsheet programming), differential equations, automata and process algebraic models, traditional regression analyses, decision tree learning techniques, Bayesian networks, machine learning, artificial neural networks and deep learning, social network analysis, and Monte Carlo methods are, unlike ABM and SD modelling, not able to explicitly simulate causal feedback among fundamentally different factors (Hulme et al., 2018a). Rather, predictive and statistical modelling, as well as sophisticated mathematical algorithms that forecast the probability of future events, are useful for understanding discrete aspects of complex systems at fixed time points and across one or more levels. These latter approaches are still computational in nature; however, they serve a purpose for a specific type of problem. Stated simply, ABM and SD modelling allows people to directly compare their mental models with the behaviours and patterns articulated through a computer-generated animation. This is why ABM and SD modelling can facilitate understanding of why things happen, how they happen, and when they might happen given that changes can be visualised dynamically over time.

19.2.1.1   Agent-Based Modelling (ABM)

ABM is a type of individual-based modelling that simulates the actions and interactions of autonomous 'agents' to assess the effects of their behaviour on the system as a whole (Bonabeau, 2002; Macal & North, 2010; Badham et al., 2018; Cassidy et al., 2019). Agents in an ABM can constitute any self-contained and goal-directed entity, including molecules, cells, pathogens, people (e.g. athletes, sports teams), animals, vehicles (e.g. autonomous cars), organisations, or entire synthetic populations (Epstein, 2009; Parker & Epstein, 2011). In the case that the agents are representative of individual people, the model user can assign demographic and lifestyle-related characteristics such as age, sex, diet, medical history, and injury or disease susceptibility, as well as cognitive rules pertaining to memory, personality, behaviour, and intelligence (Luke & Stamatakis, 2012). This means that agents can learn over time based on past experiences, update their internal states, adapt to changing environmental circumstances, and demonstrate any other characteristic or behaviour that has been explicitly defined. Based on a ground-up modelling approach, ABM can be used to explain how populations self-organise, and create patterns of global behaviour that are not predictable or programmed into each agent type a priori. For this reason, ABM is a powerful approach for exploring the mechanism(s) by which collective behaviour among individual agents gives rise to emergent phenomena (e.g. proportion or rates of sports injury occurrence). The reader is referred to several comprehensive resources covering the origins, purpose, and general use of ABM (Bonabeau, 2002; Macal & North, 2010; Luke & Stamatakis, 2012; Marshall, 2017; Tracy, Cerdá, & Keyes, 2018), including issues pertaining to the development, verification, and validation of simulations (Rand & Rust, 2011; Caro et al., 2012; GOS, 2018; Calder et al., 2018).

19.2.1.1.1   *ABM in Sports Injury Prevention Research*

A first-of-its-kind ABM in the field of sports injury prevention research was developed in the context of recreational distance running injury (Hulme et al., 2018b). The rationale behind the study was to demonstrate the use of computational modelling in a health-related field other than infectious disease epidemiology which represents a more traditional domain of application (Marshall et al., 2012). In brief, the main characteristics of complex systems described in Chapter 2 (Table 19.1) have been a recent topic of conversation in sports injury prevention research (Hulme & Finch, 2015; Bekker & Clark, 2016; Bittencourt et al., 2016; Mooney et al., 2017; Gokeler, Verhagen, & Hirschmann, 2018; Pol et al., 2018; Hulme et al., 2018a). The growing interest around complexity in the sports science domain is attributable to the fact that traditional risk factor identification methods (e.g. stepwise regression modelling) and linear cause-effect analyses fail to reflect the complex mechanisms of sports injury causation. Indeed, effective sports injury prevention requires us to understand the *complex relationships* that occur among a 'web of interacting determinants' (Bittencourt et al., 2016), rather than try to isolate the causal effect of *individual factors*. Therefore, what was lacking in the sports science literature was the application of a method that could account for complex sports injury relationships and simulate their effects to inform practice. This is because complex systems thinking was, and still is to some extent, preoccupied with theoretical descriptions (Quatman, Quatman, & Hewett, 2009; Mendiguchia, Alentorn-Geli, & Brughelli, 2011; Hulme & Finch, 2015; Bekker & Clark, 2016; Cook, 2016; Gokeler, Verhagen, & Hirschmann, 2018) and static models of complexity (Hulme et al., 2017a, 2017b, 2019) that cannot fully account for the dynamic nature of sports injury systems, including how systems change over time. Although static models are useful for depicting complexity more generally (as will be discussed later in this chapter), there remained an opportunity to promote a proof-of-concept ABM that would encourage other sports scientists to build on the underlying theory and concepts presented.

19.2.1.1.2   *The Aim and Operation of the Distance Running ABM*

The so-called 'distance running ABM' (Hulme et al., 2018b) modelled the relationship between changes to workload (i.e. weekly running distances in km) and running injury occurrence in a synthetic population of 1,000 athletes (Figure 19.1). The purpose of the model was to examine how the injury incidence proportion responded to the manipulation of various athlete management controls. The aim was to determine the ideal set of conditions that supported optimal health and performance across the running population.

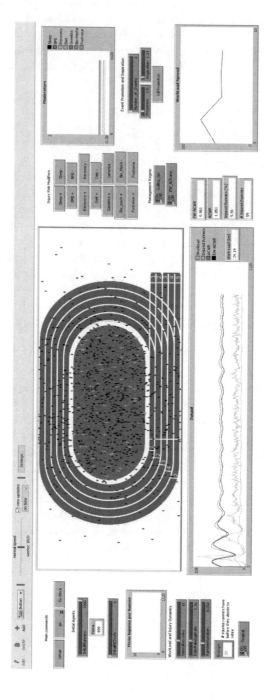

FIGURE 19.1 The distance running NetLogo interface and simulation environment. There were 1000 athletes in the simulation, represented by the figurine symbols overlaid onto the track and field visual. The distance running ABM includes several types of interface items, including buttons, switches, sliders, monitors, counts, and dynamic plots. When in operation, the runners on-screen move about their virtual track environment and change from black to red when injury is sustained.

Regarding model development, each virtual athlete in the ABM was randomly programmed to have their own personal characteristics and workload-related behaviours, such as age, body mass index, and running distance, as well as an inbuilt physiological capacity to tolerate the applied workload. These individual agent attributes were drawn from a random-normal distribution which produced population-level patterns that followed real-world sampling practices. The athletes in the distance running ABM were young and old, weak and strong, and exhibited poor or good dietary regimes. Next, the simulation was designed to support the use of various athlete management controls. These controls included how far the athletes were to run in a given week, including the rate at which workloads were applied (e.g. increases of 5% increments through to 30%). Likewise, the model incorporated an option to vary the extent to which the athlete cohort misunderstood by how much workloads should be safely increased. One way to view the manipulation of these controls is to imagine a running coach pulling different levers to see what happens under a certain set of conditions. In relation to the aforementioned five purposes of computational modelling that were outlined in the previous section (GOS, 2018; Calder et al., 2018), the distance running ABM falls into the category of theoretical modelling as the simulation is deterministic in nature. That is, a simulation supporting consistent and precisely determined outcomes without room for random variation when a threshold (e.g. physiological capacity for injury) has been surpassed.

In total, the set combination of multiple different athlete management controls produced 72 different possible conditions impacting on the running workload-injury dynamics. These conditions were repeated (model 'run') ten times resulting in a computational model containing 720,000 simulated runners managed and monitored over ~100,000 weeks. The methodological development behind the distance running ABM was exhaustive, and so a complete overview of the steps taken to develop the simulation can be viewed in the source material (Hulme et al., 2018b).

19.2.1.1.3   What Were the Results of the Distance Running ABM?

There were a number of key findings associated with the distance running ABM. Starting with the way that workloads were applied, it can be concluded that building weekly running distances over time, even if undertaken within reported 'training safe zones' (Gabbett, 2016), will eventually result in the occurrence of sports injury as athletes reach and surpass their physiological limits. Specifically, even if athletes do not rapidly increase their workloads, there will still come the point whereby a law of diminishing returns reduces the margin of error to which runners can afford to make mistakes and remain healthy. In and of itself, this finding is to be expected. However, introducing training-related error into the simulation and modelling a fixed upper ceiling of workload capacity resulted in a higher injury incidence proportion across the athlete population. Theoretically, this indicated fragility at the extremities of performance for the more serious runner who might aim to participate in competitive events. This finding has two practical implications. First, it may be advantageous for runners who wish to maintain high distances over extended periods of time to think long term about their training and refrain from regularly operating at their perceived level of peak performance (i.e. they are always on a 'knife edge'). This recommendation still stands when safe training practices are followed. Second, although external workload monitoring is essential for overall athlete health and well-being, the distance running ABM emphasised the need to also track subjective forms of performance feedback, including the monitoring of internal psychological and physiological markers of overtraining (Hulme et al., 2018b).

19.2.1.1.4 What Does the Future Hold for ABM in Sports Injury Prevention Research?

Moving forwards, it will be interesting to observe how ABM will be used to study other physiological and wider environmental factors that may impact on the work-load-injury relationship. For example, regardless of the sport in question, future ABMs could be programmed with a dynamic rather than a fixed upper physiological ceiling corresponding to the adaptation or maladaptation that occurs to an athlete's musculoskeletal capacity. This would make sense given that all athletes experience both positive and negative feedback cycles in terms of the ability to tolerate increasing workloads. Incorporating the option to modify different capacity-related exposures in the model, such as genetic endowment and other daily stressors such as employment-related obligations, is something that could be of interest. In addition, there is a need to evaluate the effect of social clustering on sports injury risk, such as how peer-to-peer interactions within and between different athletic groups could modify workload behaviours through information exchange dynamics. There is also some benefit in looking at wider environmental determinants that shape attitudes, social norms, and beliefs about training practices. These could extend to the political and regulatory environment, and cover aspects relating to the affordability of coaching education or access to community healthcare services.

Similarly, comparative analyses of rural versus built environments and how neighbourhood and city design impacts on sports participation levels is an area that could be explored with computational modelling. For example, an ABM has examined the impact of certain policies aimed to change attitudes towards walking (Yang et al., 2011), and so a similar approach to explicate the determinants of running participation is warranted. Irrespective of the changes that could be implemented, the future of ABM in sports injury prevention research is promising. The full extent and overall contribution of ABM is not yet known, prompting a requirement for further work of this kind in the sports sciences.

19.2.1.2 System Dynamics (SD) Modelling

The other computational approach that this chapter covers is SD modelling. Despite their different ways of approaching the study of complexity, both ABM and SD modelling are capable of simulating the main characteristics of complex systems as described in Table 19.2. However, in the case of SD modelling, there is no tracking of agents or factors on an individual basis. This is why SD modelling is referred to as an aggregated top-down method rather than an individual ground-up computational approach (Luke & Stamatakis, 2012; Cassidy et al., 2019). In ABM, the user has to initially define the rules of each individual agent from which the behaviour of the system as a whole emerges. Conversely, in SD modelling, the relationships between elements and the global behaviour of the system is defined from the outset and so it is the individual effects resulting from broader structural dynamics that are of interest. Therefore, SD modelling is well suited for elucidating the overall structure of complex systems and exposing optimal leverage points to test hypothesised policies and interventions.

19.2.1.2.1   The Origins and Uses of SD Modelling

SD modelling was founded by Jay Forrester at the Massachusetts Institute of Technology in the mid-1950s (Forrester, 1961, 1969). Used originally to help large corporations such as General Electric to improve understanding of how to effectively manage a range of corporate issues, SD modelling has since been extended to many other disciplines. From a methodological perspective, SD modelling is used to conceptualise the cyclical, interlocking relationships that are typically found among the different elements of complex systems. To simulate the non-linear behaviours of complex systems over time, SD modelling makes use of stocks, flows, variables, and feedback loops, all of which are underpinned by a series of differential equations and/or mathematical formulae. A key initial phase of SD modelling involves the development of Causal Loop Diagrams (CLDs) to describe the positive (i.e. reinforcing) and negative (i.e. balancing) feedback loops that underpin the behaviour of complex systems (Figure 19.2).

19.2.1.2.2   Why Use SD modelling?

Given a focus on causal feedback, SD modelling is useful for the study of emergence within complex systems at a defined population-level. This is because complex systems tend to exhibit counterintuitive behaviours, and the implementation of a health-based intervention might have unanticipated consequences that exacerbate the problem we are trying to fix. For example, the introduction of a new policy around the use of protective equipment such as headguards that purport to better prevent injuries in team sports contexts may lead athletes to engage in more risky behaviours (e.g. harder and more aggressive tackling), which in turn may lead to an increase in the number of injuries across the population. This is known as a risk compensation effect (McIntosh, 2005), which describes that sports injury countermeasures may

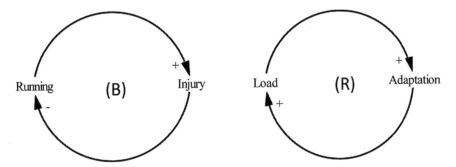

FIGURE 19.2   Simple Causal Loop Diagram (CLD) of a balancing (B) and reinforcing (R) loop. CLDs are useful for capturing hypotheses about causal dynamics, for eliciting and conceptualising mental models, and for communicating the feedback mechanisms that may be responsible for a problem. (Permission to reuse figure granted on 24 January 2020 by Rightslink Copyright Clearance Center; license # 4755170253590.)

have negative consequences that cannot always be anticipated. These effects might involve shifting the distribution of sports injury, change individual athlete behaviours due to a false sense of security, or even reduce the level of sports participation across a once active population (Hagel & Meeuwisse, 2004). As other elements and behaviours are integrated into the causal architecture of a system exhibiting risk compensation effects and complex forms of feedback, it is apparent that preventing sports injury requires a broader perspective. SD modelling can help sports injury researchers to locate and understand where in the system, and by what mechanisms, any potential unexpected outcomes could emerge.

19.2.1.2.3   An Early Working Proof-of-Concept SD Model

Figure 19.3 visualises a CLD of running injury development which has been simulated using the SD modelling software 'Sysdea' (www.sysdea.com). The proof-of-concept SD model can be viewed at: https://goo.gl/1fFj3E, and the model is presented in Figure 19.4.

As in the previous distance running ABM application, the SD model in Figure 19.4 focusses on recreational distance running injury, albeit in this case the athlete population transitions through an injury–healthcare–recovery cycle. The current simulation falls into the category of illustrative modelling as it visualises sports injury occurrence and athletic recovery without being used to predict, explain, or theorise about the described dynamics.

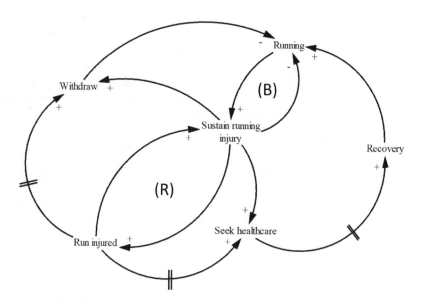

FIGURE 19.3   Causal Loop Diagram (CLD) of running injury development. A simulated version of Figure 19.3 can be found at https://goo.gl/1fFj3E, which represents the initial phase of a detailed and meticulous model building process. The parallel dashed lines indicate a time delay. (Permission to reuse figure granted on 24 January 2020 by Rightslink Copyright Clearance Center; license # 4755170253590.)

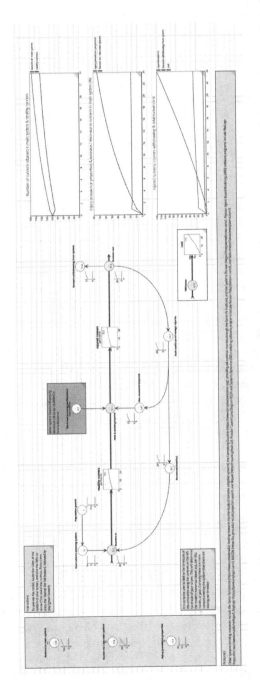

FIGURE 19.4 A screenshot of an early working distance running injury SD model. It includes stocks (i.e. the two square icons that contain the running population and fill and drain over time), flows (i.e. the thick middle arrows and circular icons that are directly linked to, and change the value of, the stocks), variables (i.e. the circular icons that define some intermediate value such as external population growth and runners entering or leaving the system), links (i.e. the thin black lines connecting variables), and dynamic graphs (i.e. the three rectangular shapes on the right of the model that plot changes over time).

There are a number of aspects to the distance running SD model that help to explain its operation. For example, the variable 'Population growth' along with the variable 'Recovered (healthy)' dictates the number of new runners who are entering the system or resuming participation following injury, respectively. The external population growth is set at an arbitrary constant value of 1.0 at each time step; however, there is room to instantiate the model with an approximated value corresponding to 'real' linear or exponential trends in population growth. This level of specificity and overall model parametrisation is required if any further explanatory or predictive power were to be sought. In a similar way to the number of new runners entering the system, the variable 'Runners out' is linked to both 'Runners withdrawing from system' and the runners who 'Seek healthcare/manage injuries'. This is because a proportion of the athlete population will cease to continue running after sustaining injury (in this case 20% is used for illustrative purposes), whereas the remaining runners will turn to therapeutic intervention with the aim of resuming participation on the basis of a successful recovery. If more runners are entering than leaving the system, there will be a net gain in the total number of runners (i.e. healthy and injured combined). Another feature of the model is the variable 'Workload (running participation) risk modifier', which determines the proportion of runners who sustain injury per month according to findings in the scientific literature (Kluitenberg et al., 2015). Although the changing nature of the risk modifier will be of interest in future model iterations, there is a need to firstly extend the boundary around the distance running injury SD model to account for the influence of other larger systems, whether they be social, healthcare, economic, and/or technological in nature. This will allow for the examination of feedback across different systems as a means to understand the dynamic causal interdependencies among a multitude of elements. In doing so it might be possible to identify optimal leverage points when thinking about the implementation of a given running injury prevention intervention. The intervention(s) might include changes to national or local policies, involve the development of new educational initiatives, change the way that a certain healthcare treatment is delivered, or relate to the implementation of athlete monitoring programmes. The key point is that without first modelling complex systems and capturing systemic relationships, it is difficult to fully understand the wider implications of a single decision or action.

In sum, the most beneficial reason for using SD modelling in sports injury research is to identify and simulate the effect of different policy interventions and examine how other elements in the system react to their implementation. This will help to answer questions such as: Is this an optimal strategy to prevent sports injury? Could there be any undesirable or unforeseen consequences resulting from this intervention? What is the 'knock-on' effect of this programme or initiative? Before getting to the point of answering such questions, there is a need to describe the system qualitatively in the form of a CLD and proceed to simulate the depicted relationships between system structures with SD modelling. A concerted multidisciplinary effort from the international community of sports scientists is required to push the SD modelling agenda forwards and further explore how it can best be used.

19.3   COMPUTATIONAL MODELLING, SYSTEMS HFE, AND SPORTS INJURY RESEARCH

This section discusses how computational modelling can complement traditional systems HFE research in sport, both from an explanatory and design-based standpoint.

19.3.1   STATIC VERSUS DYNAMIC MODELLING

Static models have no internal memory and lack the capacity to directly draw on past behaviours as a way to model future outcomes. The term 'static model' refers not to algebraic formulae or other mathematical concepts (e.g. regression analyses), but rather to the use of qualitative representations and diagrams that form the basis of many systems HFE models and outputs. For example, the benefit of applying the Systems Theoretic Accident Model and Process (STAMP) method (Leveson, 2004) in the context of recreational distance running (Hulme et al., 2017a) was its ability to describe the political, regulatory, organisational, and individual athlete systems that play a role in the management of sports injury. The scope and scale offered by STAMP was used to identify systemic opportunities or 'leverage points' to guide the implementation of sustainable running injury prevention interventions. Another static modelling approach that features in Chapter 16 of this book is Causal Loop Diagrams (CLDs). CLDs have been used extensively in other areas to identify and depict the feedback loops underpinning behaviour in complex systems, and form the input for SD modelling. In a similar manner to STAMP, CLDs are useful as they can be used to identify key leverage points which influence system behaviour. Despite their many offerings, a disadvantage of systems HFE and static models more generally, including STAMP and CLDs, is that they provide only a 'snapshot' of the system in question and do not account for the time-dependent nature of multiple decisions and behaviours that shape future performances and outcomes. The capability of ABM and SD modelling to account for historical circumstances and monitor the time-varying phenomena of complex systems is a distinct advantage of these approaches. Static models do indeed have an important place in systems HFE and sports science research, and so it is a matter of deciding when, and for what purposes, static models and computational approaches are to be applied.

19.3.2   A PROPOSED 'MODELLING CONTINUUM' TO SUPPORT SYSTEMS THINKING IN SPORT

One aspect that is common across systems HFE methods is the capacity to scale up and conceptualise multiple levels of the sports system in question (Hulme et al., 2017a, b, 2019; Clacy et al., 2017). Computational modelling approaches generally do not offer this holistic scope in quite the same way, and tend to sacrifice what has been termed 'socioecological generality' in favour of a more refined understanding of complex system behaviours and processes (Ip et al., 2013). This can enable researchers to more closely study the characteristics of complex systems (e.g. causal feedback, emergence; see Chapter 2) whilst preserving a relatively high degree of model precision, otherwise defined as the ability for a model to quantify parameters and generate numerical outputs (Ip et al., 2013). Abstracted hierarchies and large-scale diagrams such as those offered by systems HFE models represent a useful starting point to identify the structures, elements, and relationships within complex systems that can be later subjected to a more intense programme of study with computational approaches. This is because attempting to capture everything in a computational model from the outset is not recommended, as doing so can detrimentally affect explanatory and predictive accuracy, or the ability for a given model to 'fit' unknown data. This is not to say that approaches such as ABM or SD modelling cannot traverse multiple system boundaries or take a broad perspective, either. Rather, a computational model that attempts to include every moving part and relationship does not automatically make it useful, hence the reason why static modelling is a viable means to initially conceptualise phenomena on a much larger scale. Again, a useful example is the distance running STAMP model which represented an initial attempt to identify how running injury could be managed and prevented from a systems HFE perspective (Hulme et al., 2017a). From there, the distance running ABM (Hulme et al., 2018b) drilled down further into the causal schematic depicting the workload-injury mechanism at level five of STAMP, which is lowest of the system levels in the model. These two approaches were mutually supportive despite having different goals and purposes.

19.3.3　Combining Design-Based Ergonomics Methods with Computational Modelling

Any discussion about computational modelling and simulation approaches in the ergonomics space should acknowledge the work by Hettinger et al. (2015) and Salmon and Read (2019). In both contributions, the authors argue that computational approaches are required to not only understand the operation and behaviours of complex systems, but to also assist with their fundamental development and design. In the case of Hettinger et al. (2015), the concept of development and design was framed in the context of workplace safety, referring to aspects such as training needs assessment, personnel allocation, resource distribution, shared cognition, and information requirements specification. These are the aspects that many modern-day complex socio-technical systems, including complex sports systems, require during their planning and development in which to operate safely and efficiently at a later point in time.

Salmon and Read (2019) advocate for a similar direction regarding the application of computational approaches to support system design processes; however, the focus was on how the Cognitive Work Analysis (CWA) toolkit could support SD modelling. Specifically, computational approaches such as SD modelling can complement CWA as it provides a means to dynamically study the potential impact of making changes to the structure(s) or properties of complex systems in a way that provides immediate results (Salmon & Read, 2019). To formalise this way of thinking, CWA could be used to design a new strength training programme to prevent hamstring injuries in football. This is because out of all the systems ergonomics methods, CWA provides a structured approach when considering all aspects of system design, ranging from interfaces and workspaces through to regulatory systems and organisational strategies (Read, Salmon, & Lenné, 2015). Following the use of CWA, the next phase could involve applying SD modelling to examine whether the implementation of the hypothesised intervention theoretically leads to the desired result (e.g. reduced football injury incidence rates), especially in consideration of the influences from other wider systems and elements. From this it can be concluded that the use of systems HFE methods should ideally precede computational modelling, and a continuum of static and dynamic methods applications as outlined in recent literature (Salmon & Read, 2019) should be followed.

19.3.4 WHAT IS NEXT FOR COMPUTATIONAL MODELLING IN SPORT?

It is possible to identify a number of future research opportunities as well as new uses for computational modelling based on the emerging role of big data and technological growth. First, there is a need to outline a multi-method framework involving the use of both static and dynamic modelling methods specifically within the field of sports injury prevention research. Relative to other domains of study, it is pertinent to determine whether the typical problems facing sports injury research necessitate a different set or combination of ergonomics and computational approaches along a so-called 'modelling continuum'. Part of this understanding would involve exploring the utility of computational modelling approaches beyond the theoretical ABM and the purely illustrative SD modelling examples provided in this chapter. A second implication is that as computational modelling receives greater interest, there will be a requirement for training around the use of what are potentially difficult-to-use methods. For example, ABM requires expertise in computer coding and programming, and so ergonomists and sports scientists will have to either obtain these skills or collaborate with modelling experts from other, specialised domains (e.g. mathematics, computer science).

Third, the list of future needs and uses of computational modelling, similar to that proposed by Calder et al. (2018), are equally applicable in the context of sports injury control and prevention. For instance, computational models will be required to keep pace with the increasing complexity of sports systems. This growth in complexity and socio-technical interconnectedness is fuelled, in part, by the introduction and proliferation of new technologies, cutting edge scientific developments, big data possibilities, and the resulting increased knowledge that will rapidly inform and positively influence the performance, health, and well-being of athletes. As such, computational models of the future are likely going to be more sophisticated with respect to their scale, level of detail, scope, ability to function in an unsupervised manner (i.e. relying on data input only), and confidence in terms of knowing where the boundaries of uncertainty lie when formulating predictions. It could well be the case that computational modelling and simulation packages will become widespread and invaluable tools to certain groups and stakeholders, particularly those affiliated with high profile sports teams, clubs, and international athletic organisations.

19.4   CONCLUSION

Systems HFE researchers and practitioners with an interest in complexity in the sports sciences are encouraged to embrace the many possibilities associated with computational methods, particularly that of ABM and SD modelling approaches. This includes their potential capability to simulate and articulate contemporary ergonomics issues in a manner that appeals to and is congruent with a conceptual understanding of how the world works in complex sports injury settings. Computational modelling affords sports ergonomists the opportunity to test the theoretical effectiveness of different injury prevention interventions before they are implemented in complex sports settings. This could potentially save time, money, and necessary resources, especially during the initial planning and design phases. Although not without its challenge, including what could be a relatively steep learning curve and exercise in familiarisation, it will only be a matter of time before computational modelling will find its way into the sports ergonomist's methodological toolkit.

REFERENCES

Badham, J., Chattoe-Brown, E., Gilbert, N., Chalabi, Z., Kee, F., & Hunter, R. F. (2018). Developing agent-based models of complex health behaviour. *Health and Place*, 54, 170–177. doi: 10.1016/j.healthplace.2018.08.022.

Bekker, S., & Clark, A. M. (2016). Bringing complexity to sports injury prevention research: from simplification to explanation. *British Journal of Sports Medicine*. doi:10.1136/bjsports-2016-096457.

Bittencourt, N. F. N., Meeuwisse, W. H., Mendonça, L. D., Nettel-Aguirre, A., Ocarino, J. M., & Fonsesca, S. T. (2016). Complex systems approach for sports injuries. Moving from risk factor identification to injury pattern recognition – A narrative review and new concept. *British Journal of Sports Medicine*, 0, 1–7.

Bonabeau, E. (2002). Agent-based modeling: Methods and techniques for simulating human systems. *Proceedings of the National Academy of Sciences of the United States of America*, 99:3: Supplement 3, 7280–7287.

Calder, M., Craig, C., Culley, D., de Cani, R., Donnelly, C. A., Douglas, R., Edmonds, B., Gascoigne, J., Gilbert, N., Hargrove, C., Hinds, D., Lane, D. C., Mitchell, D., Pavey, G., Robertson, D., Rosewell, B., Sherwin, S., Walport, M., & Wilson, A. (2018). Computational modelling for decision-making: where, why, what, who and how. *Royal Society Open Science*, 5:6, 172096. doi: 10.1098/rsos.172096.

Caro, J. J., Briggs, A. H., Siebert, U., & Kuntz, K. M. (2012). Modeling good research practices – overview: A report of the ISPOR-SMDM modeling good research practices task force-1. *Value in Health*, 15:6, 796–803. doi: 10.1016/j.jval.2012.06.012.

Cassidy, R., Singh, N. S., Schiratti, P. R., Semwanga, A., Binyaruka, P., Sachingongu, N., Chama-Chiliba, C. M., Chalabi, Z., Borghi, J., & Blanchet, K. (2019). Mathematical modelling for health systems research: a systematic review of system dynamics and agent-based models. *BMC Health Services Research*, 19:1, 845. doi: 10.1186/s12913-019-4627-7.

Clacy, A., Goode, N., Sharman, R., Lovell, G. P., & Salmon, P. (2017). A systems approach to understanding the identification and treatment of sport-related concussion in community rugby union. *Applied Ergonomics*. doi: 10.1016/j.apergo.2017.06.010.

Claudino, J. G., de Oliveira Capanema, D., de Souza, T. V., Serrão, J. C., Pereira, A. C. M., & Nassis, G. P. (2019). Current approaches to the use of artificial intelligence for injury risk assessment and performance prediction in team sports: a systematic review. *Sports Medicine – Open*, 5:1, 28. doi: 10.1186/s40798-019-0202-3.

Cook, C. (2016). Predicting future physical injury in sports: it's a complicated dynamic system. *British Journal of Sports Medicine*. doi: 10.1136/bjsports-2016-096445.

Davis, M. C., Hughes, H. P. N., McKay, A., Robinson, M. A., & van der Wal, C. N. (2019). Ergonomists as designers: computational modelling and simulation of complex sociotechnical systems. *Ergonomics*, 1–14. doi: 10.1080/00140139.2019.1682186.

Epstein, J. M. (2009). Modelling to contain pandemics. *Nature*, 460:7256, 460–687.

Forrester, J. (1961). *Industrial Dynamics*. Cambridge, MA: MIT Press.

Forrester, J. (1969). *Urban Dynamics*. Cambridge, MA: MIT Press.

Gabbett, T. J. (2016). The training-injury prevention paradox: Should athletes be training smarter and harder? *British Journal of Sports Medicine*, 50:5, 273–280.

Gokeler, A., Verhagen, E., & Hirschmann, M. T. (2018). Let us rethink research for ACL injuries: a call for a more complex scientific approach. *Knee Surgery, Sports Traumatology, Arthroscopy*. doi: 10.1007/s00167-018-4886-6.

GOS. (2018). *Computational Modelling: Blackett Review*. https://www.gov.uk/government/publications/computational-modelling-blackett-review, accessed on 5 December.

Hagel, B., & Meeuwisse, W. (2004). Risk compensation: A 'side effect' of sport injury prevention? *Clinical Journal of Sport Medicine*, 14:4, 193–196.

Hernán, M. A., Hsu, J., & Healy, B. (2019). A second chance to get causal inference right: A classification of data science tasks. *Chance*, 32:1, 42–49. doi: 10.1080/09332480.2019.1579578.

Hettinger, L. J., Kirlik, A., Goh, Y. M., & Buckle, P. (2015). Modelling and simulation of complex sociotechnical systems: envisioning and analysing work environments. *Ergonomics*, 58:4, 600–614. doi: 10.1080/00140139.2015.1008586.

Holman, M., Walker, G., Lansdown, T., & Hulme, A. (2019). Radical systems thinking and the future role of computational modelling in ergonomics: an exploration of agent-based modelling. *Ergonomics*, 1–18. doi: 10.1080/00140139.2019.1694173.

Hulme, A., & Finch, C. F. (2015). From monocausality to systems thinking: A complementary and alternative conceptual approach for better understanding the development and prevention of sports injury. *Injury Epidemiology*, 2:1, 1–12. doi: 10.1186/s40621-015-0064-1.

Hulme, A., McLean, S., Read, G. J. M., Dallat, C., Bedford, A., & Salmon, P. M. (2019). Sports organizations as complex systems: Using cognitive work analysis to identify the factors influencing performance in an elite netball organization. *Frontiers Sports Management and Marketing*, 1:56. doi: 10.3389/fspor.2019.00056.

Hulme, A., Mclean, S., Salmon, P. M., Thompson, J., Lane, B. R., & Nielsen, R. O. (2018a). Computational methods to model complex systems in sports injury research: agent-based modelling (ABM) and systems dynamics (SD) modelling. *British Journal of Sports Medicine*. doi: 10.1136/bjsports-2018-100098.

Hulme, A., Salmon, P. M., Nielsen, R. O., Read, G. J. M., & Finch, C. F. (2017a). From control to causation: Validating a 'complex systems model' of running-related injury development and prevention. *Applied Ergonomics*, 65. doi: 10.1016/j.apergo.2017.07.005.

Hulme, A., Salmon, P. M., Nielsen, R. O., Read, G. L. M., & Finch, C. F. (2017b). Closing Pandora's Box: Adapting a systems ergonomics methodology for better understanding the ecological complexity underpinning the development and prevention of running-related injury. *Theoretical Issues in Ergonomics Science*, 18:4, 338–359. doi: 10.1080/1463922X.2016.1274455.

Hulme, A., Thompson, J., Nielsen, R., Read, G., & Salmon, P. (2018b). Towards a complex systems approach in sports injury research: Simulating running-related injury with Agent-Based Modelling. *British Journal of Sports Medicine*, 53:9.

Ip, E. H., Rahmandad, H., Shoham, D. A., Hammond, R., Huang, T. K., Wang, Y., & Mabry, P. L. (2013). Reconciling statistical and systems science approaches to public health. *Health Education and Behavior*, 40:1 Supplement, 123–131.

Kluitenberg, B., van Middelkoop, M., Diercks, R., & van der Worp, H. (2015). What are the differences in injury proportions between different populations of runners? A systematic review and meta-analysis. *Sports Medicine*, 45:8, 1143–1161.

Leveson, N. G. (2004). A new accident model for engineering safer systems. *Safety Science*, 42: 4, 237–270.

Luke, D. A., & Stamatakis, K. A. (2012). Systems science methods in public health. *Annual Review of Public Health*, 33, 357–376.

Macal, C. M., & North, M. J. (2010). Tutorial on agent-based modelling and simulation. *Journal of Simulation*, 4:3, 151–162.

Marshall, B. D. L., Paczkowski, M. N., Seemann, L., Tempalski B., Pouget, R. R., Galea, S., & Friedman, S. R. (2012). A complex systems approach to evaluate HIV prevention in metropolitan areas: Preliminary implications for combination intervention strategies. *PLoS One*, 7:9. doi: 10.1371/journal.pone.0044833.

Marshall, B. M. (2017). Agent-based modelling. In: *Systems Science and Population Health*, pp. 87–98. New York, NY: Oxford University Press.

McIntosh, A. (2005). Risk compensation, motivation, injuries, and biomechanics in competitive sport. *British Journal of Sports Medicine*, 39:1, 2–3.

Mendiguchia, J., Alentorn-Geli, E., & Brughelli, M. (2011). Hamstring strain injuries: are we heading in the right direction? *British Journal of Sports Medicine*. doi: 10.1136/bjsm.2010.081695.

Mooney, M., Charlton, P. C., Soltanzadeh, S., & Drew, M. K. (2017). Who 'owns' the injury or illness? Who 'owns' performance? Applying systems thinking to integrate health and performance in elite sport *British Journal of Sports Medicine*. doi: 10.1136/bjsports-2016-096649.

NIBIB. (2016). *Computational Modeling*. https://www.nibib.nih.gov/science-education/science-topics/computational-modeling, accessed on 5 December.

Parker, J., & Epstein, J. M. (2011). A distributed platform for global-scale agent-based models of disease transmission. *ACM Transactions on Modeling and Computer Simulation*, 22:1, 2. doi: 10.1145/2043635.2043637.

Pol, R., Hristovski, r., Medina, d., & Balague, N. (2018). From microscopic to macroscopic sports injuries. Applying the complex dynamic systems approach to sports medicine: a narrative review. *British Journal of Sports Medicine*, 53:19, 1214–1220.

Pratt, S. C., Sumpter, D. J. T., Mallon, E. B., & Franks, N. R. (2005). An agent-based model of collective nest choice by the ant *Temnothorax albipennis*. *Animal Behaviour*, 70:5, 1023–1036. doi: 10.1016/j.anbehav.2005.01.022.

Quatman, C. E., Quatman, C. C., & Hewett, T. E. (2009). Prediction and prevention of musculoskeletal injury: a paradigm shift in methodology. *British Journal of Sports Medicine*, 43:14, 1100–1107. doi: 10.1136/bjsm.2009.065482.

Rand, W., & Rust, R. T. (2011). Agent-based modeling in marketing: Guidelines for rigor. *International Journal of Research in Marketing*, 28:3, 181–193. doi: 10.1016/j.ijresmar.2011.04.002.

Ratamero, E. M. (2015). Modelling peloton dynamics in competitive cycling: A quantitative approach. In: *Sports Science Research and Technology Support*. Cham, Switzerland: Springer.

Read, G. J. M., Salmon, P. M., & Lenné, M. G. (2015). Cognitive work analysis and design: current practice and future practitioner requirements. *Theoretical Issues in Ergonomics Science*, 16:2, 154–173. doi: 10.1080/1463922X.2014.930935.

Salmon, P. M., & Read, G. J. M. (2019). Many model thinking in systems ergonomics: a case study in road safety. *Ergonomics*, 62:5, 612–628. doi: 10.1080/00140139.2018.1550214.

Tracy, M., Cerdá, M., & Keyes, K. M. 2018. Agent-based modeling in public health: Current applications and future directions. *Annual Review of Public Health*, 39:1, 77–94. doi: 10.1146/annurev-publhealth-040617-014317.

Yang, Y., Diez Roux, A. V., Auchincloss, A. H., Rodriguez, D. A., & Brown, D. G. 2011. A spatial agent-based model for the simulation of adults' daily walking within a city. *American Journal of Preventative Medicine*, 40:3, 353–361.

Section V

Future Applications of HFE in Sport

20 Summary and Future Applications of Human Factors and Ergonomics in Sport

Paul M. Salmon, Scott McLean, and Adam Hulme

CONTENTS

20.1 INTRODUCTION

The aim of this book was to communicate contemporary sports HFE research and practice from across the world, with a view to facilitating cross-disciplinary interaction between HFE and the sports sciences and stimulating new applications of HFE in sport. The chapters presented include physical, cognitive, and systems HFE applications covering a diverse set of topics across a wide range of sports, including football, running, cycling, cricket, rugby, equestrian sports, parasport, basketball, Formula One, and Australian Rules Football. Clearly HFE has much to offer, and its models and methods appear to be applicable in a broad set of sporting contexts. Whilst each chapter has its own set of findings and implications for sports research and practice, taken together the body of work presented in this book has a number of important messages. In this final chapter our attention turns to summarising these key messages, and to outlining a series of important future applications for HFE in sport.

20.2   KEY MESSAGES

20.2.1   Sport as a Complex System

Many of the chapters presented in the book emphasise the fact that sports systems, whether elite, amateur, or at the grassroots level, are complex and socio-technical in nature. In Chapter 2, Hulme et al. describe the key characteristics of complex systems and confirms their presence within sport. Part IV of the book subsequently presents a series of studies in which systems HFE methods are applied to identify factors beyond the athlete and across sports systems which interact and influence performance and injury. This message of complexity being inherent in sports systems aligns strongly with the recent sports science literature, where an increasing number of authors have emphasised the need not only to acknowledge that sports systems are complex in nature, but also to apply appropriate scientific methods which consider the requisite features of complexity (Davids et al., 2013; Bittencourt et al., 2016; Hulme et al., 2018; McGarry et al., 2002; McLean et al., 2017; Mooney et al., 2017). This book has added further to this discourse, but perhaps more importantly has showcased a number of systems HFE methods that can be used to describe and understand sport through a complex systems lens.

20.2.2   More to HFE in Sport Than Just Injury Prevention and Equipment Design

As well as covering a diverse set of sports, the contributions presented in this book showcase various purposes for which HFE can be used in sport. Whilst HFE has a long history of being used to support sports injury prevention and sports equipment design, and many of the contributions focus on these two areas, this book has covered a variety of additional important purposes for which HFE can be used. Chapters 3 (Gorman et al.) and 12 (Vickery et al.) focus on the use of HFE to support the design of training and coaching programmes for skill acquisition, whilst McLean et al. (Chapter 13) and Salmon et al. (Chapter 15) describe how HFE can be used to support more meaningful and useful performance analysis. In Chapter 16, Berber et al. use systems HFE to examine talent identification in dressage, and Kean et al. (Chapter 14) analyse environmental influences on elite and community parasport. Finally, in what is becoming an increasingly important application area for HFE in sport, Neville et al. (Chapter 10) use HFE methods to investigate the impacts of introducing new technology on performance, situation awareness, and decision making. The book therefore provides a strong message that HFE has more to offer to sport than merely injury prevention and sports equipment design. Though these will continue to be important areas in which HFE can contribute, applications for the other purposes described in this book will no doubt increase in future.

20.2.3 HFE as a Modelling Science

The need to model sports systems and their likely behaviour is a strong theme through-out the book, and it is notable that this requirement stands regardless of the purposes of the study or analysis. Part IV of the book contains various applications whereby systems HFE methods are applied for precisely this purpose, and Hulme et al. (Chapter 19) takes this a step further to simulate the dynamics of workload within a recreational running injury system. Such applications present many possibilities for HFE in sport, including new application areas, such as athlete and coach health and well-being, sports policy and governance design, and coaching curriculum development to name only a few of the broader systemic issues that could be tackled through these approaches. The importance of analyses that consider sports systems as the unit of analysis cannot be understated. It is our view that the absence of system modelling applications is a sig-nificant limitation of sports science work to date, and is perhaps where HFE can make its most powerful contribution to sports research and practice. Indeed, the lack of such studies suggest that there is much to learn regarding the behaviour and composition of complex sports systems. A critical requirement moving forward is the use of systems HFE methods that are capable of modelling the behaviour of sports systems whilst at the same time providing useful outputs for practitioners.

20.3 FUTURE APPLICATIONS OF HFE IN SPORT

There are many existing and emerging areas of focus in which HFE will have an important role to play in the future. Whilst existing areas of strength such as injury prevention and sports equipment design will continue to represent important lines of enquiry, there are a series of additional areas that represent important areas for future HFE applications.

20.3.1 Technology Insertion

There is no doubt that sports systems are becoming increasingly reliant on advanced technologies. Many sports now have technological support for officiating, and it is likely that advances in areas such as artificial intelligence, robotics, wearable athlete monitoring sensors, and workload-related quantification technologies will lead to new sporting innovations in different areas. The need to jointly optimise the human and technological elements of sports systems is perhaps more pressing than it has ever been, and this will only increase with further innovations. As the science that focuses on optimising the interactions between humans and technologies, HFE has a significant role to play in the design, evaluation, and implementation of sports technologies. As showcased by many of the contributions in this book, various HFE methods exist to support the design of new sports technologies and evaluate their likely impacts once implemented.

20.3.2 Disrupting Illicit Sports Systems

An important and emerging application area that has not been touched upon in this book is the use of HFE methods to inform the development of strategies to disrupt the functioning of systems that have been created to achieve illicit ends (Lane et al., 2019; Salmon et al., 2018). For example, Salmon et al. (2018) used Work Domain Analysis (WDA; Naikar, 2013) to identify interventions that could be used to disrupt the activities of Islamic State style terrorist cells. Lane et al. (2019) used the Event Analysis of Systemic Teamwork (EAST; Stanton et al., 2018) to analyse the processes involved when buying and selling identity credentials on DreamMarket (a darknet marketplace). Lane et al. (2019) subsequently applied the EAST-Broken Links approach (EAST-BL; Stanton & Harvey, 2017) to identify strategies for disrupting the trading of illicit goods in darknet marketplaces. Whilst the primary aim of HFE is to help optimise systems, these applications demonstrate the capacity to also disrupt and degrade performance.

Systems HFE methods are suitable for this kind of application as they support the description of entire systems, their component parts, and importantly the relationships and interactions between these parts, which in turn allows analysts to identify strategies to disrupt or prevent these interactions (Salmon et al., 2018). Specifically, by examining the outputs derived from systems HFE methods such as WDA and EAST, it is possible to identify the impacts of disrupting components – or the interactions between components. For example, Salmon et al. (2018) described how it was possible to trace the impact on terrorist cell functioning when the ability to create and use 'Propaganda' was disrupted. This use of HFE to disrupt systems, rather than optimise them, opens many new sports application areas where illicit means are used to achieve success. The obvious example is doping in sport, whereby methods such as CWA or EAST-BL could be used to model doping systems and identify avenues for disrupting access to and use of performance-enhancing drugs. However, many other application areas also exist, such as sports corruption, corporate malpractice, racism, child sexual abuse, gender equality, and sexism in sport.

20.3.3 Para-sport

Para-sport covers all sporting endeavours practised by people with physical, visual, and intellectual impairments (Vanlandewijck & Thompson, 2011). Participation and interest in para-sport are increasing substantially, with the Paralympic Games, for example, now considered to be the second-largest multisport event in the world (Patatas et al., 2018). Whilst Chapters 7 (wheelchair design) and 14 (participation in para-sport) present studies in this growing area, there is a clear lack of systems and cognitive HFE work in para-sport and indeed in disability sports generally. As with other areas of application, the potential contribution of HFE in para-sport and disability sports could span various purposes including equipment design, injury prevention, performance analysis, coaching and training design, policy and governance, and so on.

20.3.4 Risk Assessment

Risk assessment is an important area of HFE work whereby analysis methods are used to pro-actively identify risks that could hinder safety and performance. The outputs of such approaches are used to support organisations in developing appropriate and effective risk controls. For example, methods such as Systems-Theoretic Process Analysis (STPA; Leveson, 2011), EAST-BL (Stanton & Harvey, 2017), the Networked Hazard Analysis and Risk Management System (Net-HARMS; Dallat et al., 2018), and the Systematic Human Error Reduction and Prediction Approach (SHERPA; Embrey, 1986) have been used for risk assessment purposes in various domains. Recent work has explored the potential for such methods to identify the risks which could degrade optimal sports performance in elite women's road cycling (Hulme et al., 2020). Building on this, there is an opportunity to extend this work and explore the use of HFE-based risk assessment methods in other sports. Example applications include the use of methods such as STPA, EAST-BL, and Net-HARMS in both individual and team sports to identify the risks that could degrade performance.

20.3.5   HFE ACROSS THE SPORTS SYSTEM DESIGN LIFECYCLE

Proactive rather than reactive HFE is critical. That is, the benefits of HFE are best realised when it is used early and throughout the system design lifecycle from design concept and prototype testing to refinement, manufacture, and implementation. Unfortunately, HFE is often used reactively, whereby new systems are designed and implemented without HFE input, problems are encountered, and the HFE practitioner is called in to solve them with only a limited scope to make design changes. Proactive HFE is more impactful, and also provides a significant cost saving over reactive HFE (Stanton et al., 2013).

With this in mind, it is critical that HFE is not seen to be a problem-solving science in new application areas such as sport. Rather, a proactive HFE approach whereby HFE methods are applied across the sports system design lifecycle is recommended. This relates not only to sports equipment design, but also to other areas such as training and coaching design, the design of injury prevention strategies, the design of performance optimisation strategies, and even the design and creation of sports organisations including their internal processes and structures. Given the important role that HFE can play in sport, sports organisations could benefit from having HFE practitioners as collaborators, consultants, or members of performance staff. This is standard practice in safety-critical domains such as transportation (e.g. road, rail, aviation), maritime, defence, energy distribution, and nuclear power.

20.3.6   SIMULATING SPORTS SYSTEM BEHAVIOUR

As outlined by Salmon et al. (Chapter 15) and Hulme et al. (Chapter 18) there exists a range of methods that can be used to simulate the behaviour of individuals, teams, organisations, and even entire socio-technical systems. As outlined by Salmon et al. (2020), a major weakness of systems HFE methods is that they produce detailed but static outputs that are unable to demonstrate the dynamic interactions that create and influence behaviour. Whilst methods such as CWA (Vicente, 1999), EAST (Stanton et al., 2018), and STAMP (Leveson, 2004) provide the capacity to comprehensively model complex systems, they do not support the dynamic simulation of system behaviour. This can often mean that it is difficult to identify interventions that will have the desired effect on performance. Computational modelling methods such as ABM and SD have thus been proposed as a means of overcoming these limitations (Hulme et al., 2018, 2019; Salmon & Read, 2019; Salmon et al., 2020). In this sense, the work of Hulme et al. provides a clear path for future applications in which potential designs, training programmes, tactics, injury prevention interventions, etc. are examined using computational modelling methods such as ABM and SD. It is our view that both approaches potentially provide will be useful for HFE researchers and practitioners wishing to understand and optimise sports system performance.

20.4   CONCLUSION

The contributions presented in this book provide further evidence that sports HFE applications provide outputs that are useful and have practical relevance in sport and sports science. The contributions give an overview of contemporary HFE work in this area; however, there are many other issues to tackle, and likely more will emerge. Since the pioneering work of Reilly and colleagues, HFE in sport has expanded both in scope and the methods used to consider many factors beyond the athlete and their equipment, including the design and functioning of entire sports systems. As sports systems evolve, it is likely that the requirement for HFE will increase, and that we will see exciting new applications and contributions. We hope that this book will provide inspiration to both HFE practitioners and sports scientists, and that athletes, sports teams, and ultimately sports systems will benefit as a result.

REFERENCES

Bittencourt, N. F., Meeuwisse, W. H., Mendonça, L. D., Nettel-Aguiree, A., Ocarino, J. M., & Fonseca, S. T. (2016). Complex systems approach for sports injuries: moving from risk factor identification to injury pattern recognition-narrative review and new concept. *British Journal of Sports Medicine*, 50:21, 1309–1314.

Dallat, C., Salmon, P. M., & Goode, N. (2018). Identifying risks and emergent risks across socio-technical systems: The NETworked hazard analysis and risk management system (NET-HARMS). *Theoretical Issues in Ergonomics Science*, 19:4, 456–482.

Davids, K., Hristovski, R., Araujo, D., Serre, N. B., Button, C., & Passos, P. (2013). *Complex Systems in Sports*. Routledge Research in Sport and Exercise Science.

Embrey, D. E. (1986). SHERPA: A systematic human error reduction and prediction approach. In *Proceedings of the International Topical Meeting on Advances in Human Factors in Nuclear Power Systems.*

Hulme, A., Mclean, S., Dallat, C., Walker, G. H., Waterson, P., Stanton, N. A., & Salmon, P. M. (2020). Systems thinking risk assessment and prediction methods applied to elite sports performance: A comparison of STPA, EAST-BL and NET-HARMS. *Ergonomics*, in review.

Hulme A., Thompson J, Nielsen RO, Lane, B., McLean, S., Salmon P. M. (2018). Computational methods to model complex systems in sports injury research: agent-based modelling and system dynamics modelling. *British Journal of Sports Medicine*. doi: 10.1136/bjsports-2018-100098.

Hulme, A., Thompson, J., Nielsen, R. O., Read, G., & Salmon, P. (2019). Towards a complex systems approach in sports injury research: Simulating running-related injury development with agent-based modelling. *British Journal of Sports Medicine*, 53, 560–569. doi: 10.1136/bjsports-2017-098871

Lane, B. R., Salmon, P. M., Cherney, A., Lacey, D., & Stanton, N. A. (2019). Using the event analysis of systemic teamwork (EAST) broken-links approach to understand vulnerabilities to disruption in a darknet market. *Ergonomics*. doi: 10.1080/00140139.2019.1621392

Leveson, N. G. (2004). A new accident model for engineering safer systems. *Safety Science*, 42:4, 237–270.

Leveson, N. G. (2011). Applying systems thinking to analyse and learn from events. *Safety Science*, 49:1, 55–64.

McGarry, T., Anderson, D. I., Wallace, S., Hughes, M., & Franks, I. M. (2002). Sport competition as a dynamical self-organising system. *Journal of Sports Sciences*, 20, 771–781.

Mclean, S., Soloman, C., Gorman, A., & Salmon, P. M. (2017). What's in a game? A systems approach to enhancing performance analysis in football. *Plos One*, 1–15.

Mooney, M., Charlton, P. C., Soltanzadeh, S., & Drew, M. K. (2017). Who 'owns' the injury or illness? Who 'owns' performance? Applying systems thinking to integrate health and performance in elite sport. *British Journal of Sports Medicine*, 51, 1054–1055.

Naikar, N. (2013). *Work Domain Analysis: Concepts, Guidelines, and Cases*. Boca Raton, FL: CRC Press.

Patatas, J. M., De Bosscher, V., & Legg, D. (2018). Understanding parasport: an analysis of the differences between able-bodied and parasport from a sport policy perspective. *International Journal of Sport Policy and Politics*, 10:2, 235–254.

Salmon, P. M., Carden, T., & Stevens, N. J. (2018). Breaking bad systems: Using work domain analysis to identify strategies for disrupting terrorist cells. Proceedings of the Chartered Institute of Ergonomics and Human Factors Annual Meeting.

Salmon, P. M., & Read, G. J. M. (2019). Many-model thinking in systems ergonomics: a case study in road safety. *Ergonomics*, 62:5, 612–628.

Salmon, P. M., Read, G. J. M., Thompson, J., McLean, S., & McClure, R. (2020). Computational modelling and systems ergonomics: a system dynamics model of drink driving-related trauma prevention. *Ergonomics*, accepted for publication on 21 December 2019.

Stanton, N. A., & Harvey, C. (2017). Beyond human error taxonomies in assessment of risk in socio-technical systems: A new paradigm with the EAST 'broken-links' approach. *Ergonomics*, 60:2, 221–233.

Stanton, N. A., Salmon, N. A., & Walker, G. H. (2018). *Systems Thinking in Practice: The Event Analysis of Systemic Teamwork*. Boca Raton, FL: CRC Press.

Stanton, N. A., Salmon, P. M., Rafferty, L. A., Walker, G. H., Baber, C., & Jenkins, D. P. (2013). *Human Factors Methods: A Practical Guide for Engineering and Design*. 2nd ed. Surrey, UK: Ashgate.

Vanlandewijck, Y. C., & Thompson, W. R. (eds.). (2011). *Handbook of Sports Medicine and Science: The Paralympic Athlete*. John Wiley & Sons.

Vicente, K. J. (1999). *Cognitive Work Analysis: Toward Safe, Productive, and Healthy Computer-Based Work*. Mahwah, NJ: Lawrence Erlbaum Associates.

Index

355

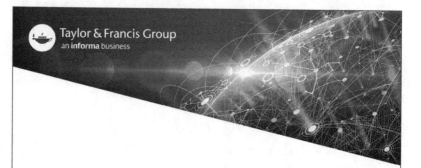